CW00811857

Neurochemical Aspects of Excitotoxicity

Neurochemical Aspects of Excitotoxicity

Akhlaq A. Farooqui
The Ohio State University
Columbus, Ohio

Wei-Yi Ong
National University of Singapore
Singapore

and

Lloyd A. Horrocks
The Ohio State University
Columbus, Ohio

Akhlaq Farooqui
Department of Molecular
and Cellular Biochemistry
The Ohio State University
3120 Herrick Road
Columbus, Ohio
USA
farooqui.1@osu.edu

Wei-Yi Ong
Department of Anatomy
Faculty of Medicine
National University of Singapore
Singapore
antongwy@nus.edu.sg

Lloyd Horrocks (deceased)

ISBN: 978-0-387-73022-6 e-ISBN: 978-0-387-73023-3

Library of Congress Control Number: 2007933089

Printed on acid-free paper

9 8 7 6 5 4 3 2 1

springer.com

Note

We regret to note that Dr. Lloyd A. Horrocks passed away on August 18, 2007. Dr. Horrocks grew up in Cincinnati, Ohio, and attended school at Ohio Wesleyan University before earning a Ph.D. in Physiological Chemistry from The Ohio State University in 1960. He served as a Project Engineer in the United States Air Force from 1955–1958, and as a Research Associate at the Cleveland Psychiatric Institute from 1960–1968 before joining the Department of Physiological Chemistry (now Molecular Cellular Biochemistry) at The Ohio State University as an Assistant Professor. He became full Professor in 1973, mentoring many graduate and post-graduate students during his tenure. Dr. Horrocks retired in 1992, serving thereafter as Professor Emeritus.

Dr. Horrocks is best known for his research on plasmalogen metabolism and phospholipases A2 in the brain. He authored over 300 research papers, book chapters, and reviews. He edited seven books, and published two monographs. He served as the Editor-in-Chief of *Neurochemical Pathology* and was a member of numerous journal editorial advisory boards, including those of the *Journal of Neurochemistry*, the *Journal of Lipid Research, Lipids*, and *Neurochemical Research*. Dr. Horrocks was dedicated to his family and the field of scientific research and writings. He was a gentleman who will be greatly missed by his friends, students, and colleagues.

Preface

About 40% of central nervous system synapses use glutamate as the neurotransmitter. It acts through ionotropic (NMDA, AMPA, and KA) and metabotropic types of glutamate receptors and plays crucial roles in developmental synaptogenesis, plasticity, long-term potentiation, and learning memory. Our understanding of the diversity, structure, and functions of various glutamate receptors has advanced significantly. Several splice variants of NMDA, AMPA, KA and metabotropic receptors exist in brain tissue. Over-stimulation of glutamate receptors produces neuronal injury or death by a mechanism called excitotoxicity. Excitotoxicity is closely associated with neurochemical and neuropathological changes involved in acute neural trauma (stroke, spinal cord trauma, and head injury) and neurodegenerative diseases such as Alzheimer disease (AD), Parkinson disease (PD), Huntington disease (HD), amyotrophic lateral sclerosis (ALS), Creutzfeldt-Jakob disease (CJD), Guam-type amyotrophic lateral sclerosis/Parkinson dementia (ALS/PDC), and multiple sclerosis (MS).

In the past decade, our understanding of the biochemistry, molecular biology, and neuropathology of the glutamate transporters and receptors has exploded. This is also true for the signal transduction mechanisms involved in the production of oxygen radicals, cytokines, and lipid mediators associated with neurodegenerative process. It is becoming increasingly evident that molecular mechanisms, which govern the transfer of the death signal from the neural cell surface to the nucleus, depend on lipid mediators and on cross talk among excitotoxicity, oxidative stress, and neuroinflammation. Thus, interactions among excitotoxicity, oxidative stress, and neuroinflammation play a major role in neuronal cell death during acute neural trauma and neurodegenerative disease. These processes may be primary initiating points in neurodegeneration or they may be the end result of the neurodegenerative process itself.

We are now empowered by technological advances in lipidomics, proteomics, and genomics. Using these techniques, investigators characterize splice variants of glutamate transporters and receptors and levels of lipid mediators and develop diagnostic tests for neurodegenerative diseases associated with glutamate-mediated toxicity. Moreover, the development of quantitative immunochemical localization techniques allows us to characterize neurodegeneration in kainate-induced model of neurotoxicity. Targeting molecular mechanisms of glutamate-mediated cell death in neurological disorders for therapeutic interventions with glutamate receptor antagonists has become a realistic objective of studies on neurodegenerative diseases, stroke, and mechanical neural trauma. This objective is supported by the constantly expanding innovative biotechnological approaches to drug design.

The purpose of this monograph is to present readers with a coherent overview of cutting edge information on glutamate metabolism in brain, the role of glutamate transporters, the involvement of glutamate receptors in the pathogenesis of acute neural trauma and neurodegenerative diseases (AD, PD, HD, and ALS), the treatment of these diseases with endogenous and exogenous antioxidants and glutamate receptor antagonists in a manner that is useful to students and teachers and also to researchers and physicians. Graduate students, neuroscientists and physicians will refer to this monograph in their professional writing. The monograph is user-friendly and designed in a way that graduate students in neuroscience will benefit to have this book as a reference. A dedicated graduate course on glutamate metabolism and toxicity in brain could use this monograph as a textbook.

The monograph has eleven chapters. The first three chapters describe the metabolism of glutamate in brain, cutting edge information on structure and function of glutamate receptors, and properties of agonists and antagonists of NMDA, AMPA, KA and metabotropic types of glutamate receptor in the central nervous system. Chapter 4 is devoted to glutamate transporters and their role in brain tissue. Chapter 5 describes the association of glutamate with neural membrane glycerophospholipid metabolism in brain. Chapter 6 covers the effect of glutamate on neurochemical parameters other than neural membrane glycerophospholipids. Chapter 7 deals with neurochemical mechanisms involved in glutamate-mediated neuronal injury. Chapter 8 describes the association of glutamate with neurological disorders (stroke, AD, PD, HD, ALS, MS, CJD, AIDS dementia, and schizophrenia). Chapter 9 is devoted to the effects of endogenous antioxidants and anti-inflammatory agents on glutamate toxicity in brain. Chapter 10 describes the use of glutamate receptor antagonists for the treatment of acute neural trauma and neurodegenerative diseases and Chapter 11 presents readers with future directions that should be followed to solve unresolved problems of glutamate-related neurological disorders and also discusses new strategies for the antagonism of the NMDA type of glutamate receptors and anti-inflammatory nutraceuticals. The chosen topics presented in this monograph are personal. They are based on our interest on glutamate metabolism in neurological disorders and in areas where major progress has been made. We have tried to ensure uniformity and mode of presentation as well as a logical progression from one topic to another and have provided extensive referencing. For the sake of simplicity and uniformity a large number of figures and line diagrams of signal transduction pathways are also included. Our attempt to integrate and consolidate knowledge of glutamate metabolism and glutamate-mediated signal transduction processes in brain will provide the basis for more dramatic advances and developments in the involvement of glutamate receptors in neurological disorders and in new strategies for the antagonism of excitotoxicity, oxidative stress, and inflammation using endogenous and exogenous glutamate receptor antagonists, antioxidants, and anti-inflammatory agents in the central nervous system.

Akhlaq A. Farooqui
Wei-Yi Ong
Lloyd A. Horrocks

Acknowledgments

We would like to thank a number of publishers who have granted us permission to reproduce figures from our earlier papers published by them. These include Springer-Verlag, Heidelberg, Germany; Elsevier Ltd, Oxford, U.K.; Blackwell Publishing, Oxford, U.K.; and Lippincott William and Wilkins, Baltimore, USA. We would also like to thank Tahira Farooqui, Siew-Mei Lim, and Marjorie Horrocks for their patience, understanding, and moral support during preparation of this monograph and Siraj A. Farooqui for drawing chemical structures of glutamate receptor agonists and antagonists.

Akhlaq A. Farooqui
Wei-Yi Ong
Lloyd A. Horrocks

Contents

Chapter 1
Glutamate and Aspartate in Brain

1.1 Introduction

L-Glutamate and L-aspartate are non-essential dicarboxylic amino acids (Fig. 1.1) synthesized mainly from 2-oxoglutarate and oxaloacetate by transamination reactions. Glutamate and aspartate have strong excitatory effects on neurons. High-affinity uptake systems move them into nerve endings. The occurrence of their receptors is known from sites where radiolabeled glutamate and aspartate selectively bind on neural cell surfaces and through the discovery of their antagonists. Both procedures identify and characterize their receptors on neuronal and glial cells. These amino acids open sodium and potassium ion channels and cause a rapid excitatory response in most neurons at a very low concentration (McGeer et al., 1987; Dingledine and McBain, 1999). The iontophoretic action of aspartate is less potent than that of glutamate. Electrical stimulation of brain slices and cultured neurons releases glutamate and aspartate in a Ca^{2+}-dependent manner. The current can be external, such as an electric shock, or internal, such as the normal current through a pre-synaptic neuron.

Enzymes responsible for the synthesis and degradation of glutamate and aspartate are in most neurons as well as the glial cells that surround them. Specialized sites containing glutamate and aspartate transporters are also on neurons and glial cells. The importance of glutamate in brain tissue can be judged by its abundance and the presence of its receptors and transporters in brain tissue. Excessive doses of glutamate and aspartate cause toxic effects on neurons and glial cells (Choi, 1988). A rapid interconversion of glutamate and aspartate occurs through aspartate aminotransferase, an enzyme found in cytosol and mitochondria (McKenna et al., 2006). This enzyme reversibly converts 2-oxoglutarate to glutamate using aspartate as the donor of the amino group (McGeer et al., 1987). This enzyme plays a role in the entry of glutamate into the citric acid cycle and resynthesis of intramitochondrial glutamate from citric acid cycle intermediates (Sonnewald et al., 1997).

In brain tissue, low concentrations of glutamate and aspartate perform as neurotransmitters, but at high concentration these amino acids act as neurotoxins. Major advances in the excitatory amino acid receptor field have come from the identification, characterization, and cloning of different families of receptors and transporters (Dingledine and McBain, 1999). These receptors and transporters are specialized

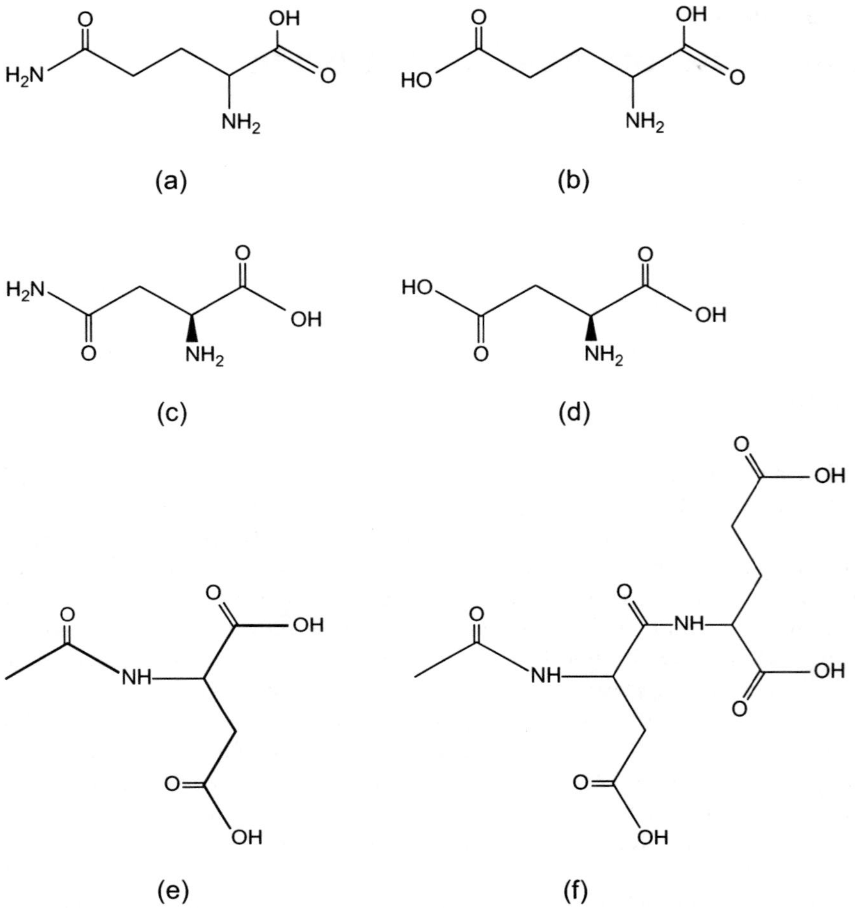

Fig. 1.1 Chemical structures of glutamate and aspartate and corresponding amines. Glutamine (a); glutamate (b); asparagine (c); aspartate (d); N-acetylaspartate (e); and N-acetylaspartyl-glutamate (f)

proteins on neural cell surfaces that interact, bind, and transport glutamate and aspartate. Therefore, these proteins modulate glutamate and aspartate concentrations in the synaptic cleft.

1.2 Glutamate Synthesis and Release in Brain

Glutamate is the main excitatory amino acid neurotransmitter in central and peripheral nervous systems. Its concentration in brain is higher than in other body tissues. In the brain, the concentration of glutamate is 3- to 4-fold greater than that of aspartate, taurine, or glutamine (McGeer et al., 1987). The most abundant amino acid

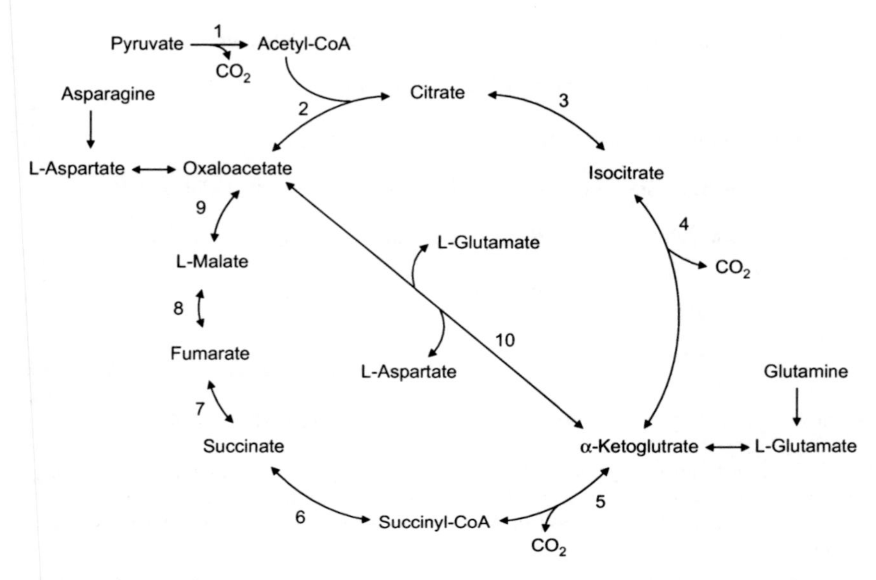

Fig. 1.2 Intermediates of the citric acid cycle showing the relationship between glutamate and aspartate. Pyruvate dehydrogenase complex (1); citrate synthase (2); aconitase (3); isocitrate dehydrogenase (4); α-ketoglutarate dehydrogenase (5); succinyl-CoA synthetase (6); fumarate (7); fumarase dehydratase (8); malate dehydrogenase (9); and aspartate aminotransferase (10)

in synaptosomes is glutamate, followed by glutamine, aspartate, γ-aminobutyric acid, and taurine. Glutamate cannot cross the blood-brain-barrier. The main source of glutamate carbon is glucose with synthesis of glutamate from glucose and other metabolites of the citric acid cycle (McGeer et al., 1987) (Fig. 1.2). The nitrogen in glutamate is from branched-chain amino acids, which are transported rapidly into the brain. Brain synthesizes glutamate, (a) from α-ketoglutarate by a transamination reaction catalyzed by aspartate aminotransferase, an enzyme found in the cytoplasm and mitochondria (McKenna et al., 2006), (b) from α-ketoglutarate and ammonium ions through the action of glutamate dehydrogenase, a pyridine nucleotide-dependent mitochondrial enzyme, and (c) by the action of mitochondrial glutaminase on glutamine. Glutamate semialdehyde, a metabolite generated by Δ^1-pyrroline-5-carboxylic acid dehydrogenase during ornithine metabolism, is also a source of glutamate. In brain tissue, proline oxidase also generates Δ^1-pyrroline-5-carboxylic acid from proline. GABA aminotransferase and alanine aminotransferase also participate in the synthesis of glutamate from α-ketoglutarate (Fig. 1.3). Although the occurrence of these reactions in brain is well established, their relative contributions to glutamate synthesis and release are still unknown. It appears however, that aspartate aminotransferase and glutaminase account for a majority of glutamate production in brain tissue (McGeer et al., 1987). Enzymes responsible for the synthesis of glutamate are in neurons as well as glial cells. A large proportion of the glutamate present in the brain is produced by astrocytes through synthesis de novo (Hertz et al., 1999), but levels of glutamate in glial cells are lower than in neurons, 2–3 mM and 5–6 mM, respectively.

Aspartate + α-Ketoglutarate $\xleftrightarrow{1}$ Oxaloacetate + Glutamate

Glutamine + H_2O $\xleftrightarrow{2}$ Glutamate + NH_3

α-Ketoglutarate + NADH + NH_4^+ $\xleftrightarrow{3}$ Glutamate + NAD^+ + H_2O

α-Ketoglutarate + GABA $\xleftrightarrow{4}$ Succinate semialdehyde + Glutamate

α-Ketoglutarate + Alanine $\xleftrightarrow{5}$ Pyruvate + Glutamate

Ornithine $\xrightarrow{6}$ Δ^1-Pyrroline 5-carboxylic acid

↓

Glutamic semialdehyde $\xrightarrow{7}$ Glutamate

Glutamine + Aspartate + ATP + H_2O $\xleftrightarrow{8}$ Glutamate + Asparagine + AMP + PPi

Fig. 1.3 Reactions showing synthesis of glutamate in brain. Aspartate aminotransferase (1); glutaminase (2); glutamate dehydrogenase (3); GABA aminotransferase (4); alanine aminotransferase (5); ornithine aminotransferase (6); Δ1-pyrroline 5-carboxylic acid dehydrogenase (7); and asparagine synthetase (8)

Glutamate plays a vital role during excitatory synaptic transmission, a process by which glutamatergic neurons communicate with each other. The specialized structures that perform this are the synapses, sometimes called nerve endings. These are asymmetrical intercellular junctions characterized by the close apposition of the presynaptic and postsynaptic plasmalemma that are separated by the synaptic cleft. During excitatory neurotransmission, glutamate-filled vesicles are docked at a specialized region of the presynaptic plasma membrane known as the active zone. Packaging and storage of glutamate into glutamatergic neuronal vesicles requires Mg^{2+}/ATP-dependent vesicular glutamate uptake systems, which utilize an electrochemical proton gradient as a driving force. Substances that disturb the electrochemical gradient inhibit this glutamate uptake into vesicles. The concentration of glutamate in vesicle reaches as high as 20–100 mM (Nicholls and Attwell, 1990).

The docked vesicles then go through a maturation process called priming to become fusion competent (Ghijsen et al., 2003; Li and Chin, 2003). In response to an action potential-induced Ca^{2+} influx in presynaptic neurons, primed vesicles undergo rapid exocytotic fusion to release glutamate. Exocytosis of synaptic vesicles occurs not only at synaptic active zones, but also at ectopic sites. Ectopic exocytosis provides a direct and rapid mechanism for neurons to communicate with glia that does not rely on transmitter spillover from the synaptic cleft (Matsui and Jahr, 2004). In the cerebellar cortex, the processes of Bergmann glia cells encase synapses between presynaptic climbing fiber varicosities and postsynaptic Purkinje cell spines, which express both AMPA receptors and electrogenic glutamate transporters. AMPA receptors expressed by Purkinje cells and Bergmann glia cells are activated predominantly by synaptic and ectopic release, respectively (Matsui and Jahr, 2004). Thus, synaptic transmission is a complex process that involves a neurotransmitter, proteins, and calcium ions both at the sending pre-synaptic and receiving post-synaptic, neurons. The released glutamate diffuses across the synaptic cleft between two neurons and stimulates the next cell in the chain by interacting with receptor proteins. The binding of glutamate to postsynaptic glutamate receptors induces an ionic flux that depolarizes the neuron.

Synaptic transmission is measured electrophysiologically as well as by imaging techniques using amphipathic molecules whose fluorescence increases on membrane binding. It can also be measured by monitoring the rise in postsynaptic calcium due to influx through transmitter-gated receptors. Glutamate binding to postsynaptic receptors also may cause metabolic changes such as the activation of second messenger systems (Dingledine and McBain, 1999). A chloride-independent membrane transport system, which is used only for re-absorbing glutamate, terminates glutamate-mediated stimulation of neurons (Attwell, 2000). Glutamate reuptake occurs through a transporter driven by Na^+-independent and

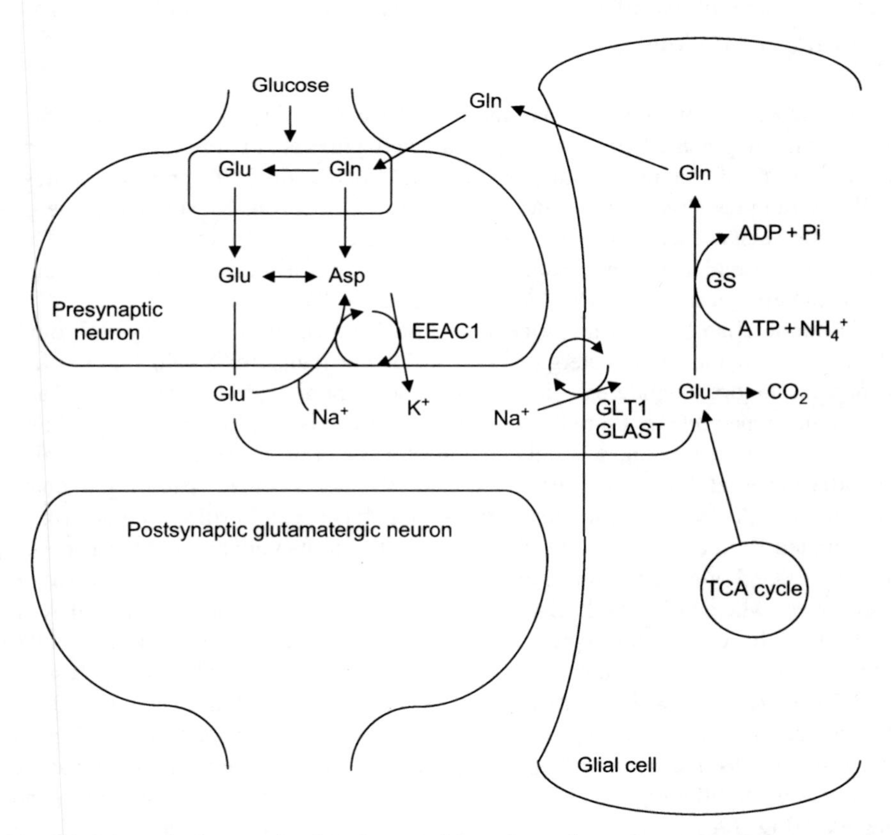

Fig. 1.4 Schematic diagram showing glutamate/glutamine cycles at the synapse. In neurons, glutamate is synthesized from glucose or glutamine. The released glutamate in the synaptic cleft interacts with postsynaptic receptor sites and is also taken up by glial cells. Astroglial cells have a higher affinity uptake system for glutamate than neurons. Astroglial cells also have glutamine synthase (GS) activity. This enzyme converts glutamate to glutamine, which has no neurotransmitter activity and can be recycled to neurons to form glutamate

Na^+-dependent mechanisms. Na^+-dependent mechanisms have a high affinity for glutamate. Sodium ion along with glutamate enters the nerve cell and potassium ion leaves the cell (Fig. 1.4). Thus, Na^+-K^+-ATPase, the sodium pump (McKenna et al., 2006), indirectly powers glutamate absorption. Synaptic vesicles also incorporate glutamate through a Na^+-independent and ATP-dependent process (Fykse and Fonnum, 1996), resulting in a high concentration of glutamate in each vesicle. Brain tissue contains two mechanisms that efficiently maintain glutamate-mediated neurotransmission, namely reuptake of glutamate from the synapse and resynthesis of the transmitter within the presynaptic site. Both mechanisms coexist within a single neuron.

1.3 Glutamate-Related Metabolic Interactions Between Neurons and Glial Cells

Neurons lack glutamine synthetase and pyruvate carboxylase. Therefore, they cannot synthesize glutamate from glucose. These enzymes are expressed in astrocytes, thus synthesis of glutamate from glucose can take place only in the presence of these cells. Astrocytes also maintain the energy status of neurons by supplying energy substrates such as lactate, intermediates of the citric acid cycle, and glutamine. This suggests that under normal conditions the trafficking of these metabolites occurs between neurons and astrocytes (Schousboe et al., 1997).

In mammalian brain, a majority of neuronal information is conveyed through the rapid excitatory glutamatergic system, which accounts for 80–90% of cortical synapses. In brain, glutamate is in a metabolic pool with α-ketoglutarate and glutamine. Once released into the synapse, glutamate is rapidly taken up by a high affinity transport system, not only in the glutamatergic nerve endings, but also by adjacent glial cells. High affinity transport systems utilize specific glutamate transporters. At least five glutamate transporters have been purified, characterized, and cloned from brain tissue (Danbolt, 2001). The glutamate aspartate transporter (GLAST) and glutamate transporter 1 (GLT-1) are predominantly located on astrocytes, whereas EAAC1, EAAT4, and EAAT5 are neuron-specific. GLAST and GLT-1 have different expression patterns in developing brain. GLAST is the major transporter for glutamate uptake during development, whereas expression of GLT-1 increases with the maturation of the nervous system (Danbolt, 2001). Glutamate transporter 1 expression follows the formation and maturation of synapses. Collectively, these studies suggest that the operation of glutamatergic synapses is tightly regulated by dynamic interactions between astrocytes and neurons (Fig. 1.4).

In isolated glial cells, the intracellular concentration of glutamate is below 1 mM, whereas in neurons it is much higher (10–12 mM). This suggests that glutamate transporter activity in glial cells may produce a substantial and rapid increase in intraglial glutamate concentration and this may act as an intracellular signal for signal processes associated with the modulation of metabolic interactions between neurons and glia. Under normal conditions, various glutamate transporters remove glutamate from the extracellular space and keep basal levels of extracellular glutamate in the range of 1–2 μM. Thus, they help to terminate glutamatergic synaptic transmission and to prevent the extracellular glutamate concentration from rising to neurotoxic values.

Treatment of astrocytes with glutamate results in an increase in astrocytic cell volume with a resulting decrease of the extracellular space. This process can alter the concentration of extracellular substances in the synaptic cleft. Many lines of evidence show that K^+ can be buffered within the astroglial gap-junction-coupled network (Tsacopoulos, 2002). These gap junctions are permeable to glutamate. All these events occur dynamically and the astroglial network has the capacity to interfere actively with neurotransmission. This contributes to a high signal-to-noise ratio for glutamatergic transmission. High-quality neuronal messages during normal physiology can then be maintained (Hertz, 2006).

Glutamate is either oxidized through the citric acid cycle located in glial cell mitochondria or converted into glutamine by glutamine synthetase (Gamberino et al., 1997; Sonnewald et al., 1997). Glutamine is rapidly discharged from glial cells into the extracellular space by channel-mediated efflux, reversal of membrane glutamate transport, or through volume-sensitive organic anion channels activated during cell swelling (Nedergaard et al., 2002). Glutamine is transferred back by low-affinity systems or by diffusion into the glutamatergic nerve endings, where it is deaminated to glutamate, which can again be used for neurotransmission. The deamination reaction is catalyzed by the enzyme glutaminase (Márquez et al., 2006). Thus, the glutamate-glutamine cycle modulates (Fig. 1.4) and completes glutamate shuttling between neurons and glial cells.

A large proportion of glutamate in astrocytes is degraded as a metabolic fuel to carbon dioxide and water through enzymes of the citric acid cycle located in mitochondria. The rate of glutamate oxidation to CO_2 is higher in astrocytes than in glutamatergic neurons. Depletion of glutamate from the glutamate-glutamine cycle induces synthesis de novo of glutamate from oxaloacetate in the citric acid cycle. At the glutamatergic synapse, 1–3 mM is the estimated level of glutamate in the cleft during a synaptic event. Diffusion or uptake by the buffering capacity of glutamate transporters removes this glutamate within 1–2 ms (Danbolt, 2001).

Glial cells play a pivotal role in glutamatergic excitatory neurotransmission in the central nervous system. The astrocytes regulate the synaptic levels of glutamate by providing glutamine through the glutamate-glutamine cycle as the sole precursor for the neurotransmitter pool of glutamate in neurons (Rao et al., 2003). The reuptake of extrasynaptic glutamate is crucial for preventing a continuous rise in glutamate concentration. High extracellular glutamate concentrations have neurotoxic effects not only on neurons, but also on oligodendrocytes (Oka et al., 1993). Besides this toxic affect, stimulation of astrocytes with glutamate increases lipid peroxidation and production of free radicals (Farooqui and Horrocks, 1991). Furthermore, increased levels of glutamate in the synaptic cleft may modify the electrical activity of the injured neurons. This may modify the expression of MHC class I molecules by neurons, which may make them a potential target for activated immune cells. Perturbation of electrical activity may also modify interactions between neurons and oligodendrocytes, in particular the phosphorylation of myelin basic protein, which may contribute to neurodegeneration.

1.4 Roles of Glutamate in Brain

L-Glutamate is present in abundance in the brain and spinal cord. Although glutamatergic neurons and synapses are widely distributed throughout the brain, they are enriched in the hippocampus, the outer layers of the cerebral cortex, and the substantia gelatinosa of the spinal cord (Dingledine and McBain, 1999). Brain has two pools of glutamate. Glutamatergic synapses represent the neurotransmitter pool with 35–45% of the brain glutamate. The remaining glutamate is in the metabolic pool. This glutamate is associated with the synthesis of γ-aminobutyric acid (GABA), general metabolism, and protein synthesis in neurons and glial cells (McGeer et al., 1987; Dingledine and McBain, 1999).

Glutamate has many important roles in the normal function of the central nervous system (CNS). Under normal conditions, glutamate is involved in rapid excitatory transmission and in the slow developing neuroplasticity changes associated with learning, memory, synaptogenesis, and neuronal development (Agranoff et al., 1999; Gibbs and Hertz, 2005). In addition, glutamate has a significant role in the regulation of bioenergetic processes. Proteins and peptides incorporate glutamate. It is involved in fatty acid synthesis. It detoxifies ammonia. It controls osmotic or anionic balance. It serves as the precursor for GABA, an inhibitory neurotransmitter (Fig. 1.5). It is a constituent of glutathione and folic acid. Glutamate induces the synthesis of cytokines in cultured cells (De et al., 2005). It also plays a key role in central pain transduction mechanisms and neural cell injury. A synergistic action of glutamate and cytokines may be involved in the pathogenesis of head and spinal cord injury (Farooqui and Horrocks, 2007a). Any disruption of normal glutamate levels in brain can have profound effects on behavior and other aspects of brain function.

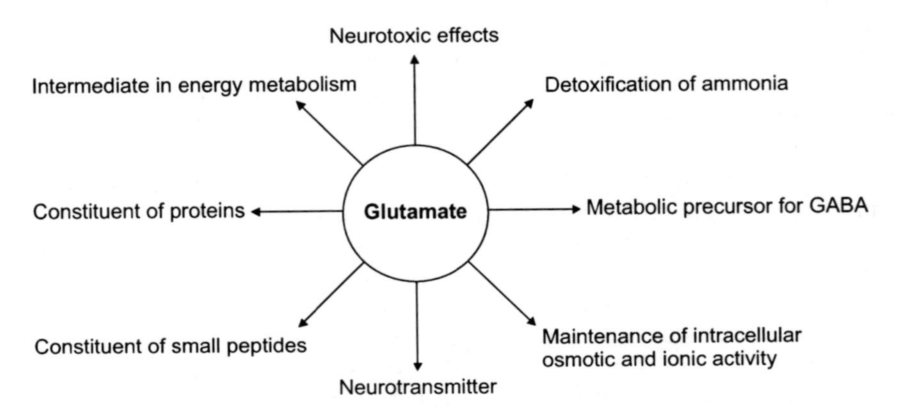

Fig. 1.5 Roles of glutamate in brain tissue

1.4.1 Glutamate and Intermediary Energy Metabolism

Glutamate is an amphibolic intermediate in the biosynthesis and degradation of amino acids. It is involved in maintenance and regulation of the bioenergetics in brain, due to its direct or indirect participation in the reactions of glycolysis, gluconeogenesis, citric acid cycle, and synthesis of ketone bodies. During the utilization of ^{14}C-labeled glucose in the brain, about 40% of the label passes through glutamate prior to the formation of $^{14}CO_2$. Between 10 and 20% of total pyruvate metabolism in brain occurs through the formation of oxaloacetate, a citric acid cycle intermediate, from pyruvate plus CO_2. This reaction is catalyzed by pyruvate carboxylase, an enzyme that occurs in astrocytes (Hertz and Hertz, 2003). Equivalent amounts of pyruvate are converted to acetyl-CoA, which condenses with oxaloacetate to form citrate. This metabolite is converted to α-ketoglutarate.

The interconversion of α-ketoglutarate to glutamate involves the malate-aspartate shuttle. This shuttle translocates α-ketoglutarate from mitochondria into the cytoplasm and then converts it to glutamate by the catalytic action of aspartate aminotransferase (McKenna et al., 2006). As part of the malate-aspartate shuttle, NADH is oxidized during reduction of oxaloacetate to malate. Malate diffuses across the outer mitochondrial membrane (Fig. 1.6). From the intermembrane space, the malate-α-ketoglutarate antiporter in the inner membrane transports malate into the matrix. For every malate molecule entering the matrix compartment, one molecule of

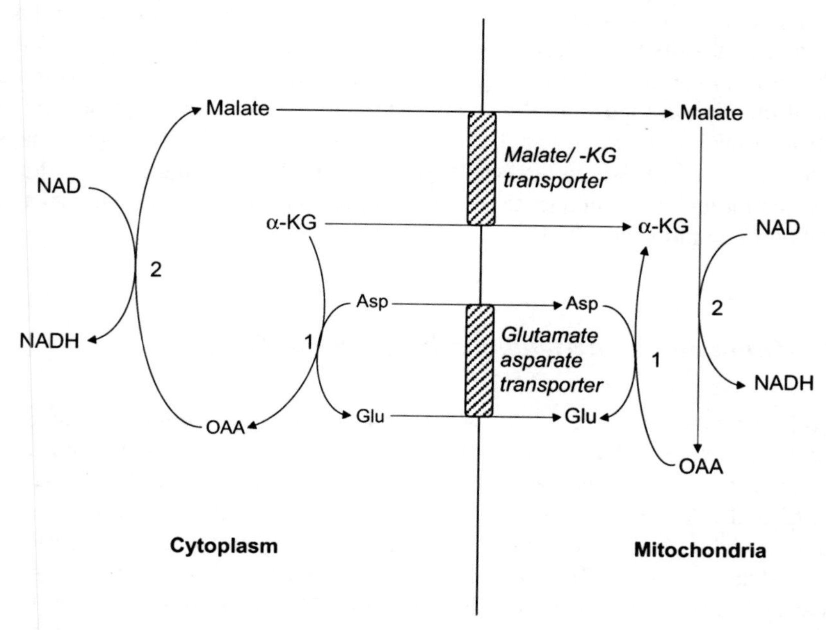

Fig. 1.6 Reactions of the malate-aspartate shuttle showing the transport of reducing equivalents from cytoplasm to mitochondria. α-KG, α-Ketoglutarate; Asp, aspartate; Glu, glutamate; OAA, oxaloacetate; aspartate aminotransferase (1); and malate dehydrogenase (2)

α-ketoglutarate is expelled. Malate is subsequently reoxidized to oxaloacetate with reduction of NAD^+ to $NADH/H^+$. In order to regenerate glutamate, oxaloacetate is transaminated to aspartate by glutamate, which forms α-ketoglutarate. The glutamate/aspartate transporter transports aspartate out of the mitochondrion in exchange for glutamate. Outside the mitochondrion, the transamination of α-ketoglutarate to glutamate converts aspartate to oxaloacetate. The full reducing equivalent of cytoplasmic NADH is thus transported into the mitochondria.

This transfer of reducing equivalents is essential for maintaining the favorable $NAD^+/NADH$ ratio required for the oxidative metabolism of glucose and synthesis of glutamate in brain (McKenna et al., 2006). The malate-aspartate shuttle is considered the most important shuttle in brain. It is particularly important in neurons. It has low activity in astrocytes. This shuttle system is fully reversible and linked to amino acid metabolism with the energy charge and citric acid cycle of neuronal cells.

1.4.2 Glutamate as a Putative Neurotransmitter

Glutamate acts as a neurotransmitter in brain. This conclusion is based on the classic criteria used for establishing the status of a neurotransmitter (Daikhin and Yudkoff, 2000). They include its content, synthesis, release, and removal from the synaptic cleft by uptake, and presence of post-synaptic receptors that are activated by exogenous glutamate and its analogs, and blockage by glutamate receptor antagonists. Glutamate is distributed widely throughout the neuraxis. Regions in which it seems particularly important include the granule cells of the cerebellum, the pyramidal cells of the hippocampus, the Betz cells of the motor strip, and the projections of the frontal lobe to the basal ganglia. Our knowledge of the glutamatergic synapse has advanced enormously in the last 20 years, primarily through the application of molecular biological techniques to the study of glutamate receptors and transporters (Dingledine and McBain, 1999).

1.4.3 Glutamate as a Metabolic Precursor of GABA

In brain, γ-aminobutyric acid (GABA) is synthesized by decarboxylation of glutamate. Pyridoxal phosphate-dependent glutamate decarboxylase catalyzes this reaction. Brain contains at least two forms of glutamate decarboxylase (GAD), namely GAD65 and GAD67. They are the products of two different genes, differing from each other in sequence, molecular weight, interaction with the cofactor pyridoxal 5′-phosphate (pyridoxal-P), and level of expression among brain regions. GAD65 is mainly localized in nerve terminals. In contrast, GAD67 is uniformly distributed throughout the cell.

The interaction of GAD with pyridoxal-P is a major factor in the short-term regulation of GAD activity. At least 50% of GAD is present in brain as the apoenzyme,

GAD without its bound cofactor; apoGAD, which serves as a reservoir of inactive GAD that can be drawn on when additional GABA synthesis is needed. GAD65 accounts for a substantial majority of apoGAD in brain, but GAD67 also contributes to the pool of apoGAD. The apparent localization of GAD65 in nerve terminals and the large reserve of apoGAD65 suggest that GAD65 is specialized to respond to short-term changes in demand for transmitter GABA. In rat neocortex GABA can also be synthesized from glutamine (Patel et al., 2001). Collective evidence suggests that glutamate and GABA are the main synaptic neurotransmitters in the hippocampus. However, their actions are not limited only to the local postsynaptic zone. These amino acids are released into the synaptic cleft by glutamate and GABA exocytosis, osmotic shock, and spillover (Sem'yanov, 2005). Glutamate and GABA receptors are also located on various parts of neurons and glial cells. Depending on the subcellular distribution of their receptors, the effects of extracellular glutamate and GABA differ considerably (Sem'yanov, 2005). Glutamate is a major excitatory transmitter whereas GABA acts as an inhibitory neurotransmitter.

1.4.4 Glutamate and Detoxification of Ammonia

Deamination of amino acids generates free ammonia in brain. Its normal concentration in brain is about 0.1–0.2 mM. High levels of ammonia are very toxic. Ammonia interferes with neurotransmission, neurotransmitter recycling, and oxidative brain metabolism (Rao and Norenberg, 2001). Thus, it must be constantly detoxified to prevent its injurious effects. In astrocytes, conjugation with glutamate removes ammonia. Glutamine synthetase, an enzyme mainly located in astrocytes, catalyzes this conjugation. When excited neurons generate and release ammonia in the extracellular space, astrocytes take it up preferentially to form glutamine. The reaction catalyzed by glutamine synthetase is a key energy-dependent process that modulates the levels of glutamate and also detoxifies ammonia in astrocytes (Dingledine and McBain, 1999; Suárez et al., 2002).

Glutamine synthetase has a high affinity for ammonia and a low affinity for glutamate. This allows ammonia at low concentrations to be utilized efficiently for the synthesis of glutamine. In addition, glutamate can control this utilization. Thus at a low ammonia concentration, the synthesis of glutamine continues until almost all available ammonia in the cell is detoxified. Under normal conditions with the ammonia concentration very low, $< 10\ \mu M$, aspartate aminotransferase and glutamate dehydrogenase reactions can utilize the available glutamate for the synthesis of other products (Fig. 1.3). The detoxification of ammonia by glutamine synthetase also controls the amounts of lactate, glutamine, and alanine generated and released by astrocytes in the extracellular space and then taken up by neurons (Tsacopoulos, 2002). Hyperammonemic states may alter functions of astrocytes, such as protein synthesis or uptake of neurotransmitters.

Normally, the induction of glutamine synthetase expression balances glutamate uptake and the increased levels of ammonia in glutamatergic areas and protects against neuronal degeneration. A decrease in glutamine synthetase expression in non-glutamatergic areas can disrupt neuron-glial metabolic interactions. Changes in glutamine synthase expression may reflect changes in astroglial function, which can affect neuronal functions. The glutamate-glutamine cycle increases the net glutamine output, reduces glutamate excitotoxicity, and avoids neuronal death. Without involvement of the astrocytic glutamate-glutamine cycle, normal glutamatergic transmission is not possible (Hertz and Zielke, 2004). Collectively these studies suggest tight coupling of the detoxification of ammonia and the supply of amino acids and lactate substrates for neurons. The interruption of this coupling and loss of interaction between neurons and astrocytes can initiate neurodegeneration, resulting in brain damage.

1.4.5 Glutamate as a Constituent of Proteins

Glutamate is the most common amino acid found in food. It is enriched 40% in plant proteins and as much as 10–15% in animal proteins. In brain under normal conditions, glutamate and glutamine are present in equal amounts in the total protein pool. However, their ratio changes considerably depending on the metabolic and physiologic state of the CNS and spinal cord. For example, in cerebral cortex more than 16% of the glutaminyl moieties are deaminated within a few minutes after the onset of ischemia. Methionine sulfoximine, but not barbiturates, can prevent this deamination. Levels of glutamate are quite high in membrane proteins and are responsible for the net negative charge. Two proteins, namely S-100 and enolase, are very rich in glutamyl residues.

1.4.6 Glutamate as a Constituent of Small Peptides

Glutathione is a tripeptide containing cysteine, glutamate, and glycine. Neural cells synthesize it in a series of steps catalyzed by specific enzymes including γ-glutamyl cysteinyl synthase and glutathione synthase. Both enzymes use ATP. Glutathione modulates the intracellular levels of glutathione by feedback inhibition of γ-glutamyl cysteinyl synthase. Glutathione peroxidase and glutathione reductase metabolize glutathione. Glutathione has many physiological functions including detoxification of reactive oxygen species in brain tissue. In non-neural mammalian cells, glutathione concentrations can be as high as 12 mM, but the levels of glutathione in brain are lower, 1–3 mM. This peptide maintains the cellular redox status and protects brain from neurotoxicity.

The glutathione contents of neural cells depend strongly on the availability of precursors for glutathione. Different types of brain cells prefer different extracellular glutathione precursors (Dringen, 2000). Glutathione is involved in the detoxification of peroxides by brain cells and in the protection against reactive oxygen species. In cocultures, astroglial cells protect other neural cell types against the toxicity of various compounds (Dringen, 2000). One mechanism for this interaction is the supply by astroglial cells of glutathione precursors to neighboring cells. Recent results confirm the prominent role of astrocytes in glutathione metabolism and the defense against reactive oxygen species in brain. These results also suggest an involvement of a compromised astroglial glutathione system in the oxidative stress reported in neurological disorders (Dringen, 2000). Besides glutathione, a number of γ-glutamyl dipeptides also occur in brain in small amounts. These peptides are intermediates in the γ-glutamyl cycle. This cycle is involved in the Na^+-independent membrane transport of a number of amino acids.

1.4.7 Glutamate and Intracellular Osmotic and Ionic Homeostasis

Many mammalian cells regulate their volume by the osmotic movement of water directed by the flux of anions and cations. Ubiquitous volume-dependent anion currents permit cells to recover volume after swelling in response to a hypotonic environment. Glutamate and Cl^- play an important role in the modulation of volume-activated anion currents (Hoffman and Garber, 2004). The Na^+,K^+-ATPase activity is markedly reduced after glutamate administration or following episodes of ischemia, hypoglycemia, or epilepsy. Under these conditions the exocytotic release of glutamate can cause depolarization of the membrane suggesting Na^+ retention that may result in osmotic swelling and possible cellular lysis (Lees, 1991).

1.4.8 Glutamate in Learning and Memory

Glutamate is essential for learning and memory processes. Memory formation takes place at the synapse, where cells communicate with each other through the release of glutamate. From a behavior viewpoint, learning is defined as the modification of behavior by experience, whereas memory is the retention of such modifications. Alterations of glutamate release, reuptake, and glutamate transport at the synapses enable animals to remember their normal activities. Glutamate receptor agonists and antagonists modulate memory formation in hippocampus-dependent tasks (Daisley et al., 1998; Daisley and Rose, 2002). In the 1-day-old chick, glutamate is released soon after training and again before memory consolidation 30 min after training (Gibbs and Hertz, 2005). Post-training injection of iodoacetate can abolish the memory consolidation. This compound inhibits glycolysis. This observation suggests that not only energy metabolism, but also pyruvate carboxylase-dependent glucose conversion to glutamate is needed for memory consolidation. Collectively these studies suggest that agonists administered post-training might amplify memory formation, whereas antagonists block this process. Furthermore, changes in glutamate transporter activity affect the strength of connections among the neurons associated with memory. This may explain memory lapses such as forgetting (Daisley et al., 1998; Daisley and Rose, 2002).

1.5 Aspartate Metabolism in Brain

L-aspartate is another non-essential dicarboxylic amino acid found in all body tissues including brain. Most L-aspartate is present as a component of neural proteins. Cerebrospinal fluid contains low amounts of free L-aspartate. The transamination of oxaloacetate synthesizes aspartate. This amino acid serves as a precursor for synthesis of proteins, oligopeptides, purines, pyrimidines, nucleic acids, and L-arginine. L-aspartate is a glycogenic amino acid that promotes energy production via its metabolism in the citric acid cycle. Like glutamate, it is released in a Ca^{2+}-dependent process from isolated neural preparations. This release is about 10% of that observed for glutamate (Nicholls and Attwell, 1990; Dingledine and McBain, 1999). L-Aspartate and glutamate are transported into neurons and glial cell by the same high-affinity transport system (Danbolt, 2001). This results in low extracellular aspartate levels in the synaptic cleft.

Aralar is a mitochondrial Ca^{2+}-regulated aspartate-glutamate carrier present in brain and skeletal muscle. It is involved in the transport of aspartate from mitochondria to cytosol and in the transfer of cytosolic reducing equivalents into mitochondria as a participant in the malate-aspartate NADH shuttle. Aralar-deficient (Aralar-/-) mice have no aralar mRNA and protein, and have no detectable malate-aspartate shuttle activity in skeletal muscle and brain mitochondria (Jalil et al., 2005). High levels of aspartate block glutamate uptake, and raise the extracellular glutamate concentration sufficiently to activate glutamate receptors and their associated ion channels. Glutamate antagonists can block the effect of a high concentration of L-aspartate. Collective evidence suggests that the effects of L-aspartate on brain tissue are indistinguishable from those of glutamate (Dingledine and McBain, 1999).

Brain tissue can convert L-aspartate to L-asparagine (Fig. 1.1). Asparagine synthetase catalyzes this reaction, which requires ATP. Rat brain synaptosomes accumulate L-asparagine with a Km value of 348 μM and a Vmax value of 3.7 nmol/mg of protein/min at 28 degrees C. L-Glutamine inhibits the uptake of L-asparagine, whereas L-asparagine can block the transport of L-glutamine. Alanine, serine, cysteine, threonine, and leucine also inhibit the uptake of L-asparagine. Other amino acids such as ornithine, lysine, arginine, and glutamate were much less effective in inhibiting L-asparagine uptake (Erecinska et al., 1991). Isomerization of L-aspartate and deamidation of L-asparagine in proteins or peptides results in the formation of L-isoaspartate by a non-enzymatic reaction via succinimide as an intermediate under physiological conditions. Isoaspartates have been identified in a variety of cellular proteins *in vivo* as well as in pathologically deposited proteins in neurodegenerative brain tissue (Shimizu et al., 2005).

D-Aspartate is also present in brain proteins at concentrations between 0.48 and 0.90 nmol/g of wet tissue, corresponding to concentrations 34 to 82 times lower than that of L-aspartate (Fisher et al., 1992). Brain does not metabolize D-aspartate so it accumulates at its uptake site. D-Aspartate competes with other glutamate agonists for binding to various glutamate receptors (see Chapter 2). The prevalence of D-aspartate in the brain suggests a possible role of this amino acid in modulating glutamate-mediated fast excitatory synaptic transmission (Gong et al., 2005).

Two aspartate derivatives, namely N-acetylaspartic acid (NAA) and N-acetyl aspartylglutamate (NAAG) (Fig. 1.1), also are in brain tissue (Baslow et al., 2000; Baslow, 2003a). The highest concentration of NAA is in cerebral cortex and the lowest in the medulla. NAA is exclusively localized in neurons in the adult brain. Its concentration in neurons of some regions reaches as high as 20 mM. This concentration is responsible for a high intracellular-extracellular gradient in brain tissue. Neuronal mitochondria synthesize it from acetyl-CoA and L-aspartate. Aspartoacylase hydrolyzes it into acetate and aspartate after its release. This enzyme is mainly located in oligodendroglial cells. Glutamate carboxypeptidase II releases NAA and glutamate from synaptic terminals. 2-Phosphonomethyl-pentanedioic acid (2-PMPA) inhibits this release (Luszczki et al., 2006). NAA also gives a strong proton magnetic resonance spectroscopic signal (MRS) (Battistuta et al., 2001), which has led to its widespread use as a neuronal marker. NAA is a putative marker of neuronal injury. Extracellular NAA is also a potential marker for monitoring interventions aimed at preserving mitochondrial function in traumatic brain injury (Belli et al., 2006). Although the precise role of NAA in brain is not known, collective evidence suggests that NAA is osmoregulatory and may be involved in the maintenance of water homeostasis (Baslow, 2002, 2003a,b).

Cycling between the anabolic L-aspartate acetylating compartment in neurons and the catabolic NAA deacetylating compartment in oligodendrocytes tightly regulates the levels of NAA in the brain. A new class of NAA-active compounds, pyrazole and pyrazole derivatives, reduce brain NAA concentrations in normal mice (Baslow et al., 2000). Abnormally low levels are detected in multiple sclerosis and Alzheimer disease and very high levels are found in Canavan disease (hyperacetylaspartemia) (Narayana, 2005; Klunk et al., 1992). Canavan disease is associated with defective activity of the NAA catabolizing enzyme in oligodendrocytes (Baslow et al., 2001; Le Coq et al., 2006). Canavan disease is an autosomal recessive disorder caused by mutation of a gene on chromosome 17, which is responsible for the production of the NAA catabolizing enzyme. In Canavan disease, a continuous efflux of NAA from neurons results in accumulation of NAA in oligodendroglial cells, which initiates demyelination. Increased levels of NAA exert osmotic pressure at the axon-oligodendrocyte paranodal junction resulting in the rupture of the axon-oligodendrocyte junctional seals (Baslow, 1999, 2002; Baslow et al., 2000) causing demyelination. In Aralar -/- mouse brain, the disturbed NAA-mediated water balance results in demyelination. Defective NAA production is also observed in cell extracts from primary neuronal cultures derived from Aralar -/- mouse embryos (Jalil et al., 2005).

NAAG, another endogenous acidic peptide found in brain and spinal cord in millimolar concentrations, is localized in cortical and hippocampal pyramidal neurons, neurites, and synaptic vesicles. NAAG is synthesized from N-acetylaspartate (NAA) and glutamate. NAA availability may limit the rate of NAAG synthesis. It is released from rat brain slices and synaptosomes by a Ca^{2+}-dependent depolarization process (Zollinger et al., 1994). Upon neurostimulation, NAAG is exported to astrocytes where it activates a specific metabotropic Glu surface receptor (mGluR3), and is then hydrolyzed by an astrocyte-specific enzyme, NAAG peptidase, liberating glutamate, which can then be taken up by the astrocyte. Thus NAAG acts as a neurotransmitter and also modulates glutamatergic transmission via interaction with mGluR3 receptors (Koenig et al., 1994; Neale et al., 2000).

The synaptic release of NAAG initiates two processes. Firstly, the activation of glutamate receptors and secondly, the release of glutamate in the synaptic cleft by the action of extracellular NAAG peptidase (Fuhrman et al., 1994). This glutamate not only activates glutamate receptors, but also is taken up by astrocytes and recycled to neurons through the glutamine cycle. NAAG peptidase is a membrane-bound enzyme that is expressed in the adult rat brain exclusively in astroglial cells (Berger et al., 1999). An interactive laser cytometer quantitates changes in intraneuronal Ca^{2+} in individual neurons. NAAG promotes a rapid increase in intraneuronal Ca^{2+}. NMDA receptor and channel antagonists block these effects (Neale et al., 2000). Acting through NMDA and metabotropic glutamate receptor type 3, NAAG reduces cyclic AMP levels, decreases voltage-dependent Ca^{2+} conductance, modulates acetylcholine release, and influences long-term potentiation and long-term depression (Baslow, 2003a). NAAG also regulates the expression of GABA subunits and blocks GABA release from cortical neurons (Baslow, 2000, 2003a; Neale et al., 2000). NAAG also suppresses neurotransmitter release through selective activation of presynaptic group II metabotropic glutamate receptors (Zhong et al., 2006). Treatment of rats with NAAG peptidase inhibitors increases extracellular NAAG levels and reduces extracellular levels of glutamate following ischemic and

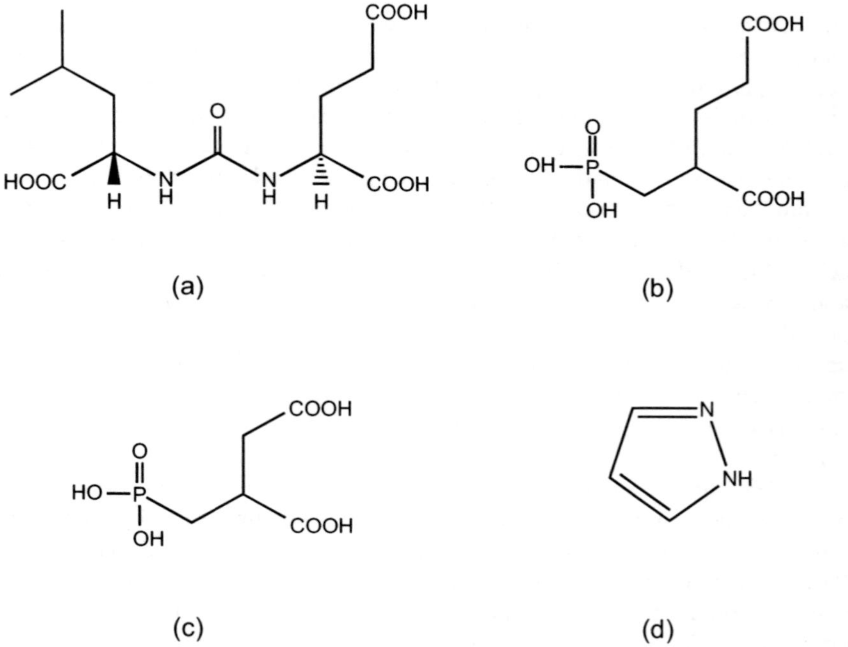

Fig. 1.7 Chemical structures of NAAG peptidase inhibitors. ZJ-43, a urea-based NAAG analog, Ki 0.8 nM (a); PMPA (2-(phosphonomethyl)pentanedioic acid), Ki 280 pM (b); PMSA (2-(phosphonomethyl)succinic acid), Ki 2.6 μM (c); and pyrazole (d)

traumatic brain injuries. These inhibitors include ZJ-43, a urea-based NAAG analog, 2-(phosphonomethyl) pentanedioic acid (PMPA), and 4,4'-phosphinicobis(butane-1,3-dicarboxylic acid) (PBDA) (Fig. 1.7). These inhibitors also block activities of glutamate carboxypeptidase II and glutamate carboxypeptidase III (Bacich et al., 2002). These zinc metallopeptidases inactivate NAAG following its release from synapses and their inhibition reduces toxic effects of glutamate following ischemic and traumatic brain injury (Zhong et al., 2006). Inhibition of glutamate carboxypeptidase II decreases the perception of chronic pain (Yamamoto et al., 2001).

NAAG also mimics the effect of nerve fiber stimulation on the glia. Although glutamate has a similar effect on glial cell membrane potential, NAAG is suggested as the primary axon-to-glia signaling agent. When the unstimulated nerve fiber is treated with cysteate, 2-amino-3-sulphopropionate, a glutamate reuptake blocker, there is a small hyperpolarization of the glial cell that can be substantially reduced by pretreatment with PMPA before addition of cysteate. A similar effect of cysteate is seen during a 50 Hz/5 s stimulation. Glutamate derived from NAAG hydrolysis appears in the periaxonal space and may contribute to the glial hyperpolarization (Gafurov et al., 2001).

References

Agranoff B. W., Cotman C. W., and Uhler M. D. (1999). Learning and memory. In: Siegel G. J., Agranoff B. W., Albers R. W., Fisher S. K., and Uhler M. D. (eds.), *Basic Neurochemistry, Molecular, Cellular, and Medical Aspects*. Lippincott-Raven Publishers, Philadelphia and New York, pp. 1027–1050.

Attwell D. (2000). Brain uptake of glutamate: food for thought. J. Nutr. 130:1023S–1025S.

Bacich D. J., Ramadan E., O'Keefe D. S., Bukhari N., Wegorzewska I., Ojeifo O., Olszewski R., Wrenn C. C., Bzdega T., Wroblewska B., Heston W. D., and Neale J. H. (2002). Deletion of the glutamate carboxypeptidase II gene in mice reveals a second enzyme activity that hydrolyzes N-acetylaspartylglutamate. J. Neurochem. 83:20–29.

Baslow M. H. (1999). Molecular water pumps and the aetiology of Canavan disease: a case of the sorcerer's apprentice. J. Inherit. Metab Dis. 22:99–101.

Baslow M. H. (2000). Functions of N-acetyl-L-aspartate and N-acetyl-L-aspartylglutamate in the vertebrate brain: role in glial cell-specific signaling. J. Neurochem. 75:453–459.

Baslow M. H. (2002). Evidence supporting a role for N-acetyl-L-aspartate as a molecular water pump in myelinated neurons in the central nervous system. An analytical review. Neurochem. Int. 40:295–300.

Baslow M. H. (2003a). Brain N-acetylaspartate as a molecular water pump and its role in the etiology of Canavan disease: a mechanistic explanation. J. Mol. Neurosci. 21:185–190.

Baslow M. H. (2003b). N-acetylaspartate in the vertebrate brain: metabolism and function. Neurochem. Res. 28:941–953.

Baslow M. H., Suckow R. F., and Hungund B. L. (2000). Effects of ethanol and of alcohol dehydrogenase inhibitors on the reduction of N-acetylaspartate levels of brain in mice *in vivo*: a search for substances that may have therapeutic value in the treatment of Canavan disease. J. Inherit. Metab Dis. 23:684–692.

Baslow M. H., Suckow R. F., Berg M. J., Marks N., Saito M., and Bhakoo K. K. (2001). Differential expression of carnosine, homocarnosine and N-acetyl-L-histidine hydrolytic activities in cultured rat macroglial cells. J. Mol. Neurosci. 17:351–359.

Battistuta J., Bjartmar C., and Trapp B. D. (2001). Postmortem degradation of N-acetyl aspartate and N-acetyl aspartylglutamate: an HPLC analysis of different rat CNS regions. Neurochem. Res. 26:695–702.

Belli A., Sen J., Petzold A., Russo S., Kitchen N., Smith M., Tavazzi B., Vagnozzi R., Signoretti S., Amorini A. M., Bellia F., and Lazzarino G. (2006). Extracellular N-acetylaspartate depletion in traumatic brain injury. J. Neurochem. 96:861–869.

Berger U. V., Luthi-Carter R., Passani L. A., Elkabes S., Black I., Konradi C., and Coyle J. T. (1999). Glutamate carboxypeptidase II is expressed by astrocytes in the adult rat nervous system. J. Comp. Neurol. 415:52–64.

Choi D. W. (1988). Glutamate neurotoxicity and diseases of the nervous system. Neuron 1: 628–634.

Daikhin Y. and Yudkoff M. (2000). Compartmentation of brain glutamate metabolism in neurons and glia. J. Nutr. 130:1026S–1031S.

Daisley J. N. and Rose S. P. (2002). Amino acid release from the intermediate medial hyperstriatum ventrale (IMHV) of day-old chicks following a one-trial passive avoidance task. Neurobiol. Learn. Mem. 77:185–201.

Daisley J. N., Gruss M., Rose S. P., and Braun K. (1998). Passive avoidance training and recall are associated with increased glutamate levels in the intermediate medial hyperstriatum ventrale of the day-old chick. Neural Plast. 6:53–61.

Danbolt N. C. (2001). Glutamate uptake. Prog. Neurobiol. 65:1–105.

De A., Krueger J. M., and Simasko S. M. (2005). Glutamate induces the expression and release of tumor necrosis factor-α in cultured hypothalamic cells. Brain Res. 1053:54–61.

Dingledine R. and McBain C. J. (1999). Glutamate and aspartate. In: Siegel G. J., Agranoff B. W., Albers R. W., Fisher S. K., and Uhler M. D. (eds.), *Basic Neurochemistry. Molecular, Cellular, and Medical Aspects*. Lippincott-Raven Publishers, Philadelphia and New York, pp. 315–333.

Dringen R. (2000). Metabolism and functions of glutathione in brain. Prog. Neurobiol. 62: 649–671.

Erecinska M., Zaleska M. M., Chiu L., and Nelson D. (1991). Transport of asparagine by rat brain synaptosomes: an approach to evaluate glutamine accumulation. J. Neurochem. 57: 491–498.

Farooqui A. A. and Horrocks L. A. (1991). Excitatory amino acid receptors, neural membrane phospholipid metabolism and neurological disorders. Brain Res. Rev. 16:171–191.

Farooqui A. A. and Horrocks L. A. (2007a). Glutamate and cytokine-mediated alterations of phospholipids in head injury and spinal cord trauma. In: Banik N. (ed.), *Brain and Spinal Cord Trauma.* Handbook of Neurochemistry. Lajtha, A. (ed.). Springer, New York, in press.

Farooqui A. A. and Horrocks L. A. (2007b). *Glycerophospholipids in the Brain: Phospholipases A_2 in Neurological Disorders*, pp. 1–394. Springer, New York.

Fisher G. H., D'Aniello A., Vetere A., Cusano G. P., Chavez M., and Petrucelli L. (1992). Quantification of D-aspartate in normal and Alzheimer brains. Neurosci. Lett. 143: 215–218.

Fuhrman S., Palkovits M., Cassidy M., and Neale J. H. (1994). The regional distribution of N-acetylaspartylglutamate (NAAG) and peptidase activity against NAAG in the rat nervous system. J. Neurochem. 62:275–281.

Fykse E. M. and Fonnum F. (1996). Amino acid neurotransmission: dynamics of vesicular uptake. Neurochem. Res. 21:1053–1060.

Gafurov B., Urazaev A. K., Grossfeld R. M., and Lieberman E. M. (2001). N-acetylaspartylglutamate (NAAG) is the probable mediator of axon-to-glia signaling in the crayfish medial giant nerve fiber. Neuroscience 106:227–235.

Gamberino W. C., Berkich D. A., Lynch C. J., Xu B., and LaNoue K. F. (1997). Role of pyruvate carboxylase in facilitation of synthesis of glutamate and glutamine in cultured astrocytes. J. Neurochem. 69:2312–2325.

Ghijsen W. E., Leenders A. G., and Lopes da Silva F. H. (2003). Regulation of vesicle traffic and neurotransmitter release in isolated nerve terminals. Neurochem. Res. 28:1443–1452.

Gibbs M. E. and Hertz L. (2005). Importance of glutamate-generating metabolic pathways for memory consolidation in chicks. J. Neurosci. Res. 81:293–300.

Gong X. Q., Frandsen A., Lu W. Y., Wan Y., Zabek R. L., Pickering D. S., and Bai D. (2005). D-aspartate and NMDA, but not L-aspartate, block AMPA receptors in rat hippocampal neurons. Br. J. Pharmacol. 145:449–459.

Hertz L. (2006). Glutamate, a neurotransmitter-And so much more. A synopsis of Wierzba III. Neurochem. Int. 48:416–425.

Hertz L. and Hertz E. (2003). Cataplerotic TCA cycle flux determined as glutamate-sustained oxygen consumption in primary cultures of astrocytes. Neurochem. Int. 43:355–361.

Hertz L. and Zielke H. R. (2004). Astrocytic control of glutamatergic activity: astrocytes as stars of the show. Trends Neurosci. 27:735–743.

Hertz L., Dringen R., Schousboe A., and Robinson S. R. (1999). Astrocytes: glutamate producers for neurons. J. Neurosci. Res. 57:417–428.

Hoffman M. M. and Garber S. S. (2004). Volume-dependent glutamate permeation depends on transmembrane ionic strength and extracellular Cl^-. J. Membr. Biol. 197:193–202.

Jalil M. A., Begum L., Contreras L., Pardo B., Iijima M., Li M. X., Ramos M., Marmol P., Horiuchi M., Shimotsu K., Nakagawa S., Okubo A., Sameshima M., Isashiki Y., Del Arco A., Kobayashi K., Satrustegui J., and Saheki T. (2005). Reduced N-acetylaspartate levels in mice lacking aralar, a brain- and muscle-type mitochondrial aspartate-glutamate carrier. J. Biol. Chem. 280:31333–31339.

Klunk W. E., Panchalingam K., Moossy J., McClure R. J., and Pettegrew J. W. (1992). N-acetyl-L-aspartate and other amino acid metabolites in Alzheimer's disease brain: a preliminary proton nuclear magnetic resonance study. Neurology 42:1578–1585.

Koenig M. L., Rothbard P. M., DeCoster M. A., and Meyerhoff J. L. (1994). N-acetyl-aspartyl-glutamate (NAAG) elicits rapid increase in intraneuronal Ca^{2+} *in vitro*. NeuroReport 5: 1063–1068.

Le Coq J., An H. J., Lebrilla C., and Viola R. E. (2006). Characterization of human aspartoacylase: the brain enzyme responsible for Canavan disease. Biochemistry 45:5878–5884.

Lees G. J. (1991). Inhibition of sodium-potassium-ATPase: a potentially ubiquitous mechanism contributing to central nervous system neuropathology. Brain Res Brain Res Rev. 16:283–300.

Li L. and Chin L. S. (2003). The molecular machinery of synaptic vesicle exocytosis. Cell Mol. Life Sci. 60:942–960.

Luszczki J. J., Mohamed M., and Czuczwar S. J. (2006). 2-phosphonomethyl-pentanedioic acid (glutamate carboxypeptidase II inhibitor) increases threshold for electroconvulsions and enhances the antiseizure action of valproate against maximal electroshock-induced seizures in mice. Eur. J. Pharmacol. 531:66–73.

Márquez J., López de la Oliva A. R., Matés J. M., Segura J. A., and Alonso F. J. (2006). Glutaminase: A multifaceted protein not only involved in generating glutamate. Neurochem. Int. 48:465–471.

Matsui K. and Jahr C. E. (2004). Differential control of synaptic and ectopic vesicular release of glutamate. J. Neurosci. 24:8932–8939.

McGeer P. L., Eccles J. C., and McGeer E. G. (1987). Putative excitatory neurons: Glutamate and aspartate. In: McGeer P. L., Eccles J. C., and McGeer E. G. (eds.), *Molecular Neurobiology of the Mammalian Brain.* Plenum Press, New York.

McKenna M. C., Hopkins I. B., Lindauer S. L., and Bamford P. (2006). Aspartate aminotransferase in synaptic and nonsynaptic mitochondria: Differential effect of compounds that influence transient hetero-enzyme complex (metabolon) formation. Neurochem. Int. 48:629–636.

Narayana P. A. (2005). Magnetic resonance spectroscopy in the monitoring of multiple sclerosis. J. Neuroimaging 15:46S–57S.

Neale J. H., Bzdega T., and Wroblewska B. (2000). N-Acetylaspartylglutamate: the most abundant peptide neurotransmitter in the mammalian central nervous system. J. Neurochem. 75: 443–452.

Nedergaard M., Takano T., and Hansen A. J. (2002). Beyond the role of glutamate as a neurotransmitter. Nat. Rev. Neurosci. 3:748–755.

Nicholls D. and Attwell D. (1990). The release and uptake of excitatory amino acids. Trends Pharmacol. Sci. 11:462–468.

Oka A., Belliveau M. J., Rosenberg P. A., and Volpe J. J. (1993). Vulnerability of oligodendroglia to glutamate: pharmacology, mechanisms, and prevention. J. Neurosci. 13:1441–1453.

Patel A. J., Lazdunski M., and Honore E. (2001). Lipid and mechano-gated 2P domain K^+ channels. Current Opinion in Cell Biology 13:422–427.

Rao K. V. and Norenberg M. D. (2001). Cerebral energy metabolism in hepatic encephalopathy and hyperammonemia. Metab Brain Dis. 16:67–78.

Rao T. S., Lariosa-Willingham K. D., and Yu N. (2003). Glutamate-dependent glutamine, aspartate and serine release from rat cortical glial cell cultures. Brain Res. 978:213–222.

Schousboe A., Westergaard N., Waagepetersen H. S., Larsson O. M., Bakken I. J., and Sonnewald U. (1997). Trafficking between glia and neurons of TCA cycle intermediates and related metabolites. Glia 21:99–105.

Sem'yanov A. V. (2005). Diffusional extrasynaptic neurotransmission via glutamate and GABA. Neurosci. Behav. Physiol 35:253–266.

Shimizu T., Matsuoka Y., and Shirasawa T. (2005). Biological significance of isoaspartate and its repair system. Biol. Pharm. Bull. 28:1590–1596.

Sonnewald U., Westergaard N., and Schousboe A. (1997). Glutamate transport and metabolism in astrocytes. Glia 21:56–63.

Suárez I., Bodega G., and Fernández B. (2002). Glutamine synthetase in brain: effect of ammonia. Neurochem. Int. 41:123–142.

Tsacopoulos M. (2002). Metabolic signaling between neurons and glial cells: a short review. J. Physiol. (Paris) 96:283–288.

Yamamoto T., Nozaki-Taguchi N., and Sakashita Y. (2001). Spinal N-acetyl-alpha-linked acidic dipeptidase (NAALADase) inhibition attenuates mechanical allodynia induced by paw carrageenan injection in the rat. Brain Res. 909:138–144.

Zhong C., Zhao X., Van K. C., Bzdega T., Smyth A., Zhou J., Kozikowski A. P., Jiang J., O'Connor W. T., Berman R. F., Neale J. H., and Lyeth B. G. (2006). NAAG peptidase inhibitor increases dialysate NAAG and reduces glutamate, aspartate and GABA levels in the dorsal hippocampus following fluid percussion injury in the rat. J. Neurochem. 97:1015–1025.

Zollinger M., Brauchli-Theotokis J., Gutteck-Amsler U., Do K. Q., Streit P., and Cuenod M. (1994). Release of N-acetylaspartylglutamate from slices of rat cerebellum, striatum, and spinal cord, and the effect of climbing fiber deprivation. J. Neurochem. 63:1133–1142.

Chapter 2
Excitatory Amino Acid Receptors in Brain

Glutamate, the main excitatory amino acid transmitter in the vertebrate brain, is involved in the dynamic changes in protein repertoire that underlie synaptic plasticity. Differential expression dependent on the activity of glutamate-dependent processes occurs in neurons and in glial cells. Glutamate serves as the neurotransmitter at more than 40–45% of synapses in brain. It plays crucial roles in developmental synaptogenesis and plasticity as well as in long-term potentiation and memory (Monaghan et al., 1989). It interacts with certain receptors on the cell surface. These receptors are the glutamate receptors or excitatory amino acid receptors. Based on their transduction mechanisms, excitatory amino acid receptors are classified into two subclasses, ionotropic and metabotropic. Cation channels are coupled directly to ionotropic excitatory amino acid receptors. They mediate fast excitatory synaptic responses at various synapses throughout the central nervous system. Metabotropic excitatory amino acid receptors are coupled to the second messenger system through a GTP-binding protein (G protein) and are involved in the generation of slow synaptic responses and the modulation of neuronal excitability (Monaghan et al., 1989; Schoepp et al., 1990).

Ionotropic excitatory amino acid receptors are further classified according to their preferential agonist into (a) N-methyl-D-aspartate (NMDA) receptor, (b) α-amino-3-hydroxy-5-methyl-4-isoxazole propionate (AMPA) receptor, and kainate (KA) receptor. The metabotropic receptors are represented by *trans*-1-amino-cyclopentyl-1, 3-dicarboxylate (*trans*-ACPD) receptors. AMPA and kainate receptors mediate fast excitatory synaptic transmission, and are permeable to Na^+ and K^+. NMDA receptors, however, are stimulated and desensitized more slowly by glutamate and are permeable to Ca^{2+}. The ionotropic receptors are also distinguished on the basis of electrophysiological studies indicating differences in cell firing patterns and membrane conductance (Mayer and Westbrook, 1987; Monaghan et al., 1989).

Regional distribution studies indicate that the highest density of kainate receptor occurs in the stratum lucidum of the hippocampus (mossy fiber system) and in the inner and outer layers of the neocortex. The highest density of NMDA receptors is found in the hippocampus, stratum radiatum, and in the striatum, thalamus, and cerebral cortex. The distribution of AMPA receptors is similar to that of NMDA receptors, but in the cerebellum AMPA receptors predominate in the molecular layer and NMDA receptors predominate in the granule cell layer (Managhan et al., 1983; Olverman et al., 1986). Although NMDA and AMPA receptors are coexpressed at most synapses, they make a minimal contribution to basal neurotransmission. They are recruited when cells are depolarized. The response to AMPA always readily self-desensitizes, whereas the response to KA usually shows little or no desensitization.

2.1 Ionotropic Receptors

2.1.1 NMDA Receptors

The discovery of potent and selective agonists and antagonists (Figs. 2.1, 2.2, and 2.3) has resulted in extensive information on the NMDA receptor-channel complex (Wood et al., 1990; Fagg and Baud, 1988). It consists of four domains: (1) the transmitter recognition site with which NMDA and L-glutamate interact;

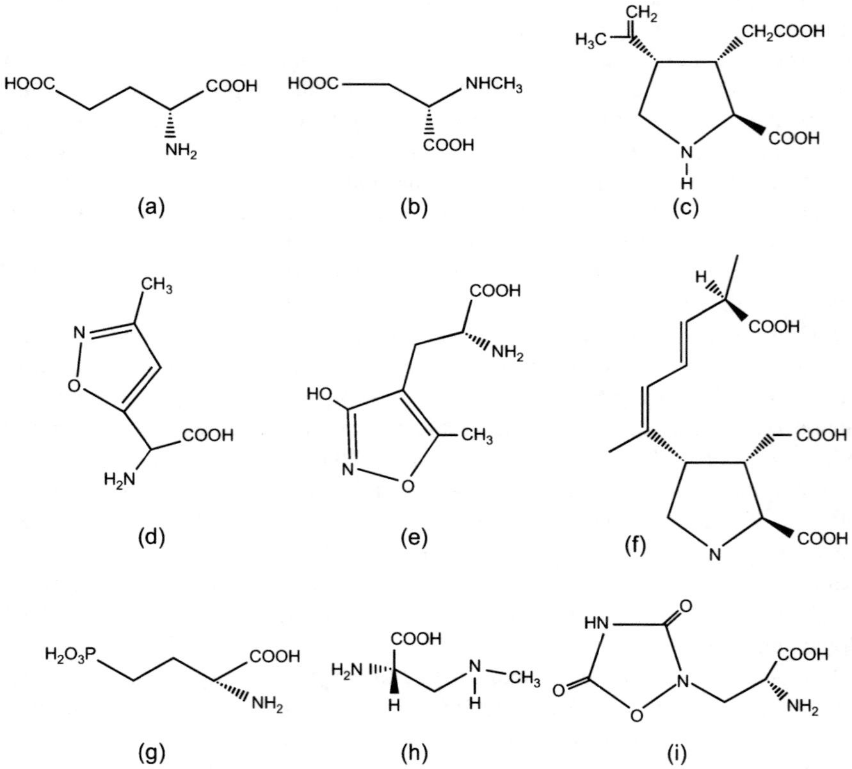

Fig. 2.1 Structures of glutamate and its analogs. L-Glutamate (a); NMDA, N-methyl-D-aspartate (b); kainate (c); ibotenate (d); AMPA (e); domoate (f); L-AP4 (g); BMAA (h); and quisqualate (i)

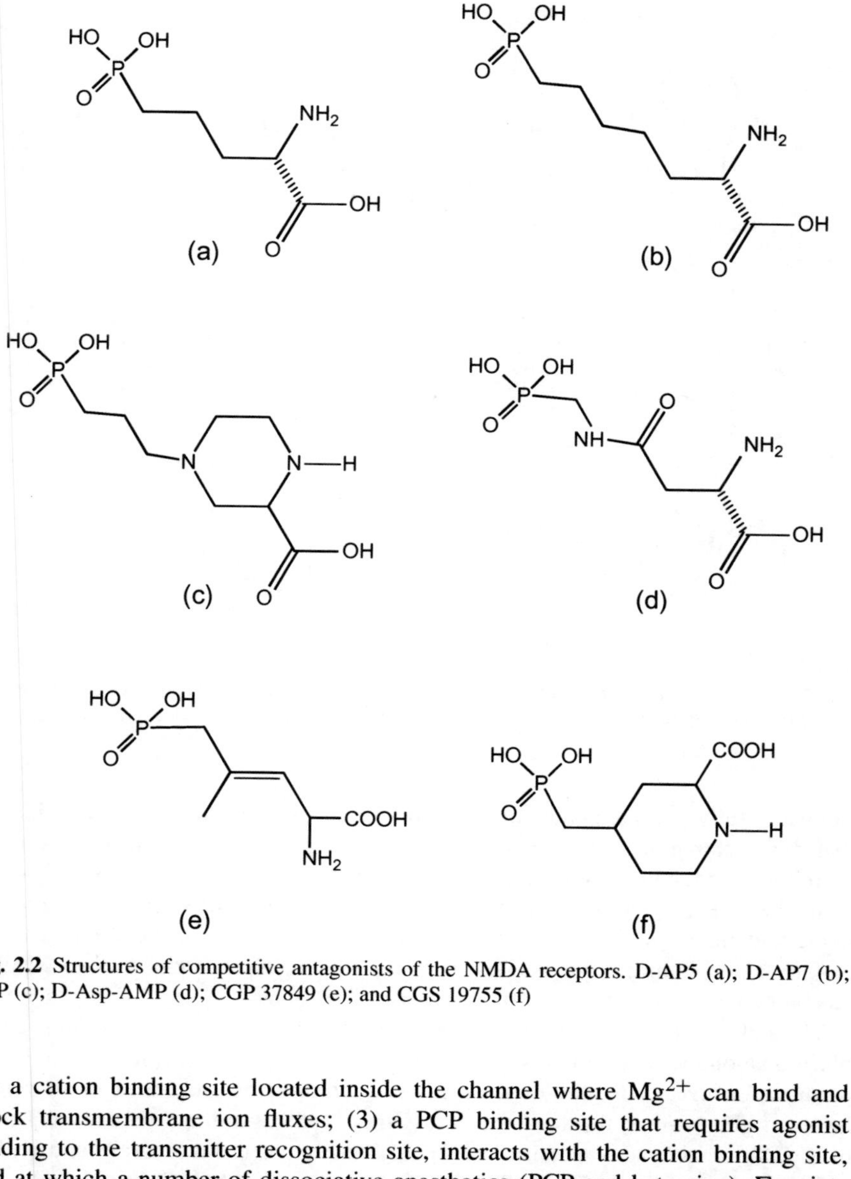

Fig. 2.2 Structures of competitive antagonists of the NMDA receptors. D-AP5 (a); D-AP7 (b); CPP (c); D-Asp-AMP (d); CGP 37849 (e); and CGS 19755 (f)

(2) a cation binding site located inside the channel where Mg^{2+} can bind and block transmembrane ion fluxes; (3) a PCP binding site that requires agonist binding to the transmitter recognition site, interacts with the cation binding site, and at which a number of dissociative anesthetics (PCP and ketamine), Σ-opiate N-allylnormetazocine (SKF-10047), and MK-801 bind and function as open channel blockers; and (4) a glycine binding site that appears to allosterically modulate the interaction between the transmitter recognition site and the PCP binding site (Fagg and Baud, 1988). NMDA is allosterically modulated by glycine, a coagonist whose presence is an absolute requirement for receptor activation.

A polyamine-binding site is in the NMDA receptor complex (Singh et al., 1990; Ransom and Stec, 1988). Like the glycine-binding site, the polyamine modulatory

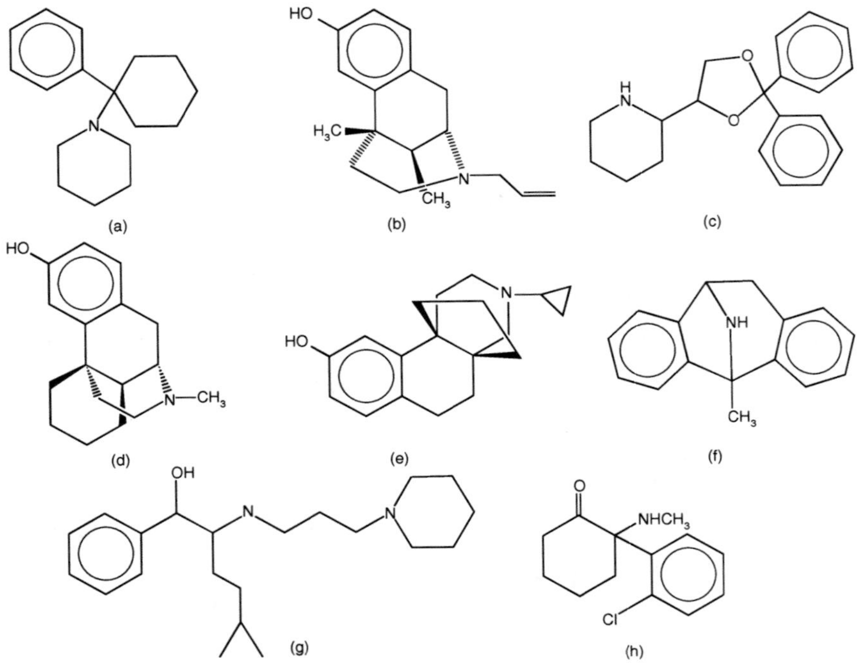

Fig. 2.3 Structures of noncompetitive antagonists of the NMDA receptor. PCP (a); SKF 10,047 (b); dexoxadrol (c); dextrophan (d); LY 154045 (e); MK 801 (f); MLV 5860 (g); and ketamine (h)

site is not fully activated under normal circumstances. In addition there is evidence that Zn^{2+}, acting at a separate site near the mouth of the ion channel, acts as an inhibitory modulator of channel function (Westbrook and Mayer, 1987; Guilarte et al., 1995; Huang, 1997). The NMDA receptor complex also contains an arachidonic acid binding site. The amino acid sequences of this binding site resemble those of fatty acid binding proteins (Petrou et al., 1993). Other modulators of NMDA receptors include sulfhydryl redox reagents and H^+ ions.

Normal functioning of the NMDA receptor complex depends on a dynamic equilibrium among various domain components. Loss of equilibrium during membrane perturbation may cause the entire system to malfunction and result in abnormal levels of glutamate in the synaptic cleft (Olney, 1989). An important consequence of NMDA receptor activation is the influx of Ca^{2+} into neurons (Murphy and Miller, 1988; Holopainen et al., 1989, 1990; MacDermott et al., 1986). Collective evidence suggests that when the membrane is depolarized, the Mg^{2+} block is relieved and the receptor can be activated by glutamate. Activation of the NMDA receptor therefore requires the association of two synaptic events: membrane depolarization and glutamate release. This associative property provides the logic for the role of the NMDA receptor in sensory integration, memory function, coordination and programming of motor activity (Collingridge and Bliss, 1987; Lester et al., 1988) associated with synaptogenesis, and synaptic plasticity.

2.1.2 NMDA Receptor Agonists and Antagonists

NMDA receptors are selectively antagonized by α-amino-ω-phosphonocarboxylic acids, including 2-amino-5-phosphonovalerate (APV). Other antagonists for NMDA receptors include 5-[3-(4-benzylpiperidin-1-yl)prop-1-ynyl]-1,3-dihydrobenzoimidazol-2-one, IC50 value, 5.3 nM; and EMD-95885 (6-[3-[4-(4-fluorobenzyl) piperidino]propionyl]-3H-benzoxazol-2-one, IC50, 3.9 nM (Roger et al., 2004). Neurons regulate excitatory transmission by altering the number and composition of excitatory amino acid receptors at the postsynaptic plasma membrane. The dynamic trafficking of excitatory amino acid receptors to and from synaptic sites involves a complex series of events including receptor assembly, trafficking through secretory compartments, membrane insertion, and endocytic cycling. The mechanisms that control the trafficking of NMDA-type excitatory amino acid receptors are only now beginning to be understood. Although previous studies have considered NMDA receptors as immobile receptors, tightly anchored to the postsynaptic membranes, recent evidence has challenged this view. NMDA receptor trafficking dynamically regulates the plasticity of neural circuits (Pérez-Otaño and Ehlers, 2004).

2.1.3 Kainic Acid Receptors

Kainic acid (KA) is a cyclic and nondegradable analog of glutamate (Fig. 2.1). It is 30- to 100-fold more potent than glutamate as a neuronal excitant. The exact molecular basis of KA-induced neurotoxicity and epileptogenesis is poorly understood (Lerma et al., 2001). KA produces a loss of neurons in specific striatal and hippocampal areas of the brain after intraventricular and intracerebral injections (Lerma, 1997; Ben-Ari and Cossart, 2000). Neuronal axons and nerve terminals are more resistant to the destructive effects of KA than the cell soma. KA produces its effects by interacting with specific receptors, the KA receptors (KAR). KAR modulate synaptic transmission by pre-and postsynaptic mechanisms (Lerma, 1997; Chittajallu et al., 1999). KA generates large ion fluxes, particularly the influx of Na^+ ion through receptor-gated ion channels, accompanied by the passive movement of Cl^- and water molecules into neurons and glial cells. KAR are multimeric associations of specific subunits that form ligand-gated ion channels.

Electrophysiological studies on rat hippocampal neurons show two types of KAR with different current responses to KA. In the KA1 response, KA causes little increase in membrane Ca^{2+} permeability and an outward rectification in the current-voltage plots. The KA2 response is characterized by prominent Ca^{2+} permeability and an inward rectification in the current voltage plots (Ozawa et al., 1991).

Using hybridization in situ with oligonucleotide probes, expression patterns of the five known kainate-type glutamate receptor subunit genes, KA1, KA2 and GluR5 to 7, have been discovered in brain tissue of adult and developing rat brain (Michaelis, 1998; Lerma et al., 2001). In situ hybridization studies also indicate that neurons showing the KA1 response are almost exclusively restricted to the CA3 region whereas KA2 message is found in all regions of brain tissue (Michaelis, 1998; Lerma et al., 2001).

2.1.4 Agonists and Antagonists of KAR

Initially it was difficult to distinguish between AMPA and KA receptors on the basis of electrophysiological and pharmacological studies (Paternain et al., 2000; Michaelis, 1998). However, the discovery of 2, 3-benzodiazepines that specifically block AMPA receptors has now made it easier to study the pharmacology and role of KAR in brain tissue. KAR display a large variety of physiological functions in synaptic transmission and in their regulation. At some synapses, KAR mediate a small and slow component of the glutamatergic synaptic response, whereas at other synapses, KAR act as presynaptic modulators of synaptic transmission. A major physiological characteristic of KAR is their rapid desensitization in response to KA and glutamate. This rapid desensitization plays an important role in determining the kinetics of KA receptor channels and is useful in studying the properties of KAR on neurons containing mixed populations of glutamate receptors. Concanavalin A prevents desensitization of native and recombinant KAR (Sperk, 1994; Lerma, 1997; Chittajallu et al., 1999).

Selective agonists of KAR include SYM 2081, (2S,4R)-4-methylglutamic acid; 5-I-Will, 5-iodowillardiine; and 2S,4R isomers of 4-cinnamyl and 4-allylglutamate (Fig. 2.4) (Chittajallu et al., 1999; Pedregal et al., 2000). AMOA, (RS)-2-amino-3-[3-(carboxymethoxy)-5-methylisoxazol-4-yl]propionic acid; AMNH, (RS)-2-amino-3-[2-(3-hydroxy-5-methylisoxazol-4-yl)-methyl-5-methyl-3-oxoisoxazolin-4-yl]propionic acid; LY294486, (3SR,4aRS,6SR,8aRS)-6-[(1H-tetrazol-5-yl) methoxymethyl]-1,2,3,4,4a,5,6,7,8,8a-decahydroisoquinoline-3-carboxylic acid; and LY382884, (3S, 4aR, 6S, 8aR)-6-((4-carboxyphenyl)methyl) 1,2,3,4,4a,5,6,7,8, 8a-decahydroisoquinoline-3-carboxylic acid, are selective antagonists for KAR (Frandsen et al., 1990; Bortolotto et al., 1999a). KA-mediated neurotoxicity is partially blocked by CNQX, 6-cyano-7-nitroquinoxaline-2,3-dione; DNQX, 6,7-dinitroquinoxaline-2,3-dione; NBQX, 2.3-dihydroxy-6-nitro-7-sulfamoyl-quinoxaline 2,3-dione; and GYKI 53655, (±)-1-(4-aminophenyl) 3-methylcarbamyl -4-methyl-3,4-dihydro-7,8-methylenedioxy-5H-2,3-benzodiazepine (Fig. 2.5).

In hippocampal slices, KA depresses GABA-mediated synaptic inhibition and increases the firing rate of interneurons. However, the sensitivity to agonists of these responses differs, suggesting that the presynaptic and somatic KAR have distinct molecular compositions. Hippocampal interneurons express several distinct KAR subunits that can assemble into heteromeric receptors with a variety of pharmacological properties and that, in principle, could fulfill different roles. To address which receptor types mediate each of the effects of kainate in interneurons, new compounds and mice deficient for specific KAR subunits have been used. In a recombinant assay, 5-carboxyl-2,4-di-benzamido-benzoic acid (NS3763) acted exclusively on homomeric glutamate receptor subunit 5 (GluR5), whereas 3S,4aR,6S,8aR-6-((4-carboxyphenyl)methyl)

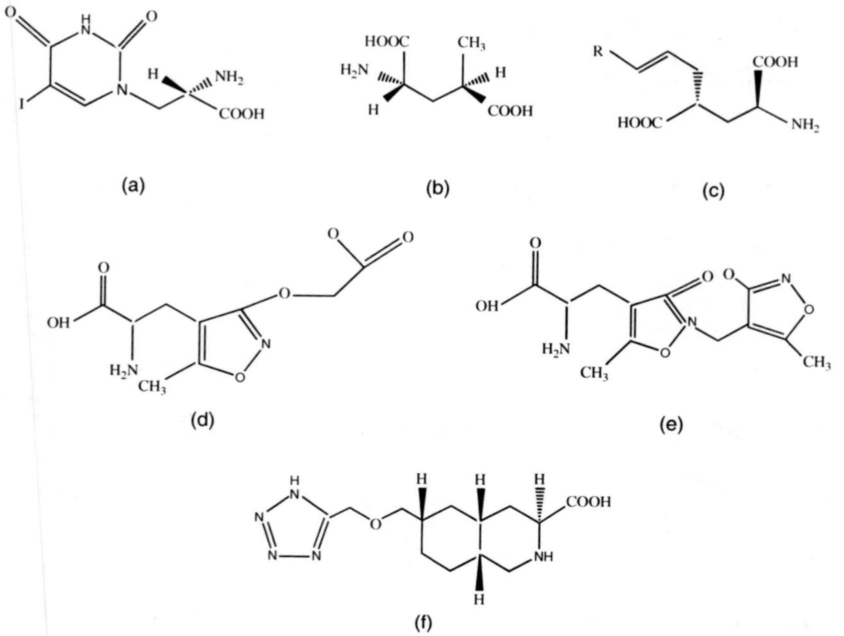

Fig. 2.4 Structures of agonists and antagonists of kainate receptors. 5-I Will, 5-iodowillardiine (a); SYM 2081, (25,4R)-4-methylglutamate (b); 2S,AR isomer of 4-cinnamylglutamic acid (c); AMOA, (RS)-2-amino-3-(3-carboxymethoxy)-5-methylisoxazol-4-yl]propionic acid (d); AMNH, (RS)-2-amino-3-[2-(3-hydroxy-5-methylisoxazol-4-yl)-methyl-5-methyl-3-oxoisoxazolin-4-yl)]propionic acid (e); and LY 294486, (3SR,4aRS,6SR,8aRS)-6-[(1H-tetrazol-5-yl)methoxymethyl]-1,2,3,4,4a,5,6,7,8,8a-decahydroisoquinoline-3-carboxylic acid (f)

1,2,3,4,4a,5,6,7,8,8a-decahydroisoquinoline-3-carboxylic acid (LY382884) antagonized homomeric GluR5 and any heteromeric combination containing GluR5 subunits (Fig. 2.5).

In hippocampal slices, LY382884, but not NS3763, prevents KA-induced depression of evoked IPSC. The selectivity of these compounds is seen additionally in knock-out mice, such that they are inactive in GluR5-/- mice, but completely effective in GluR6-/- mice. These studies indicate that in wild-type mice, CA1 interneurons express heteromeric GluR6 -KA2 receptors in their somatic compartments and GluR5-GluR6 or GluR5-KA2 at presynaptic terminals. However, functional compensation is known to take place in the null mutants (Christensen et al., 2004). A kainate autoreceptor that depresses synaptic transmission has also been described (Kidd et al., 2002). This autoreceptor is present at developing thalamo-cortical synapses in the barrel cortex, specifically regulates transmission at frequencies corresponding to those observed *in vivo* during whisker activation, and is developmentally down-regulated during the first postnatal week. Therefore this receptor may limit the transfer of high-frequency activity to the developing cortex, the loss of which mechanism may be important for the maturation of sensory processing.

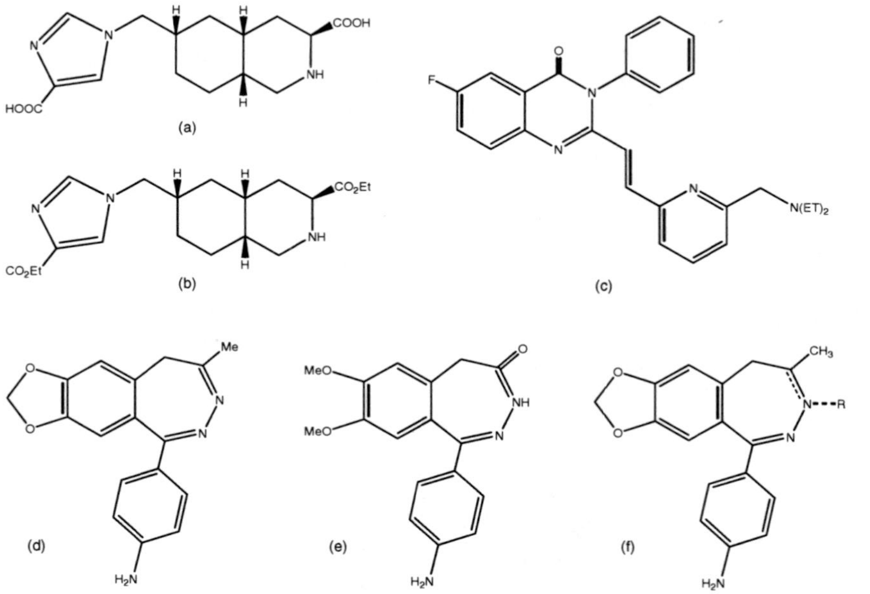

Fig. 2.5 Structures of kainate and AMPA receptor antagonists. LY382884 (a) and LY382884 diethyl ester (b) are kainate receptor antagonists; CP-465022 (c), GYKI 52466 (d); CFM-2 (e); and GYKI 53655 (R= CONHMe) (f) are AMPA receptor antagonists

2.1.5 AMPA Receptors

AMPA and kainate receptors share similar pharmacological and biophysical properties in that they are activated by common agonists and display rapid activation and desensitization characteristics. AMPA receptors are densely distributed in the mammalian brain. They are primarily involved in mediating fast excitatory synaptic transmission. AMPA receptors are made up of GluR1, GluR2, and GluR4 subunits. AMPA receptor ion channels consist of various combinations of the subunits GluR1-GluR4, which bestow certain properties. For example, AMPAR that lack GluR2 are highly permeable to Ca^{2+} and generate inwardly rectifying currents. 1-Naphthylacetyl spermine (NAS), a selective antagonist of Ca^{2+}-permeable AMPAR, inhibits AMPAR-mediated currents in subsets of interneurons and principal cells in cultures and slices in a dose-dependent manner.

2.1.6 Agonists and Antagonists of AMPA Receptors

1-Aryl-3,5-dihydro-7,8-methylenedioxy-4H-2,3-benzodiazepin-4-ones and GYKI 52466, well-known noncompetitive AMPA receptor antagonists. GYKI 53784 and LY303070 [(-)1-(4-aminophenyl)-4-methyl-7,8-methylenedioxy-4,5-dihydro-

3-methylcarbamoyl-2,3-benzodiazepine] are selective and non-competitive antagonists of AMPA receptors (Fig. 2.5). These compounds belong to homophthalazines family which include the original GYKI-52466 (1-4-aminophenyl-methyl-7,8-methylenedioxy-5H-2,3-benzodiazepine), its more potent derivative GYKI 53655 and the active isomer of the latter, GYKI 53784. Unlike the quinoxalinedione family compounds (NBQX, CNQX, and NBQX), that block both AMPA and kainate receptors, GYKI 53784 does not block the activation of kainate receptors. Furthermore, GYKI 53784 does not act at the same receptor site as positive AMPA modulators (i.e., cyclothiazide, BDP-12, 1-BCP or aniracetam). These observations suggest that GYKI 53784 is a powerful neuroprotective agent in both in vitro and *in vivo* models of AMPA receptor-mediated neurotoxicity (Ruel et al., 2002).

AMPA receptors are phosphorylated on GluR1, GluR2, and GluR4 subunits. All phosphorylation sites reside at serine, threonine, or tyrosine on the intracellular C-terminal domain. Several key protein kinases, such as protein kinase A, protein kinase C, Ca^{2+}/calmodulin-dependent protein kinase II, and tyrosine kinase (Trk) receptors or non-receptor family Trk are associated with the regulation of the AMPA receptor. Other glutamate receptors, N-methyl-D-aspartate receptors and metabotropic glutamate receptors, also regulate AMPA receptors through a protein phosphorylation mechanism. Emerging evidence indicates that as a rapid and short-term mechanism, dynamic protein phosphorylation directly modulates the electrophysiological, morphological (externalization and internalization trafficking and clustering), and biochemical (synthesis and subunit composition) properties of the AMPA receptor, as well as protein-protein interactions between the AMPA receptor subunits and various intracellular interacting proteins (Wang et al., 2005). We propose that these modulations underlie the major molecular mechanisms that ultimately affect many forms of synaptic plasticity.

Various combinations of the subunits GluR1 to GluR4 modulate properties of AMPA receptors. For example, AMPAR that lack GluR2 are highly permeable to Ca^{2+} and generate inwardly rectifying currents. Application of 1-naphthylacetyl spermine (NAS), a selective antagonist of Ca^{2+}-permeable AMPAR, prevents AMPAR-mediated currents in subsets of interneurons and principal cells in cultures and slices. The addition of spermine to the electrode produces inwardly rectifying current-voltage plots in some cells. Bath application of NAS with d,l-2-amino-5-phosphonovaleric acid (AP5) inhibits the amplitudes of synaptic responses compared with responses obtained using AP5 alone. These results suggest that AMPAR-mediated Ca^{2+} entry at synapses may be associated with facilitation of synaptic vesicle fusion.

2.2 Metabotropic Glutamate Receptors

At least eight metabotropic glutamate receptors (mGluR1-8) occur in brain tissue. They are members of the group C family of G-protein-coupled receptors, GPCR (Bonsi et al., 2005). Based on sequence homology, agonist pharmacology, and coupling to intracellular transduction mechanisms, metabotropic receptors are classified into three groups (Endoh, 2004). Group I consists of mGluR1 and mGluR5, including their splice variants. These receptors are coupled to an inositol phosphate/Ca^{2+}intracellular signaling pathway. These receptors are found on postsynaptic membranes (Endoh, 2004). They are involved in the modulation of the permeability of Na^+ channels and K^+ channels (Chu and Hablitz, 2000). Their action can be excitatory, increasing conductance, causing more glutamate to be released from the presynaptic cell, but they also increase inhibitory postsynaptic potentials (Chu and Hablitz, 2000). They can also inhibit glutamate release and can modulate voltage dependent Ca^{2+} channels (Endoh, 2004).

Group II comprises mGluR2 and mGluR3, and Group III comprises mGluR4, mGluR6, mGluR7 and mGluR8. Groups II and III inhibit the formation of cAMP by activating a G protein that inhibits the enzyme adenylate cyclase that forms cAMP from ATP (Chu and Hablitz, 2000; Bonsi et al., 2005). They are found on both pre- and postsynaptic membranes. These receptors are involved in presynaptic inhibition (Endoh, 2004), and do not appear to affect the postsynaptic membrane potential by themselves. Receptors in groups II and III reduce the activity of postsynaptic potentials, both excitatory and inhibitory, in the cortex (Chu and Hablitz, 2000). Group II and III receptors mediate presynaptic actions of glutamate in several brain areas, but these receptors may also mediate postsynaptic effects in other brain regions (Chu and Hablitz, 2000). Similarly, there is evidence that Group I receptors are found at presynaptic locations in some cases. Within the thalamus of adults, in situ hybridization studies have shown particularly prominent expression of mGluR1, mGluR4 and

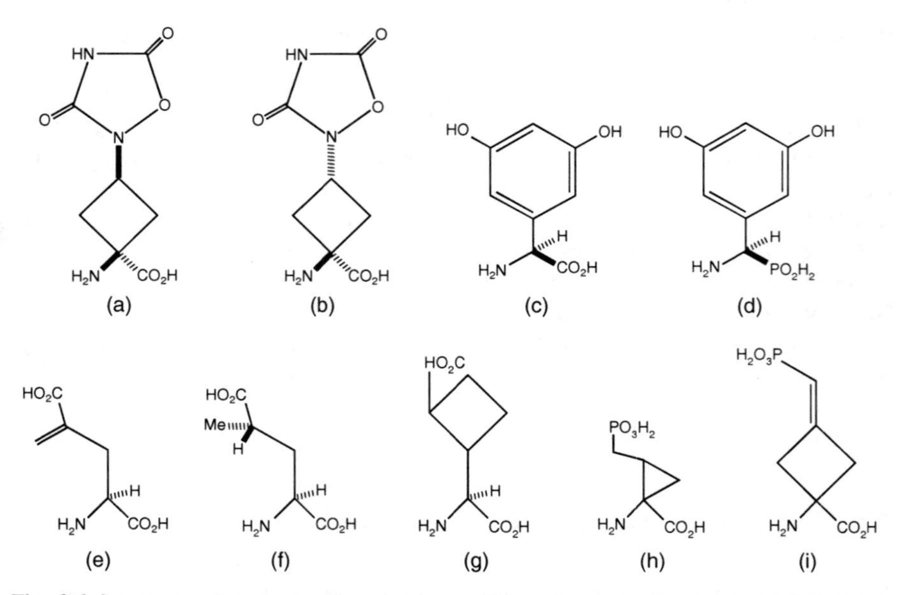

Fig. 2.6 Structures of group I mGlu agonists. ZCBQA (a); E-CBQA (b); S-3,5-DHPG (c); DHPMP (d); S-4MeGlu (e); S,S-4MG (f); L-CBG-1 (g); Z-cyclopropyl-AP4 (h); and cyclobutylene AP5 (i)

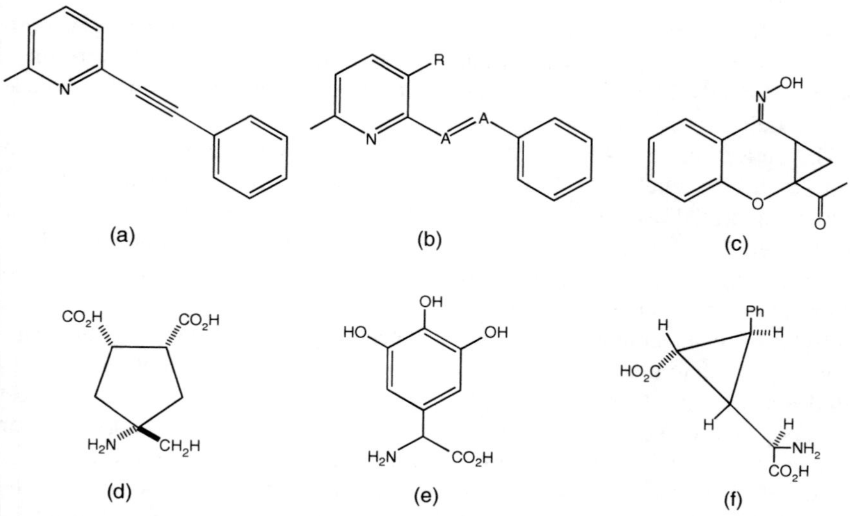

Fig. 2.7 Structures of selective group I, group II, and group III mGlu receptor antagonists. MPEP (a); SIB-1893 where R=H and A=CH (b); CPCCOEt where R=OEt (c); ACPT-II (d); THPG (e); and PCCG-13 (f)

mGluR7 (Bates et al., 2002). mGluR3 mRNA is highly expressed in neurons of the thalamic reticular nucleus. Metabotropic effects of some excitatory amino acid agonists do not fit into the currently known classification scheme of metabotropic receptors. It is likely that this is due to unknown metabotropic receptors that remain to be discovered and classified.

(1S,3R)-ACPD acts as an agonist for most of the known metabotropic receptors with a varying degree of potency. That makes it very difficult to draw conclusions about the physiological role(s) of the various metabotropic receptors when agonists are applied to complex neural systems, which almost certainly contain a variety of receptors at different pre- and postsynaptic loci. These difficulties are exacerbated by the lack of selective competitive antagonists to the metabotropic receptors. Several agonists and antagonists for group I, II, and III metabotropic receptors have been discovered (Figs. 2.6 and 2.7). The pharmacological and neurochemical properties of group I metabotropic receptors and the role of metabotropic receptors in synaptic function are studied with these compounds. In general, metabotropic glutamate receptors (mGluR) regulate neuronal excitability and synaptic strength (Bortolotto et al., 1999b).

2.3 Glutamate Receptors and Glutamate-Mediated Neural Cell Death

The mechanisms underlying glutamate and its analog-induced neuronal damage are quite complex. The molecular sequence of these events is not fully understood. The mechanisms include inhibition of mitochondrial function and increases in intracellular calcium ion, due to depolarization, and free radical generation and the formation of 4-hydroxynonenal (4-HNE). These processes result in apoptotic as well as necrotic cell death during glutamate and its analog-induced neurotoxicity.

The activation of KAR produces membrane depolarization. The result is an increase in intracellular Ca^{2+} concentration (Meldrum and Garthwaite, 1990). In neuronal and glial cells, glutamate and its analogs induce cell death. This is abolished when Ca^{2+} is removed from the culture medium, indicating that calcium ion influx triggers the excitotoxic cascade (Choi, 1988; Weiss and Sensi, 2000). Glutamate and its analogs induce Na^{+} loading and reversal of the membrane Na^{+}/Ca^{2+} exchanger in neurotoxicity mediated by glutamate and its analogs. Thus, an influx of calcium and sodium ions is required to trigger the neuronal and glial cell death cascade (Brorson et al., 1994). Nuclear microscopic studies also indicate a steady increase in calcium ion at 1, 2, and 3 weeks post-KA or NMDA injections (Ong et al., 1999). This increase in calcium ion may be responsible for the stimulation of many Ca^{2+}-dependent enzymes including phospholipase A_2 (Sandhya et al., 1998), calpain II (Ong et al., 1997), and endonucleases during KA-induced neurotoxicity.

Glutamate and its analogs also produce an increase in nitric oxide as measured by increased levels of citrulline (Farooqui et al., 2004). This increase in citrulline precedes the decrease in high-energy metabolites. Pretreatment with N-*tert*-butyl-α-phenylnitrone and vitamin E prevents the changes in citrulline and energy levels mediated by glutamate analogs. Collective evidence thus suggests that neurodegeneration mediated by glutamate and its analogs is brought about by prolonged and repeated depolarization resulting in changes in membrane permeability and irreversible alterations in ion homeostasis.

Cholesterol and 7-ketocholesterol levels also markedly increase during neurodegeneration induced by glutamate and its analogs (Ong et al., 2003). Cholesterol is a multifaceted molecule. It is an essential membrane component, a cofactor for signaling molecules, and a precursor of steroid hormones. Alterations in cholesterol and its metabolites may not only influence neural membrane properties, but also affect synapse development and stability (Pfrieger, 2003). Collectively, alterations in cholesterol metabolism mediated by glutamate and its analogs may have major effects on neural cell survival.

References

Bates B., Xie Y., Taylor N., Johnson J., Wu L., Kwak S., Blatcher M., Gulukota K., and Paulsen J. E. (2002). Characterization of mGluR5R, a novel, metabotropic glutamate receptor 5-related gene. Brain Res. Mol. Brain Res. 109:18–33.

Ben-Ari Y. and Cossart R. (2000). Kainate, a double agent that generates seizures: two decades of progress. Trends Neurosci. 23:580–587.

Bonsi P., Cuomo D., De Persis C., Centonze D., Bernardi G., Calabresi P., and Pisani A. (2005). Modulatory action of metabotropic glutamate receptor (mGluR) 5 on mGluR1 function in striatal cholinergic interneurons. Neuropharmacology 49(Suppl 1):104–113.

Bortolotto Z. A., Clarke V. R., Delany C. M., Parry M. C., Smolders I., Vignes M., Ho K. H., Miu P., Brinton B. T., Fantaske R., Ogden A., Gates M., Ornstein P. L., Lodge D., Bleakman D., and Collingridge G. L. (1999a). Kainate receptors are involved in synaptic plasticity. Nature 402:297–301.

Bortolotto Z. A., Fitzjohn S. M., and Collingridge G. L. (1999b). Roles of metabotropic glutamate receptors in LTP and LTD in the hippocampus. Curr. Opin. Neurobiol. 9:299–304.

Brorson J. R., Manzolillo P. A., and Miller R. J. (1994). Ca^{2+}entry via AMPA/KA receptor and excitotoxicity in cultured cerebellar Purkinje cells. J. Neurosci. 14:187–197.

Chittajallu R., Braithwaite S. P., Clarke V. R. J., and Henley J. M. (1999). Kainate receptors: subunits, synaptic localization and function. Trends Pharmacol. Sci. 20:26–35.

Choi D. W. (1988). Glutamate neurotoxicity and diseases of the nervous system. Neuron 1: 628–634.

Christensen J. K., Paternain A. V., Selak S., Ahring P. K., and Lerma J. (2004). A mosaic of functional kainate receptors in hippocampal interneurons. J. Neurosci. 24:8986–8993.

Chu Z. and Hablitz J. J. (2000). Quisqualate induces an inward current via mGluR activation in neocortical pyramidal neurons. Brain Res. 879:88–92.

Collingridge G. L. and Bliss T. V. P. (1987). NMDA receptors-their role in long-term potentiation. Trends Neurosci. 10:288–293.

Endoh T. (2004). Characterization of modulatory effects of postsynaptic metabotropic glutamate receptors on calcium currents in rat nucleus tractus solitarius. Brain Res. 1024:212–224.

Fagg G. E. and Baud J. (1988). Characterization of NMDA receptor-ionophore complexes in the brain. In: Lodge D. (ed.), *Excitatory Amino Acids in Health and Disease*. John Wiley & Sons, New York, pp. 63–90.

Farooqui A. A., Ong W. Y., and Horrocks L. A. (2004). Neuroprotection abilities of cytosolic phospholipase A_2 inhibitors in kainic acid-induced neurodegeneration. Curr. Drug Targets Cardiovasc. Haematol. Disord. 4:85–96.

Frandsen A., Krogsgaard-Larsen P., and Schousboe A. (1990). Novel glutamate receptor antagonists selectively protect against kainic acid neurotoxicity in cultured cerebral cortex neurons. J. Neurochem. 55:1821–1823.

Guilarte T. R., Miceli R. C., and Jett D. A. (1995). Biochemical evidence of an interaction of lead at the zinc allosteric sites of the NMDA receptor complex: effects of neuronal development. Neurotoxicology. 16:63–71.

Holopainen I., Enkvist M. O. K., and Akerman K. E. O. (1989). Glutamate receptor agonists increase intracellular Ca^2+ independently of voltage-gated Ca^2+ channels in rat cerebellar granule cells. Neurosci. Lett. 98:57–62.

Holopainen I., Louve M., Enkvist M. O. K., and Akerman K. E. O. (1990). Coupling of glutamatergic receptors to changes in intracellular Ca^2+ in rat cerebellar granule cells in primary culture. J. Neurosci. Res. 25:187–193.

Huang E. P. (1997). Metal ions and synaptic transmission: think zinc. Proc. Natl. Acad. Sci. USA 94:13386–13387.

Kidd F. L., Coumis U., Collingridge G. L., Crabtree J. W., and Isaac J. T. (2002). A presynaptic kainate receptor is involved in regulating the dynamic properties of thalamocortical synapses during development. Neuron 34:635–646.

Lerma J. (1997). Kainate reveals its targets. Neuron 19:1155–1158.

Lerma J., Paternain A. V., Rodriguez-Moreno A., and Lopez-Garcia J. C. (2001). Molecular physiology of kainate receptors. Physiol. Rev. 81:971–998.

Lester R. A. J., Herron C. E., Coan E. J., and Collingridge G. L. (1988). The role of NMDA receptors in synaptic plasticity and transmission in the hippocampus. In: Lodge D. (ed.), *Excitatory Amino Acids in Health and Disease*. John Wiley & Sons Ltd., New York, pp. 275–295.

MacDermott A. B., Mayer M. L., Westbrook G. L., Smith S. J., and Barker J. L. (1986). NMDA-receptor activation increases cytoplasmic calcium concentration in cultured spinal cord neurones. Nature 321:519–522.

Managhan D. T., Holets V. R., Toy D. W., and Cotman C. W. (1983). Anatomical distribution of four pharmacologically distinct 3H-L-glutamate binding sites. Nature 306:176–179.

Mayer M. L. and Westbrook G. L. (1987). The physiology of excitatory amino acids in the vertebrate central nervous system. Prog. Neurobiol. 28:197–276.

Meldrum B. and Garthwaite J. (1990). Excitatory amino acid neurotoxicity and neurodegenerative disease. Trends Pharmacol. Sci. 11:379–387.

Michaelis E. K. (1998). Molecular biology of glutamate receptors in the central nervous system and their role in excitotoxicity, oxidative stress and aging. Prog. Neurobiol. 54:369–415.

Monaghan D. T., Bridges R. J., and Cotman C. W. (1989). The excitatory amino acid receptors: Their classes, pharmacology, and distinct properties in the function of the central nervous system. Annu. Rev. Pharmacol. Toxicol. 29:365–402.

Murphy S. N. and Miller R. J. (1988). A glutamate receptor regulates Ca^2+ mobilization in hippocampal neurons. Proc. Natl. Acad. Sci. USA 85:8737–8741.

Olney J. W. (1989). Excitotoxicity and N-methyl-D-aspartate receptors. Drug Dev. Res. 17: 299–319.

Olverman H. J., Monaghan D. T., Cotman C. W., and Watkins J. C. (1986). [3H]CPP, a new competitive ligand for NMDA receptors. Eur. J. Pharmacol. 131:161–162.

Ong W. Y., He Y., Suresh S., and Patel S. C. (1997). Differential expression of apolipoprotein D and apolipoprotein E in the kainic acid-lesioned rat hippocampus. Neuroscience 79:359–367.

Ong W. Y., Ren M. Q., Makjanic J., Lim T. M., and Watt F. (1999). A nuclear microscopic study of elemental changes in the rat hippocampus after kainate-induced neuronal injury. J. Neurochem. 72:1574–1579.

Ong W. Y., Goh E. W. S., Lu X. R., Farooqui A. A., Patel S. C., and Halliwell B. (2003). Increase in cholesterol and cholesterol oxidation products, and role of cholesterol oxidation products in kainate-induced neuronal injury. Brain Path. 13:250–262.

Ozawa S., Iino M., and Tsuzuki K. (1991). Two types of kainate response in cultured rat hippocampal neurons. J. Neurophysiol. 66:2–11.

Paternain A. V., Herrera M. T., Nieto M. A., and Lerma J. (2000). GluR5 and GluR6 kainate receptor subunits coexist in hippocampal neurons and coassemble to form functional receptors. J. Neurosci. 20:196–205.

Pedregal C., Collado I., Escribano A., Ezquerra J., Dominguez C., Mateo A. I., Rubio A., Baker S. R., Goldsworthy J., Kamboj R. K., Ballyk B. A., Hoo K., and Bleakman D. (2000). 4-Alkyl- and 4-cinnamylglutamic acid analogues are potent GluR5 kainate receptor agonists. J. Medicinal Chem. 43:1958–1968.

Pérez-Otaño I. and Ehlers M. D. (2004). Learning from NMDA receptor trafficking: clues to the development and maturation of glutamatergic synapses. Neurosignals 13:175–189.

Petrou S., Ordway R. W., Singer J. J., and Walsh J. V. (1993). A putative fatty acid-binding domain of the NMDA receptor. Trends Biochem. Sci. 18:41–42.

Pfrieger F. W. (2003). Role of cholesterol in synapse formation and function. Biochim. Biophys. Acta Biomembr. 1610:271–280.

Ransom R. W. and Stec N. L. (1988). Cooperative modulation of [3H]-MK-801 binding to the N-methyl-D-aspartate receptor-ion channel complex by L-glutamate, glycine, and polyamines. J. Neurochem. 51:830–836.

Roger G., Dollé F., de Bruin B., Liu X., Besret L., Bramoullé Y., Coulon C., Ottaviani M., Bottlaender M., Valette H., and Kassiou M. (2004). Radiosynthesis and pharmacological evaluation of $[^{11}C]$EMD-95885: a high affinity ligand for NR2B-containing NMDA receptors. Bioorg. Med. Chem. 12:3229–3237.

Ruel J., Guitton M. J., and Puell J. L. (2002). Negative allosteric modulation of AMPA-preferring receptors by the selective isomer GYKI 53784 (LY303070), a specific non-competitive AMPA antagonist. CNS Drug Rev. 8:235–254.

Sandhya T. L., Ong W. Y., Horrocks L. A., and Farooqui A. A. (1998). A light and electron microscopic study of cytoplasmic phospholipase A_2 and cyclooxygenase-2 in the hippocampus after kainate lesions. Brain Res. 788:223–231.

Schoepp D., Bockaert J., and Sladeczek F. (1990). Pharmacological and functional characteristics of metabotropic excitatory amino acid receptors. Trends Pharmacol. Sci. 11:508–515.

Singh L., Oles R., and Woodruff G. (1990). *In vivo* interaction of a polyamine with the NMDA receptor. Eur. J. Pharmacol. 180:391–392.

Sperk G. (1994). Kainic acid seizures in the rat. Prog. Neurobiol. 42:1–32.
Wang Q., Yu S., Simonyi A., Sun G. Y., and Sun A. Y. (2005). Kainic acid-mediated excitotoxicity as a model for neurodegeneration. Mol. Neurobiol. 31:3–16.
Weiss J. H. and Sensi S. L. (2000). Ca^{2+}-Zn^{2+} permeable AMPA or kainate receptors: possible key factors in selective neurodegeneration. Trends Neurosci. 23:365–371.
Westbrook G. L. and Mayer M. L. (1987). Micromolar concentrations of Zn++ antagonize NMDA and GABA responses of hippocampal neurones. Nature 328:640.
Wood P. L., Rao T. S., Iyengar S., Lanthorn T., Monahan J., Cordi A., Sun E., Vazquez M., Gray N., and Contreras P. (1990). A review of the in vitro and *in vivo* neurochemical characterization of the NMDA/PCP/glycine/ion channel receptor macrocomplex. Neurochem. Res. 15:217–230.

Chapter 3
Multiplicity of Glutamate Receptors in Brain

Glutamate is the major excitatory neurotransmitter in the central nervous system (CNS) and mediates its actions via activation of both ionotropic and metabotropic receptor families (Fig. 3.1). The agonist-binding site of glutamate receptors (GluR) consists of amino acid residues in the extracellular S1 and S2 domains in the N-terminal and M3-M4 loop regions, respectively. The interaction of glutamate and its analogs with receptors can be characterized by two fundamental properties, affinity and efficacy. Affinity defines how tightly glutamate and its analogs associate with its receptor, and efficacy measures the ability of the bound glutamate and its analog to activate the receptor. Although affinity and efficacy are independent properties, the binding and activation processes that they describe are tightly coupled. This strong coupling has complicated the interpretation of concentration-response phenotypes caused by receptor mutations. The development of selective ligands, including competitive agonists and antagonists and positive and negative allosteric modulators, has enabled investigation of the functional roles of glutamate receptor family members.

3.1 Structure and Distribution of NMDA Receptor Subunits in Brain

NMDA receptors are multimeric associations of NMDAR1 (NR1), NMDAR2, and NMDAR3 (NR2A-NR2D and NR3A-B) subunits that form ligand-gated ion channels (Ibrahim et al., 2000; Schrattenholz and Soskic, 2006). These subunits are essential for the formation of functional channels. The amino acid sequence of C-terminal regions of the NR2A and NR2B subunits resemble each other with 30% identity. The amino acid sequence of C-terminal region of NR1 shows poor homology with NR2. NR1 and NR2 subunits have been cloned and characterized physicochemically as well as pharmacologically (Moriyoshi et al., 1991; Monyer et al., 1992; Ishii et al., 1993).

NR1 is a component of all native NMDA receptors. The NRI subunit is the glycine binding subunit. It exists as 8 splice variants of a single gene. C-terminal deletion of the NR1 subunit causes NMDAR inactivation, changes in downstream

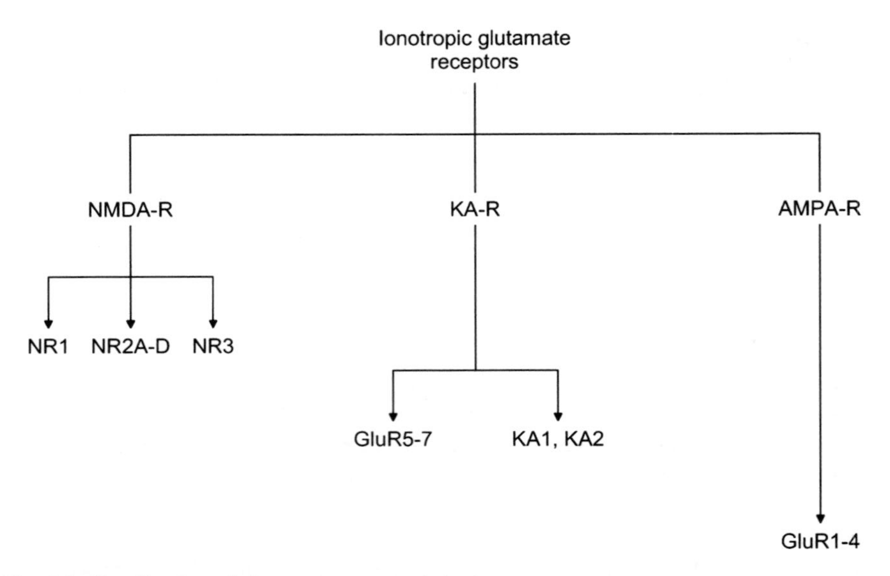

Fig. 3.1 Classification of glutamate receptors in brain

signaling, and gene expression. This deletion has no effect on global increases in intracellular Ca^{2+}. This results in increased Ca^{2+} throughout the neuron. NMDAR-mediated signaling requires coupling through the NR1 C-terminus. The gene encoding the NR1 subunit contains 21 exons, three of which are alternatively spliced to generate at least seven NR1 splice variants, NR1a-NR1g) (Baethmann, 1990). They are formed from the insertion or deletion of three short exon cassettes in the NH_2-terminal (N1), and COOH-terminal (C1, C2) regions of the NR1 molecules. Pharmacologically, NR1 splice variants differ from each other in their spatial and temporal expression patterns and sensitivity to phorbol esters, polyamines, protons, and Zn^{2+} (Baethmann, 1990; Laurie and Seeburg, 1994; Ehlers et al., 1996). Both the gene sequence and the exon-intron organization of the human NR1 gene are similar to those of the rat gene, with exception of the deletion of rat exon #3 (Hynd et al., 2001). Exon #4 encodes a 21-amino acid cassette (N1) within the amino-terminal domain, while exons #20 and #21 encode two consecutive 37- ('C1') and 38-amino acid ('C2') cassettes at the carboxy terminus (Hynd et al., 2001). Deletion of the cassette potentiates responses to polyamines and Zn^{2+}.

Calmodulin (CaM), a Ca^{2+}-binding protein, binds directly to the NR1 subunit of the NMDA receptor in a Ca^{2+}-dependent manner. CaM binding occurs at two distinct sites in the COOH –terminal region of NR1 molecule. One CaM binding site is contained within the spliced C1 exon cassette, whereas the other is located in a region common to all NR1 splice variants (Ehlers et al., 1996). Collective evidence suggests that NMDA receptor function can be modulated by CaM binding to the NR1 subunit and this process may be related to activity-dependent feedback inhibition and Ca^{2+}-dependent inactivation of NMDA receptors.

NR2 is the glutamate binding subunit. This subunit also contains the polyamine-binding site. It is generated as the product of four distinct genes (Yamakura and Shimoji, 1999). The NR2 subunit is responsible for most of the structural heterogeneity in NMDA receptors (Lynch and Guttmann, 2001). When NR2 is co-expressed with NR1, it forms native NMDA-receptor-channel complexes (Yamakura and Shimoji, 1999). The different NR2 subunits appear to confer different physiological and pharmacological properties on the receptors. For example, NR1-NR2C channels are more sensitive to Mg^{2+} blockade and display the highest affinity sites for glycine binding compared to other heteromeric channels, whereas the NR1-NR2A channel differs from the others in its response to reducing agents. The NR1 subunit is distantly related to the NR2 subunits with 18% sequence homology (Monyer et al., 1992). Collectively these studies suggest that at the structural level NMDA receptors have an enormous flexibility due to seven genes, alternative splicing, RNA-editing, and extensive posttranslational modifications, including phosphorylation and glycosylation (Schrattenholz and Soskic, 2006). Differential localization and functional regulation of NMDA receptors may depend on neuregulins and receptor tyrosine kinases in cholesterol-rich membrane domains (lipid rafts), Ca^{2+}-related mitochondrial feedback-loops and subsynaptic structural elements like PSD-95, a post-synaptic density protein of 95 kDa (Schrattenholz and Soskic, 2006).

NMDA receptors are clustered at postsynaptic sites where cytoplasmic adapter proteins anchor them to the cytoskeleton. Some adapter proteins contain a PDZ domain while others lack this. The PDZ domain is a protein-protein interaction motif of approximately 90 amino acids, which in most cases, binds their target proteins at the C-terminus ends. For example a PDZ domain protein, PSD-95, interacts with the C-terminus of the NR2 subunit. Its interaction with tubulin-based cytoskeletal protein is mediated by a novel postsynaptic protein called CRIPT (Niethammer et al., 1998). Thus NR1 and NR2 subunits and their related proteins are crucial for efficient gating of NMDA receptor channels.

The NR1 subunit is ubiquitous throughout brain tissue, whereas there is a differential distribution of NR2 subunits. NR2C expression levels are high in the granule cell layer of the cerebellum, but low elsewhere. NR2A and NR2B are found in the thalamus, although NR2A is distributed more prominently in the lateral thalamic nuclei, especially the ventrobasal complex, and NR2D is expressed early during development rather than in the adult (Ibrahim et al., 2000). During brain development, the NR2 subunit heterogeneity of NMDA receptors results from differences in the structure of ligand binding regions, as well as structural differences between subtypes in a modulatory region called the LIVBP-like domain. This region in NR1 and NR2B controls the action of NR2B-selective drugs like ifenprodil (Borza and Domany, 2006), whilst this domain in receptors containing the NR2A subunit controls the action of NR2A-selective modulators such as Zn^{2+} (Frederickson et al., 2004). The proximal amino-terminal domains (ATD) of NMDA receptors constitute many modulatory binding sites that may serve as potential drug targets (Wang et al., 2005). There are few biochemical and structural data on the ATD of NMDA receptors, as it is difficult to produce the functional proteins. Several methods have been optimized to reconstitute the insoluble recombinant ATD of the NMDA receptor NR2B subunit (ATD2B) through productive refolding of 6xHis-ATD2B protein from inclusion bodies. Circular dichroism and dynamic light scattering characterizations revealed that the solubilized and refolded 6xHis-ATD2B adopted well-defined secondary structures and monodispersity. More significantly, the soluble

6xHis-ATD2B specifically bound ifenprodil to saturation (Wang et al., 2005). These studies provide information not only on the ATD domain of the NR2B subunit as a promising therapeutic target but also provide a framework for designing structurally novel NR2B-selective antagonists.

The mRNA encoding the NR3 subunit of the NMDA receptor was identified in rat brain (Sun et al., 1998). It contains a 60-bp insertion at the nucleotide position 3007 in the intracellular domain of the C-terminal. The NR3 mRNA exists in at least two variant forms—with the insert, NR3-long; NR3-l, and without the insert, NR3-short; NR3-s. The NR3-l variant is expressed throughout the adult rat brain. Moreover, this variant predominates in the occipital and entorhinal cortices, thalamus, and cerebellum. Analysis of NR3-l development indicates that it is regulated in a region-specific manner. The NR3 subunit can be coexpressed with NR1/NR2 subunits. 100 mM ethanol inhibits all NMDA receptor subunit combinations. The inhibition of NR1/NR2B receptors by the NR2B subunit-selective antagonist, ifenprodil, is not altered by co-expression of the NR3 subunit. Overall, these results suggest that the NR3A subunit is not a determinant of ethanol sensitivity in recombinant NMDA receptors (Smothers and Woodward, 2003).

The movement of the transmembrane domain M3 within NMDAR subunits may be a structural determinant linking agonist binding to channel gating. Covalent modification of NR1-A652C or the analogous mutation in NR2A, -2B, -2C, or -2D by methanethiosulfonate ethylammonium (MT-SEA) occurs only in the presence of glutamate and glycine. That modification potentiates recombinant NMDA receptor currents (Yuan et al., 2005). The modified channels remain open even after removing glutamate and glycine from the external solution. The degree of potentiation depends on the identity of the NR2 subunit (NR2A < NR2B < NR2C and D) inversely correlating with previous measurements of channel open probability. MTSEA-induced modification of channels is associated with increased glutamate potency, increased mean single-channel open time, and slightly decreased channel conductance (Yuan et al., 2005). Modified channels are insensitive to the competitive antagonists D-2-amino-5-phosphonovaleric acid (APV) and 7-Cl-kynurenic acid, as well as allosteric modulators of gating, extracellular protons, and Zn^{2+}. However, channels remain fully sensitive to Mg^{2+} blockade and partially sensitive to pore block by (+)MK-801, (–)MK-801, ketamine, memantine, amantadine, and dextrorphan. The partial sensitivity to (+)MK-801 may reflect its ability to stimulate agonist unbinding from MT-SEA-modified receptors. These results suggest that the transmembrane domain M3 is a conserved and critical determinant of channel gating in all NMDA receptors (Yuan et al., 2005).

3.2 Structure and Distribution of KA Receptor Subunits in Brain

Kainate receptors (KAR) are heteromeric ionotropic glutamate receptors that play a variety of roles in the regulation of synaptic network activity. KAR are composed of five different subunits designated as GluR5, GluR6, GluR7, KA1, and KA2. These subunits have molecular masses of 100 kDa. The primary structure of KA receptor subunits shows 40% homology with AMPA receptors and 20% homology with NMDA receptors. GluR5, GluR6, and GluR7 show 75–80% homology with each other. KA1 and KA2 have 68% homology. Channels formed by the expression of GluR6 alone are stimulated by KA, but not by α-amino-3-hydroxy-5-methyl-4-isoxazole propionate (AMPA). Homomers of GluR5 form channels that are activated by domoate, KA, glutamate, and AMPA. GluR7 and KA1 and KA2 do not form channels when each protein is expressed alone, but contribute to the formation of heteromeric assemblies of KAR when expressed with GluR5 and GluR6. Co-expression of the GluR5-7 and KA1 and KA2 proteins in different combinations leads to the formation of channels similar to KAR (Mulle et al., 1998, 2000; Chittajallu et al., 1999).

Thus GluR5, GluR6, and GluR7 correspond to the low-affinity KAR whereas KA1 and KA2 correspond to high-affinity KAR. The cDNA for GluR5, GluR6 and GluR7 have 35–40% homology to the AMPA receptor subunits and are components of high affinity KA receptors. It is difficult to distinguish between AMPA and KA receptors on the basis of electrophysiological and pharmacological studies (Michaelis, 1998; Paternain et al., 2000). However, the discovery of 2, 3-benzodiazepines (GYK1 compounds) that specifically block AMPA receptors has now made it easier to study the pharmacology and role of KAR in brain tissue. These studies are also supported by results obtained from KAR knockout mice indicating that KAR are different from AMPA receptors. Thus KAR knockout mice are useful tools for validating the pharmacological and electrophysiological studies on these receptors.

The function of glutamate receptors is highly dependent on their surface density in specific neuronal domains. Alternative splicing is known to regulate surface expression of GluR5 and GluR6 subunits. The KAR subunit GluR7 exists under different splice variant isoforms in the C-terminal domain (GluR7a and GluR7b) (Jaskolski et al., 2005). Trafficking studies on GluR7 splice variants in cultured hippocampal neurons from wild-type and KAR mutant mice indicate that alternative splicing regulates surface expression of GluR7-containing KAR. GluR7a and GluR7b differentially traffic from the endoplasmic reticulum to the plasma membrane. GluR7a is highly expressed at the plasma membrane, and its trafficking is dependent on a stretch of positively charged amino acids also found in GluR6a (Jaskolski et al., 2005). In contrast, GluR7b is detected at the plasma membrane at a low level and retained mostly in the endoplasmic reticulum (ER). The RXR motif of GluR7b does not act as an ER retention motif, at variance with other receptors and ion channels, but might be involved during the assembly process. Like GluR6a, GluR7a promotes surface expression of ER-retained subunit splice variants when assembled in heteromeric KAR. However, positive regulation of KAR trafficking is limited by the ability of different combinations of subunits to form heteromeric receptor assemblies (Jaskolski et al., 2005). These studies further define the complex rules that govern membrane delivery and subcellular distribution of KAR in brain tissue.

According to in situ hybridization studies, KAR are widely expressed in brain. However, KAR subunits display heterogeneous expression patterns throughout the brain tissue. Thus the GluR5 transcript is mainly associated with DRG neurons, subiculum, septal nuclei, piriform and the cingulated cortices, and Purkinje cells of the cerebellum. GluR6 transcript is abundantly expressed in the cerebellar granule cells, dentate gyrus, and the CA3 region of the hippocampus and striatum. GluR7 mRNA is expressed at low levels throughout the brain tissue. KA1 is exclusively expressed in the CA3 region, dentate gyrus, amygdala, and entorhinal cortex. The KA2 message is expressed throughout the mammalian brain. The highly heterogeneous distribution of KAR in various regions of mammalian brain supports the view that these receptors play an important role not only in neurotransmitter release but also in synaptic plasticity.

KAR contribute to presynaptic as well as postsynaptic signaling (Ben-Ari and Cossart, 2000; Kamiya, 2002). The mechanism and physiological function of presynaptic KAR have been studied at the hippocampal mossy fiber-CA3 synapse including their activation modulating neurotransmitter release bi-directionally. Thus weak activation enhances glutamate release, whereas strong activation leads to inhibition (Kamiya, 2002). This observation explains the possible ionotropic action of KA that leads to axonal depolarization, which in turn modulates several voltage-dependent channels associated with action potential-dependent Ca^{2+}entry. This process may be involved in the induction of long-term potentiation (LTP) and long-term depression (LTD) (Nicoll and Malenka, 1995; Kobayashi et al., 1996). Activation of postsynaptic KAR depolarizes interneurons to induce high-frequency firing and thereby inducing massive release of GABA. This process in turn depresses evoked GABAergic transmission indirectly by acting on the presynaptic GABA receptors (Kamiya, 2002). Collectively these studies suggest that presynaptic and postsynaptic KAR contribute to the modulation of synaptic efficacy in the brain tissue.

Hippocampal mossy fibers, the axons of dentate granule cells, form powerful excitatory synapses onto the proximal dendrites of CA3 pyramidal cells. It is well known that KAR containing both GluR6 and KA2 subunits are involved in KAR-mediated EPSCs at mossy fiber synapses on CA3 pyramidal cells (Schmitz et al., 2001). Endogenous glutamate, by activating KAR, reversibly inhibits the slow Ca^{2+}-activated K^{+}current I(sAHP) and increases neuronal excitability through a G-protein-coupled mechanism. Using KAR knockout mice, KA2 is essential for the inhibition of I(sAHP) in CA3 pyramidal cells by low nanomolar concentrations of kainate, in addition to GluR6. In GluR6(-/-) mice, both ionotropic synaptic transmission and inhibition of I(sAHP) by endogenous glutamate released from mossy fibers is lost. In contrast, inhibition of I(sAHP) is absent in KA2(-/-) mice despite the preservation of KAR-mediated EPSCs. These data indicate that the metabotropic action of KAR does not rely on the activation of a KAR-mediated inward current. Biochemical analysis of knockout mice revealed that KA2 is required for the interaction of KAR with Gα (q/11)-proteins known to be involved in I(sAHP) modulation. Ionotropic and metabotropic actions of KAR at mossy fiber synapses are differentially sensitive to the competitive glutamate receptor ligands, kainate and kynurenate. Based on these observations a model has been proposed in which KAR operate in two modes at mossy fiber synapses: through a direct ionotropic action of GluR6, and through an indirect G-protein-coupled mechanism requiring the binding of glutamate to KA2 (Ruiz et al., 2005). Collectively, these studies suggest that KAR containing GluR6 directly increase excitability of CA1 pyramidal cells and help explain the propensity for seizure activity following KAR activation.

3.3 Structure and Distribution of AMPA Receptor Subunits in Brain

AMPA-type glutamate receptors are tetrameric ion channels that mediate fast excitatory synaptic transmission in the mammalian brain. The assembly of GluR-A, -B, -C, and -D (GluR1, GluR2, GluR3, and GluR4) subunits into homo- and heteromeric channels generates AMPA receptors (Fig. 3.2). The GluR-B subunit is dominant in determining functional properties of heteromeric AMPA receptors (Burnashev et al., 1992). This subunit exists in developmentally distinct edited and unedited forms, GluR-B(R) and GluR-B(Q), which differ in a single amino acid in transmembrane segment TM2 (Q/R site). Homomeric GluR-B(R) channels expressed in neural and non-neural cells display a low divalent permeability, whereas homomeric GluR-B(Q) and GluR-D channels exhibit a high divalent permeability. Mutational analysis reveals that both the positive charge and the size of the amino acid side chain located at the Q/R site control the divalent permeability of homomeric channels (Burnashev et al., 1992). Coexpression of Q/R site arginine- and glutamine-containing subunits generates cells with varying divalent permeabilities depending on the amounts of expression vectors used for cell transfection. Intermediate divalent permeabilities can be traced to the presence of both divalent permeant homomeric and impermeant heteromeric channels. The positive charge contributed by the arginine of the edited GluR-B(R) subunit may determine low divalent permeability in heteromeric GluR

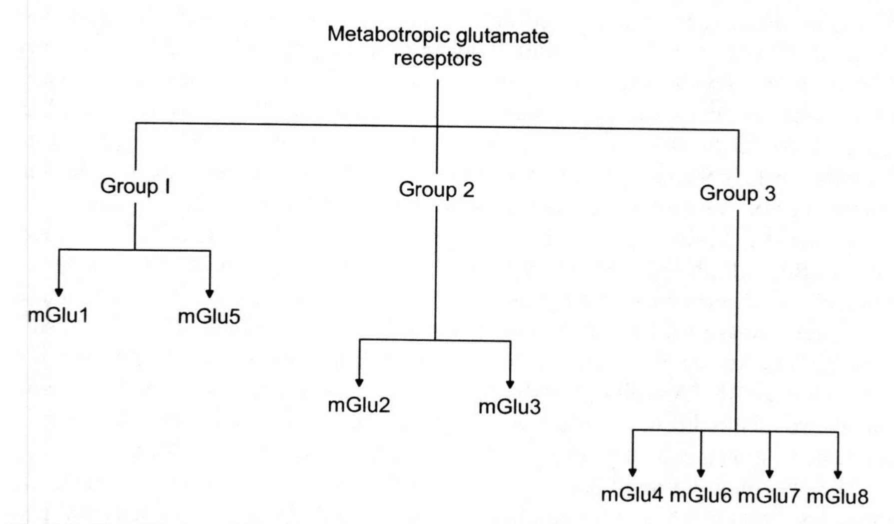

Fig. 3.2 Classification of metabotropic glutamate receptors in brain

channels and changes in GluR-B(R) expression may regulate the AMPA receptor-dependent divalent permeability of a cell (Burnashev et al., 1992).

The differential expression of AMPA receptor subunit mRNA has been studied in the adult rat brain and spinal cord using quantitative RT-PCR with laser capture microdissection. Rat brain expresses the mRNA of all AMPA receptor subunits. The mRNA expression level for GluR2 is the highest in all brain areas and neuronal subsets (Sun et al., 2005). In motor neurons, the levels of GluR2 mRNA and its expression are lowest relative to the other subunits. The unique AMPA receptor expression profile of motoneurons may render them selectively vulnerable to AMPA receptor-mediated excitotoxicity. mRNA encoding GluR 1 and 2 are localized to dendrites of hippocampal neurons and are regulated by paradigms that alter synaptic efficacy. NMDAR activation depletes dendritic levels of AMPAR mRNA. Reduction in mRNA occurs through intracellular Ca^{2+} elevation, activation of extracellular signal-regulated kinase/mitogen-activated protein kinase signaling, and transcriptional arrest at the level of the nucleus (Grooms et al., 2006). The reduction in mRNA is accompanied by a long-lasting reduction in synaptic AMPAR number. This is consistent with reduced synaptic efficacy (Grooms et al., 2006).

Subunit composition determines the function and trafficking of AMPAR (Greger et al., 2006). Mechanisms underlying channel assembly are thus central to the efficacy and plasticity of glutamatergic synapses containing AMPAR. RNA editing at the Q/R site of the GluR2 subunit contributes to the assembly of AMPAR heteromers by attenuating formation of GluR2 homotetramers (Greger et al., 2006). The function of the Q/R may site depend on subunit contacts between adjacent ligand binding domains (LBDs). Changes of LBD interface contacts alter GluR2 assembly properties, forward traffic, and expression at synapses. Interestingly, developmentally regulated RNA editing within the LBD (at the R/G site) produces analogous effects. Collectively these studies show that editing to glycine reduces the self-assembly competence of this critical subunit and slows GluR2 maturation in the endoplasmic reticulum (ER). Therefore, RNA editing sites, located at strategic subunit interfaces, shape AMPA-R assembly and trafficking in a developmentally regulated manner (Greger et al., 2006). AMPA receptor trafficking and channel gating is modulated by a family of small transmembrane AMPA receptor regulatory proteins (TARP). TARP provide the first example of auxiliary subunits of AMPAR that plays an important role in the turnover and life cycle of AMPAR (Nicoll et al., 2006). Recent studies in both heterologous expression systems and cultured neurons have indicated that the AMPAR can be phosphorylated on their subunits (GluR1, GluR2, and GluR4). All phosphorylation sites reside at serine, threonine, or tyrosine on the intracellular C-terminal domain. Several key protein kinases, such as protein kinase A, protein kinase C, Ca^{2+}/calmodulin-dependent protein kinase II, and tyrosine kinases (Trk; receptor or nonreceptor family Trk) are involved in the site-specific regulation of AMPAR phosphorylation. Other glutamate receptors, NMDAR and metabotropic glutamate receptors, also regulate AMPA receptors through a protein phosphorylation mechanism (Simonyi et al., 2005; Vanhoose et al., 2006).

The phosphorylation of serine 845 (Ser-845) in the GluR1 subunit of AMPAR indicates a relationship between phospho-Ser-845, GluR1 surface expression, and synaptic strength in hippocampal neurons (Oh et al., 2006). About 15% of surface

AMPAR in cultured neurons are phosphorylated at Ser-845 basally, whereas chemical potentiation (forskolin/rolipram treatment) persistently increases this to 60%. In contrast, NMDA treatment (chemical depression) decreases it to 10%. These changes in Ser-845 phosphorylation are paralleled by corresponding changes in the surface expression of AMPAR in both cultured neurons and hippocampal slices. For every 1% increase in net phospho-Ser-845, there is a 0.75% increase in the surface fraction of GluR1. Phosphorylation of Ser-845 correlates with a selective delivery of AMPAR to extrasynaptic sites, and their synaptic localization requires coincident synaptic activity. Furthermore, increasing the extrasynaptic pool of AMPA receptors induces a stronger theta burst LTP. These results support a two-step model for delivery of GluR1-containing AMPAR to synapses during activity-dependent LTP, in which Ser-845 phosphorylation can traffic AMPAR to extrasynaptic sites for subsequent delivery to synapses during LTP (Oh et al., 2006).

The phosphorylation of subunits forming AMPAR is regulated by constitutive phospholipase A_2 (PLA_2) activity in rat brain sections. Incubation of rat brain sections with bromoenol lactone (BEL) increases phosphorylation on the Ser-831 residue of the AMPA receptor GluR1 subunit in synaptosomal P2 fractions, whereas $AACOCF_3$ causes increased phosphorylation on residues Ser-880/891 of GluR2/3 subunits. These effects are restricted to the AMPA receptor subtype as no changes in phosphorylation were elicited on the NMDA receptor NR1 subunit. The effects of BEL and $AACOCF_3$ are not occluded during blockade of protein phosphatases since AMPA receptor phosphorylation is also observed in the presence of okadaic acid, indicating that the PLA_2 inhibitor-induced increase in AMPA receptor phosphorylation does not rely on a decrease in dephosphorylation reactions. However, pretreatment of rat brain sections with a cell-permeable protein kinase C (PKC) inhibitor blocks BEL- and $AACOCF_3$-induced phosphorylation on the Ser831 and Ser880/891 sites of GluR1 and GluR2/3 subunits, respectively. Constitutive $cPLA_2$and $iPLA_2$ systems may differentially influence AMPA receptor properties and function in the rat brain through mechanisms involving PKC activity. As a rapid and short-term mechanism, dynamic protein phosphorylation directly modulates the electrophysiological, morphological (externalization and internalization trafficking and clustering), and biochemical (synthesis and subunit composition) properties of the AMPAR, as well as protein-protein interactions between the AMPAR subunits and various intracellular interacting proteins. These modulations underlie the major molecular mechanisms that not only affect many forms of synaptic plasticity, but also play a critical role in synaptic expression of AMPAR, channel properties, and synaptic plasticity (Song and Huganir, 2002; Wang et al., 2005; Vanhoose et al., 2006).

Very little information is available on the molecular mechanisms underlying differences in the ligand binding properties of AMPA, KA, and NMDA subtype glutamate receptors. Crystal structures of the GluR5 and GluR6 kainate receptor ligand binding cores in complexes with glutamate, 2S,4R-4-methylglutamate, KA, and quisqualate (Mayer, 2005) reveal that the ligand binding cavities are 40% (GluR5) and 16% (GluR6) larger than for GluR2. The binding of AMPA- and GluR5-selective agonists to GluR6 is prevented by steric occlusion, which also interferes with the high-affinity binding of 2S,4R-4-methylglutamate to AMPAR.

Strikingly, the extent of domain closure produced by the GluR6 partial agonist KA is only 3 degrees less than for glutamate and 11 degrees greater than for the GluR2 kainate complex. Furthermore, mutations in GluR6 designed to disrupt these interactions reduce agonist apparent affinity, speed up receptor deactivation, and increase the rate of recovery from desensitization. Conversely, introduction of mutations in GluR2 not only causes additional interdomain interactions in the agonist-bound state, increases agonist apparent affinity 15-fold, and slows both deactivation and recovery from desensitization (Weston et al., 2006). These observations together with extensive interdomain contacts between domains 1 and 2 of GluR5 and GluR6, absent from AMPAR, may be responsible for the high stability of GluR5 and GluR6 kainate complexes (Mayer, 2005). It is also proposed that interdomain interactions have evolved as a distinct mechanism that contributes to the unique kinetic properties of AMPAR and KAR.

KAR require not only the neurotransmitter L-glutamate (L-Glu), but also external sodium and chloride ions for activation (Wong et al., 2005). Removal of external ions traps KAR in a novel inactive state that binds L-Glu with picomolar affinity. Moreover, occupancy of KAR by L-Glu precludes external ion binding, demonstrating crosstalk between ligand- and ion-binding sites. AMPAR function normally in the absence of external ions, revealing that even closely related AMPAR subfamilies operate by distinct gating mechanisms. This behavior is interchangeable via a single amino acid residue that operates as a molecular switch to confer AMPA receptor behavior onto KAR. These findings indicate the existence of a novel allosteric site that singles out KAR from all other ligand-gated ion channels (Wong et al., 2006).

3.4 Structure and Distribution of Metabotropic Glutamate Receptor Subunits in Brain

Metabotropic glutamate receptors (mGluR) are divided into three classes based on sequence homologies, mechanisms of signal transduction, as well as pharmacological characteristics. Generally, group I mGluR mediate neuronal excitation whereas activation of groups II and III mGluR decreases synaptic transmission. Group I receptors are coupled to phospholipase C (PLC), which hydrolyzes phosphoinositides and releases Ca^{2+} from intracellular stores, whereas groups II and III mGluR are negatively linked to adenylate cyclase, which catalyzes the production of cyclic adenosine monophosphate (cAMP). Classes of mGlu receptor subtypes include group I, mGlu(1) and mGlu(5); group II, mGlu(2) and mGlu(3); and group III, mGlu(4), mGlu(7) and mGlu(8) receptors. Each metabotropic glutamate receptor is composed of seven transmembrane-spanning domains, similar to other members of the superfamily of metabotropic receptors, which includes noradrenergic, muscarinic acetylcholinergic, dopaminergic, serotonergic, and γ-aminobutyric acid (GABA) type B receptors. The role of mGluR in brain is not fully understood. However, mGluR have multiple actions on neuronal excitability through G-protein-linked modifications of enzymes and ion channels. They act presynaptically to modify glutamatergic and GABAergic transmission and can contribute to long-term changes in synaptic function. It is proposed that these changes specifically modulate excitability within crucial brain structures involved in anxiety states (Swanson et al., 2005).

Immunohistochemical studies also indicate that high levels of PLCβ4 are detected in the somatodendritic domain of neuronal populations expressing the metabotropic glutamate receptor (mGluR) type 1α, including olfactory periglomerular cells, neurons in the bed nucleus anterior commissure, thalamus, substantia nigra, inferior olive, and unipolar brush cells and Purkinje cells in the cerebellum (Nakamura et al., 2004). Low to moderate levels can be detected in many other mGluR1α-positive neurons and in a few mGluR1α-negative neurons. In Purkinje cells, immunogold electron microscopic studies localize PLCβ4 to the perisynapse, at which mGluR1α is concentrated, and to the smooth endoplasmic reticulum in dendrites and spines, an intracellular Ca^{2+} store gated by Ins-P_3 receptors. In the cerebellum, immunoblots demonstrate a concentrated distribution in the post-synaptic density and microsomal fractions, where mGluR1α and type 1 Ins-P_3 receptor were also greatly enriched. Furthermore, PLCβ4 forms co-immunoprecipitable complexes with mGluR1α, type 1 Ins-P_3 receptor, and Homer 1. These results suggest that PLCβ4 is preferentially localized in the perisynapse and smooth endoplasmic reticulum as a component of the physically linked phosphoinositide signaling complex (Nakamura et al., 2004).

Age-dependent changes in the expression of group I and II metabotropic glutamate (mGlu) receptors occur in thalamic nuclei and the hippocampal CA3 region (Simonyi et al., 2005). However, a slight decrease in mGlu1a receptor mRNA expression in individual Purkinje neurons and a decline in cerebellar mGlu1a receptor protein levels were detected in aged animals. In contrast, mGlu1b receptor mRNA levels are increased in the cerebellar granule cell layer. Although mGlu5 receptor mRNA expression decreases in many regions, its protein expression remained unchanged during aging. Compared to the small changes in mGlu2 receptor mRNA levels, mGlu3 receptor mRNA levels show substantial age differences. An increased mGlu2/3 receptor protein expression has found in the frontal cortex, thalamus, hippocampus, and corpus callosum in aged animals. These results demonstrate region- and subtype-specific, including splice variant specific changes in the expression of mGlu receptors in the brain with increasing age (Simonyi et al., 2005).

Collective evidence indicates that mGluR play multiple roles in synaptic plasticity. LTP induction, potentiation of NMDA receptor activity in CA1 neurons, and hippocampal NMDA receptor-dependent LTD involve mGlu receptors.

References

Baethmann A. (1990). Pathophysiology of acute brain damage following epilepsy. Acta Neurochir. Suppl. (Wien.) 50:14–18.

Ben-Ari Y. and Cossart R. (2000). Kainate, a double agent that generates seizures: two decades of progress. Trends Neurosci. 23:580–587.

Borza I. and Domany G. (2006). NR2B selective NMDA antagonists: the evolution of the ifenprodil-type pharmacophore. Curr. Top. Med. Chem. 6:687–695.

Burnashev N., Monyer H., Seeburg P. H., and Sakmann B. (1992). Divalent ion permeability of AMPA receptor channels is dominated by the edited form of a single subunit. Neuron 8:189–198.

Chittajallu R., Braithwaite S. P., Clarke V. R. J., and Henley J. M. (1999). Kainate receptors: subunits, synaptic localization and function. Trends Pharmacol. Sci. 20:26–35.

Ehlers M. D., Zhang S., Bernhadt J. P., and Huganir R. L. (1996). Inactivation of NMDA receptors by direct interaction of calmodulin with the NR1 subunit. Cell 84:745–755.

Frederickson C. J., Maret W., and Cuajungco M. P. (2004). Zinc and excitotoxic brain injury: a new model. Neuroscientist 10:18–25.

Greger I. H., Akamine P., Khatri L., and Ziff E. B. (2006). Developmentally regulated, combinatorial RNA processing modulates AMPA receptor biogenesis. Neuron 51:85–97.

Grooms S. Y., Noh K. M., Regis R., Bassell G. J., Bryan M. K., Carroll R. C., and Zukin R. S. (2006). Activity bidirectionally regulates AMPA receptor mRNA abundance in dendrites of hippocampal neurons. J. Neurosci. 26:8339–8351.

Hynd M. R., Scott H. L., and Dodd P. R. (2001). Glutamate$_{NMDA}$ receptor NR1 subunit mRNA expression in Alzheimer's disease. J. Neurochem. 78:175–182.

Ibrahim H. M., Healy D. J., Hogg A. J., Jr., and Meador-Woodruff J. H. (2000). Nucleus-specific expression of ionotropic glutamate receptor subunit mRNAs and binding sites in primate thalamus. Brain Res. Mol. Brain Res. 79:1–17.

Ishii T., Moriyoshi K., Sugihara H., Sakurada K., Kadotani H., Yokoi M., Akazawa C., Shigemoto R., Mizuno N., Masu M., and Nakanashi S. (1993). Molecular characterization of the family of the N-methyl-D-aspartate receptor subunits. J. Biol. Chem. 268:2836–2843.

Jaskolski F., Normand E., Mulle C., and Coussen F. (2005). Differential trafficking of GluR7 kainate receptor subunit splice variants. J. Biol. Chem. 280:22968–22976.

Kamiya H. (2002). Kainate receptor-dependent presynaptic modulation and plasticity. Neurosci. Res. 42:1–6.

Kobayashi K., Manabe T., and Takahashi T. (1996). Presynaptic long-term depression at the hippocampal mossy fiber-CA3 synapse. Science 273:648–650.

Laurie D. J. and Seeburg P. H. (1994). Regional and developmental heterogeneity in splicing of the rat brain NMDAR1 mRNA. J. Neurosci. 14:3180–3194.

Lynch D. R. and Guttmann R. P. (2001). NMDA receptor pharmacology: perspectives from molecular biology. Curr. Drug Targets 2:215–231.

Mayer M. L. (2005). Crystal structures of the GluR5 and GluR6 ligand binding cores: molecular mechanisms underlying kainate receptor selectivity. Neuron 45:539–552.

Michaelis E. K. (1998). Molecular biology of glutamate receptors in the central nervous system and their role in excitotoxicity, oxidative stress and aging. Prog. Neurobiol. 54:369–415.

Monyer H., Sprengel R., Schoepfer R., Herb A., Higuchi M., Lomeli H., Burnashev N., Sakmann B., and Seeburg P. (1992). Heteromeric NMDA receptors: Molecular and functional distinction of subtypes. Science 256:1217–1221.

Moriyoshi K., Masu M., Ishii T., Shigemoto R., Mizuno N., and Nakanishi S. (1991). Molecular cloning and characterization of the rat NMDA receptor. Nature 354:31–37.

Mulle C., Sailer A., Perez-Otano I., Dickinson-Anson H., Castillo P. E., Bureau I., Maron C., Gage F. H., Mann J. R., Bettler B., and Heinemann S. F. (1998). Altered synaptic physiology and reduced susceptibility to kainate-induced seizures in GluR6-deficient mice. Nature 392:601–605.

Mulle C., Sailer A., Swanson G. T., Brana C., O'Gorman S., Bettler B., and Heinemann S. F. (2000). Subunit composition of kainate receptors in hippocampal interneurons. Neuron 28:475–484.

Nakamura M., Sato K., Fukaya M., Araishi K., Aiba A., Kano M., and Watanabe M. (2004). Signaling complex formation of phospholipase Cβ4 with metabotropic glutamate receptor type 1α and 1,4,5-trisphosphate receptor at the perisynapse and endoplasmic reticulum in the mouse brain. Eur. J. Neurosci. 20:2929–2944.

Nicoll R. A. and Malenka R. C. (1995). Contrasting properties of two forms of long-term potentiation in the hippocampus. Nature 377:115–118.

Nicoll R. A., Tomita S., and Bredt D. S. (2006). Auxiliary subunits assist AMPA-type glutamate receptors. Science 311:1253–1256.

Niethammer M., Valtschanoff J. G., Kapoor T. M., Allison D. W., Weinberg T. M., Craig A. M., and Sheng M. (1998). CRIPT, a novel postsynaptic protein that binds to the third PDZ domain of PSD-95/SAP90. Neuron 20:693–707.

Oh M. C., Derkach V. A., Guire E. S., and Soderling T. R. (2006). Extrasynaptic membrane trafficking regulated by GluR1 serine 845 phosphorylation primes AMPA receptors for long-term potentiation. J. Biol. Chem. 281:752–758.

Paternain A. V., Herrera M. T., Nieto M. A., and Lerma J. (2000). GluR5 and GluR6 kainate receptor subunits coexist in hippocampal neurons and coassemble to form functional receptors. J. Neurosci. 20:196–205.

Ruiz A., Sachidhanandam S., Utvik J. K., Coussen F., and Mulle C. (2005). Distinct subunits in heteromeric kainate receptors mediate ionotropic and metabotropic function at hippocampal mossy fiber synapses. J. Neurosci. 25:11710–11718.

Schmitz D., Mellor J., Frerking M., and Nicoll R. A. (2001). Presynaptic kainate receptors at hippocampal mossy fiber synapses. Proc. Natl. Acad. Sci. USA 98:11003–11008.

Schrattenholz A. and Soskic V. (2006). NMDA receptors are not alone: dynamic regulation of NMDA receptor structure and function by neuregulins and transient cholesterol-rich membrane domains leads to disease-specific nuances of glutamate signalling. Curr. Top. Medicinal Chem. 6:663–686.

Simonyi A., Ngomba R. T., Storto M., Catania M. V., Miller L. A., Youngs B., DiGiorgi-Gerevini V., Nicoletti F., and Sun G. Y. (2005). Expression of groups I and II metabotropic glutamate receptors in the rat brain during aging. Brain Res. 1043:95–106.

Smothers C. T. and Woodward J. J. (2003). Effect of the NR3 subunit on ethanol inhibition of recombinant NMDA receptors. Brain Res. 987:117–121.

Song I. and Huganir R. L. (2002). Regulation of AMPA receptors during synaptic plasticity. Trends Neurosci. 25:578–588.

Sun L., Margolis F. L., Shipley M. T., and Lidow M. S. (1998). Identification of a long variant of mRNA encoding the NR3 subunit of the NMDA receptor: its regional distribution and developmental expression in the rat brain. FEBS Lett. 441:392–396.

Sun H., Kawahara Y., Ito K., Kanazawa I., and Kwak S. (2005). Expression profile of AMPA receptor subunit mRNA in single adult rat brain and spinal cord neurons in situ. Neurosci. Res. 52:228–234.

Swanson C. J., Bures M., Johnson M. P., Linden A. M., Monn J. A., and Schoepp D. D. (2005). Metabotropic glutamate receptors as novel targets for anxiety and stress disorders. Nat. Rev. Drug Discov. 4:131–144.

Vanhoose A. M., Clements J. M., and Winder D. G. (2006). Novel blockade of protein kinase A-mediated phosphorylation of AMPA receptors. J. Neurosci. 26:1138–1145.

Wang J. Q., Arora A., Yang L., Parelkar N. K., Zhang G., Liu X., Choe E. S., and Mao L. (2005). Phosphorylation of AMPA receptors: mechanisms and synaptic plasticity. Mol. Neurobiol. 32:237–249.

Weston M. C., Gertler C., Mayer M. L., and Rosenmund C. (2006). Interdomain interactions in AMPA and kainate receptors regulate affinity for glutamate. J. Neurosci. 26:7650–7658.

Wong E., Ng F. M., Yu C. Y., Lim P., Lim L. H., Traynelis S. F., and Low C. M. (2005). Expression and characterization of soluble amino-terminal domain of NR2B subunit of N-methyl-D-aspartate receptor. Protein Sci. 14:2275–2283.

Wong A. Y., Fay A. M., and Bowie D. (2006). External ions are coactivators of kainate receptors. J. Neurosci. 26:5750–5755.

Yamakura T. and Shimoji K. (1999). Subunit- and site-specific pharmacology of the NMDA receptor channel. Prog. Neurobiol. 59:279–298.

Yuan H., Erreger K., Dravid S. M., and Traynelis S. F. (2005). Conserved structural and functional control of N-methyl-D-aspartate receptor gating by transmembrane domain M3. J. Biol. Chem. 280:29708–29716.

Chapter 4
Glutamate Transporters and Their Role in Brain

Glial and neuronal transport proteins regulate the extracellular glutamate concentration in brain. Glutamate transporters regulate extracellular glutamate concentrations so as to maintain dynamic and high fidelity cell signaling processes in the brain tissue. Some glutamate transporters depend on an electrochemical gradient of sodium ions, whereas other glutamate transporters are Na^+-gradient independent (Danbolt, 2001). Na^+-independent glutamate transporters have a low affinity for glutamate. This poorly characterized uptake system supplies brain tissue with amino acids for metabolic purposes (Erecinska and Silver, 1990). Na^+-dependent glutamate transporters have a high affinity for glutamate and are distributed throughout the brain tissue (Danbolt, 2001).

Five glutamate transporters were identified, characterized, and cloned from brain tissue (Saier, Jr., 1999; Slotboom et al., 1999; Sims and Robinson, 1999) (Fig. 4.1). They include EAAT1 (GLAST), EAAT2 (GLT1), EAAC1 (EAAT3), EAAT4, and EAAT5. EAAT1 and EAAT2 are primarily astrocytic, whereas EAAT3, EAAT4, and EAAT5 are neuronal. EAAT3 is localized to the presynaptic cleft and non-synaptic area of glutamatergic and GABAergic neurons (Rothstein et al., 1994; Conti et al., 1998). EAAT4 is expressed primarily in the cerebellum and EAAT5 is present in retina. EAAT2 predominates quantitatively and is responsible for most of the glutamate uptake activity in juvenile and adult brain tissue. These transporters are homotrimeric complexes. Trimerization of monomers occurs efficiently after synthesis of the individual subunits resulting in stable trimers (Gendreau et al., 2004).

Glutamate transporters possess six to ten transmembrane domains, which may form an aqueous transmembrane pore (Seal and Amara, 1998). Among the five known mammalian EAAT subtypes, the glial carriers, EAAT1 and EAAT2, have the greatest impact on clearance of glutamate released during neurotransmission. These transporters are essential for terminating synaptic transmission as well as for maintaining extracellular glutamate concentration below neurotoxic levels. In addition, brain tissue also contains a cystine-glutamate antiporter (Erecinska and Silver, 1990). This antiporter acts as a cystine transporter that uses the transmembrane gradient of glutamate as a driving force. This transporter may protect neurons against oxidative stress by providing a suitable amount of cysteine to produce glutathione. Glutamate does not cross the blood-brain barrier. Most of the glutamate present

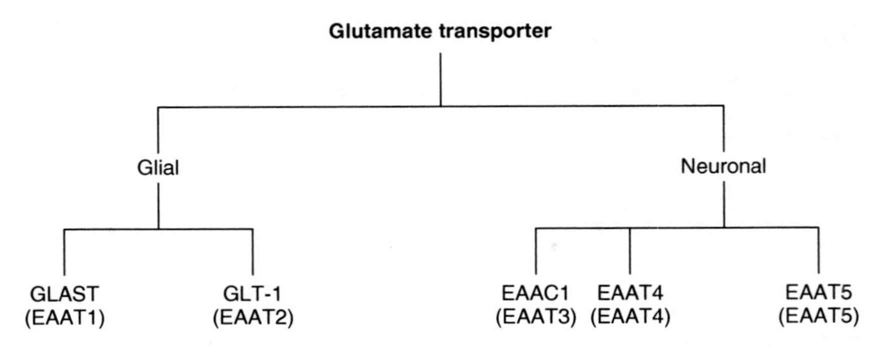

Fig. 4.1 Classification of glutamate transporters in brain. Names in parentheses refer to human glutamate transporters

in the brain is synthesized de novo by astrocytes that have lower levels of cytosolic glutamate than neurons. Astrocytes contain glutamine synthase, an enzyme that converts glutamate into glutamine (Hertz et al., 1999).

These glutamate transport mechanisms provide glutamate to brain tissue for many metabolic activities including synthesis of GABA, glutathione, proteins, and energy production. Glutamate transport mechanisms are modulated by cytokines and growth factors (Danbolt, 2001). Collective evidence suggests that glutamate transporters represent a major mechanism for removal of glutamate from the extracellular fluid (Fig. 4.2). Their activities are important for the long-term maintenance of low and non-toxic concentration of glutamate at the synapse. Although the interactions of glutamate with its transporters decrease the levels of free glutamate in the synaptic cleft, these interactions also slow down the diffusion of glutamate away from the release site. A high density of glutamate transporters on glial cells may provide an opportunity for glutamate to re-enter the synaptic cleft upon dissociation from the glutamate transporter (Danbolt, 2001).

4.1 Astrocytic Glutamate Transporters

EAAT1 and EAAT2 are selective markers for astrocytic plasma membranes. The density of EAAT2 is much higher in hippocampal astrocytes than in cerebellar

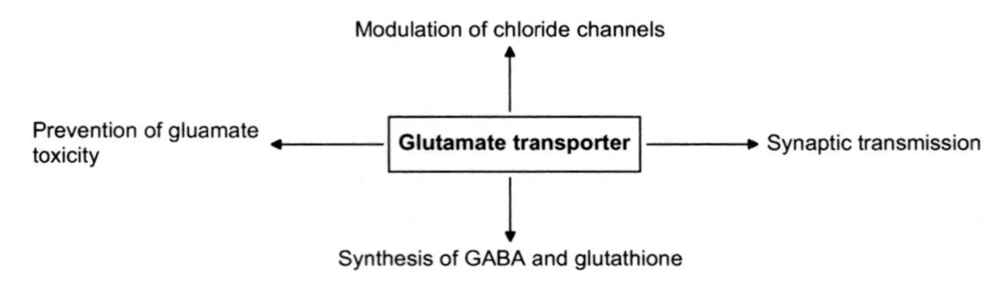

Fig. 4.2 Roles of glutamate transporters in brain

astrocytes (Chaudhry et al., 1995). Astroglial membrane EAAT1 densities are ranked as follows: Bergmann > cerebellar granular layer ≈ hippocampus > cerebellar white matter. Astrocytic membranes facing capillaries, pia, or stem dendrites show a lower density of glutamate transporters than those facing nerve terminals, axons, and spines. Parallel fiber glutamatergic boutons synapsing on interneuron dendritic shafts are surrounded by lower transporter densities than those synapsing on Purkinje cell spines. These results suggest the localization of glutamate transporters is carefully regulated by developmental processes (Chaudhry et al., 1995; Suárez et al., 2000).

EAAT2 antibodies label a broad heterogeneous band of proteins with a maximum density around 73 kDa, whereas the antibody against EAAT1 labels a similarly broad band around 66 kDa in the cerebellum and a few kDa lower in other brain regions (Lehre et al., 1995). EAAT2 is expressed at the highest concentrations in the hippocampus, lateral septum, cerebral cortex, and striatum, whilst EAAT1 is preferentially expressed in the molecular layer of the cerebellum. However, both transporters are present throughout the brain, and have roughly parallel distributions in the cerebral hemispheres and brainstem.

Both EAAT1 and EAAT2 are restricted to astrocytes, but their proportions are different in different regions of brain tissue (Lehre et al., 1995). EAAT1 and EAAT2 have a specific pattern of expression in embryonic and postnatal rat brain (Furuta et al., 1997). EAAT1 immunoreactivity is low prenatally but is enriched in cerebellar Bergmann glia early postnatally and then is also present in forebrain later postnatally. The post-translational modification of EAAT1 is unique among the subtypes; glycosylated EAAT1 is increased with maturation, whereas non-glycosylated EAAT1 is decreased in abundance postnatally. EAAT2 is present in fetal brain and spinal cord, with expression progressively increasing to adult levels throughout the rat neuraxis by postnatal day 26. Transient expression of EAAT2 immunoreactivity along axonal pathways is observed prenatally, in contrast to the exclusive localization of EAAT2 to astrocytes in the adult CNS (Furuta et al., 1997). EAAT2 is regulated by phosphorylation (Casado et al., 1993).

Immunocytochemical studies also indicate the presence of EAAT1 and EAAT2 in microglial cell cultures (Nakajima and Kohsaka, 2004). The EAAT2 activity of microglial cells can be suppressed by dihydrokainic acid, a specific EAAT2 inhibitor. In general, the immunocytochemical distribution of EAAT1 and EAAT2 agrees with the distribution of mRNA for EAAT1 and EAAT2 (Storck et al., 1992; Torp et al., 1994). EAAT1 and EAAT2 may be expressed in the same cells, but in different proportions in different regions of brain tissue. Collective evidence suggests that during CNS development the expression of glutamate transporter subtypes is differentially regulated, regionally segregated, and coordinated in astrocytes and microglial cells. EAAT1 and EAAT2 in astrocytes and microglia protect neurons from glutamate toxicity (Nakajima and Kohsaka, 2004).

Pure astroglial cultures express only EAAT1, whereas astrocytes grown in the presence of cerebellar granule neurons or cortical neurons express both EAAT1 at increased levels and EAAT2 (Gegelashvili et al., 1997). The induction of EAAT2 protein and its mRNA can be reproduced in pure cortical astroglial cultures supplemented with conditioned media from cortical neuronal cultures or from mixed neuron-glia cultures. This treatment does not change the levels of EAAT1. These results suggest that soluble neuronal factors differentially regulate the expression of EAAT2 and EAAT1 in cultured astroglia (Gegelashvili et al., 1997). Further elucidation of the molecular nature of the secreted neuronal factors and corresponding signaling pathways regulating the expression of the astroglial glutamate transporters in vitro may reveal mechanisms important for the understanding and treatment of neurological diseases.

Glutamate transporters maintain a 10,000-fold gradient between extracellular glutamate, 3–10 mM, and intracellular glutamate, 0.3–1.0 μM, which is driven by the ionic gradient. Furthermore, the potential inside neural cells is negative, and glutamate bears a net negative charge. Thus, energy is needed to pump glutamate into neural cells up its electrochemical gradient. This energy is not provided by ATP, but comes from the co-transport of ions moving down their electrochemical gradients. Thus in astrocytes, high affinity glutamate transport is an electrogenic process in which glutamate transporters are coupled to the co-transport of 2–3 sodium ions, 1 proton, and the counter-transport of a potassium ion (Barbour et al., 1988; Kanai et al., 1995; Zerangue and Kavanaugh, 1996) (Fig. 4.3). Thus, at least one net positive charge enters the neural cell per molecule of glutamate transported. This is compensated by an uncoupled flux of chloride ion (Wadiche et al., 1995). The

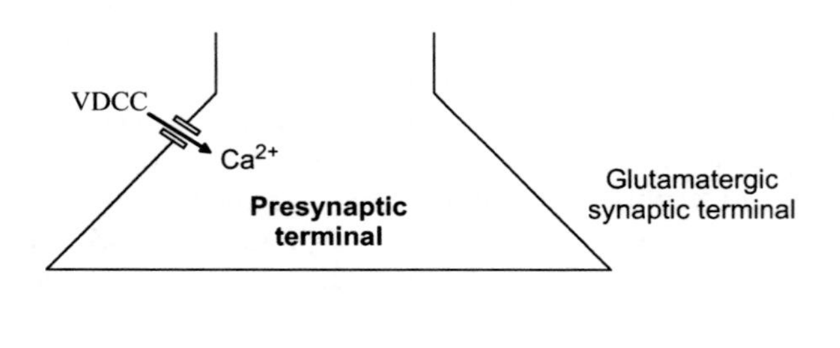

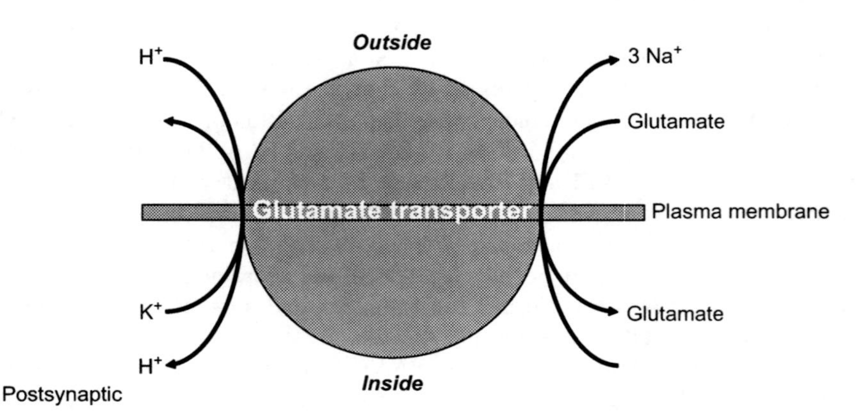

Fig. 4.3 Uptake of glutamate is associated with ion transport across neural membranes. Ion transport results in entry of 3 sodium ions and 1 proton in the cell whilst 1 potassium ion is transported out (modified from Attwell, 2000)

relative magnitude of Cl^- conductance varies with each cloned glutamate receptor subtype; for EAAT1 and EAAT2, the magnitude of the Cl^-current at physiological membrane potentials is similar to that of the electrogenic co-transport current, but the currents generated by EAAT4 (see below) are almost entirely due to the flux of chloride ions. During glutamate transport, Na^+,K^+-ATPase generates an inwardly directed electrochemical Na^+ gradient that is utilized by the transporter to drive the uptake of glutamate. An increase in intracellular Na^+ is accompanied by the stimulation of Na^+,K^+-ATPase, oxygen consumption, and glucose utilization in astrocytes (Eriksson et al., 1995; Stanimirovic et al., 1997). Similarly, ouabain-sensitive stimulation of glycolysis occurs in astrocytes as a result of Na^+-dependent glutamate uptake (Pellerin and Magistretti, 1994). Hence glutamate release from neurons causes an increase in metabolic activity in the surrounding astrocytes. This suggests that neuronal-astrocyte interactions are important for glutamate transport and metabolism in brain tissue.

Little is known about the signaling pathways associated with the expression of EAAT1 and EAAT2. Treatment of astrocytic cultures with dibutyryl-cAMP, epidermal growth factor (EGF), or other growth factors induces the expression of EAAT2 and increases the expression of EAAT1 in astrocytic cultures (Li et al., 2006). The induction of EAAT2 correlates with morphological and biochemical changes that are consistent with astrocytic maturation. Phosphatidylinositol 3-kinase (PtdIns-3K) and the nuclear transcription factor-kappaB (NF-κB) may be involved in the induction of EAAT2 expression (Li et al., 2006). In several signaling systems, Akt (protein kinase B, PKB) functions downstream to PtdIns-3K. Using lentiviral vectors engineered to express dominant-negative (DN), constitutively active (CA), or null variants of Akt, the expression of DN-Akt can attenuate the EGF-dependent induction of EAAT2. Expression of CA-Akt produces a dose- and time-dependent increase in EAAT2 protein, increased EAAT2 mRNA levels, increased dihydrokainate-sensitive (presumably EAAT2 mediated) transport activity, and a change in astrocyte morphology to a more stellate shape, but has no effect on EAAT1 protein levels. Finally, the expression of CA-Akt up-regulates the expression of a reporter construct containing a putative promoter fragment from the human homolog of EAAT2. On the basis of these studies, Akt may induce the expression of EAAT2 through increased transcription and regulate EAAT2 expression without increasing EAAT1 expression in astrocytes (Li et al., 2006).

EAAT2c is a novel carboxyl-terminal splice-variant of the glutamate transporter EAAT2. In rat brain low levels of protein and message are detected, protein expression being restricted to the end feet of astrocytes apposed to blood vessels or to some astrocytes adjacent to the ventricles (Rauen et al., 2004). Conversely, within the retina, this variant is selectively and heavily expressed in the synaptic terminals of both rod and cone photoreceptors in both humans and rats. Double-immunolabeling with antibodies to the carboxyl region of EAAT2b/EAAT2v, which is strongly expressed in apical dendrites of bipolar cells and in cone photoreceptors, reveals that EAAT2c is co-localized with EAAT2b/EAAT2v in rat cone photoreceptors, but not with EAAT2b/EAAT2v in rat bipolar cells. EAAT2c expression is developmentally regulated, appearing only around postnatal day 7 in the rat retina, when photoreceptors first exhibit a dark current (Rauen et al., 2004). Since the glutamate transporter

EAAT5 (see below) is also expressed in terminals of rod photoreceptor terminals, rod photoreceptors express two glutamate transporters with distinct properties. Similarly, cone photoreceptors express two glutamate transporters. Differential usage of these transporters by rod and cone photoreceptors may influence the kinetics of glutamate transmission by these neurons (Rauen et al., 2004).

The functional differences between EAAT1 and EAAT2 illustrate the complexity of the glutamate synapse. For example, EAAT2, but not EAAT1, is inhibited by dihydrokainate, suggesting that anatomic or circuit-dependent differences in glutamate transporter pharmacology may reflect differences in the functional nature of a given glutamate synapse or circuit (Dunlop et al., 1999; Robinson, 1998). The variability in effects of transporter antisense or knockout experiments is consistent with this notion. EAAT2 knockout animals develop lethal seizures, whereas EAAT1 knockout animals have subtler defects, including difficulties with coordination (Rothstein et al., 1996; Tanaka et al., 1997; Watase et al., 1998).

In rats traumatic brain injury is accompanied by alterations in EAAT1 and EAAT2 activity (van Landeghem et al., 2001). Western blot analyses indicate that levels of EAAT1 and EAAT2 are decreased by 40–54% and 42–49% between 24 and 72 h post-trauma. By 8 h CSF glutamate levels are increased to 10.5 μM vs. 2.56 μM in controls; $P < 0.001$, reaching maximum values by 48 h. A significant increase in ramified microglia expressing EAAT1 and EAAT2 de novo can be seen within 4 h. This reaches a stable level by 48 h, and remains high up to 72 h after traumatic injury. Furthermore, traumatic injury is accompanied by a decrease in the expression of EAAT1/EAAT2 that reaches minimum levels within 8 h. This reduction of expression can be due either to protein down-regulation or loss of astrocytes (van Landeghem et al., 2001).

Similarly, cortical lesions also decrease striatal concentrations of the astrocytic glutamate transporter proteins, EAAT1 and EAAT2 (Levy et al., 1995). Because GABA uptake activity is not decreased and glial fibrillary acidic protein is increased, it is proposed that the cortical lesion-mediated decrease in EAAT2 and EAAT1 is not caused by a general impairment of neuronal or glial function. The observed reduction in the two astrocytic glutamate transporters after corticostriatal nerve terminal degeneration indicates dependence of their expression levels on glutamatergic innervation (Levy et al., 1995). Collective evidence indicates that traumatic injury and cortical lesions cause a reduction in astrocytic EAAT1 and EAAT2 protein levels and may contribute to secondary injury, which is accompanied by the stimulation of cytosolic phospholipase A_2, release of arachidonic acid, and generation of eicosanoids (Farooqui et al., 1994) (Fig. 4.4). At a low concentration, arachidonic acid stimulates the glutamate transporter, but at a high concentration it strongly inhibits the transport of glutamate (Nieoullon et al., 2006).

An up-regulation of EAAT1 and EAAT2 immunoreactivities in the CA1 region was observed from 30 min to 12 h after ischemia-reperfusion injury. Their expression begins to decrease at 24 h (Kim et al., 2006; Kosugi and Kawahara, 2006). In contrast, EAAT3 immunoreactivity is transiently reduced in the CA1 region at 30 min after ischemia injury, re-enhanced from 3 to 12 h after ischemia, and re-reduced at 24 h after ischemia. Malfunctions of plasma membrane glutamate transporters may play an important role in the elevation of extracellular glutamate concentrations

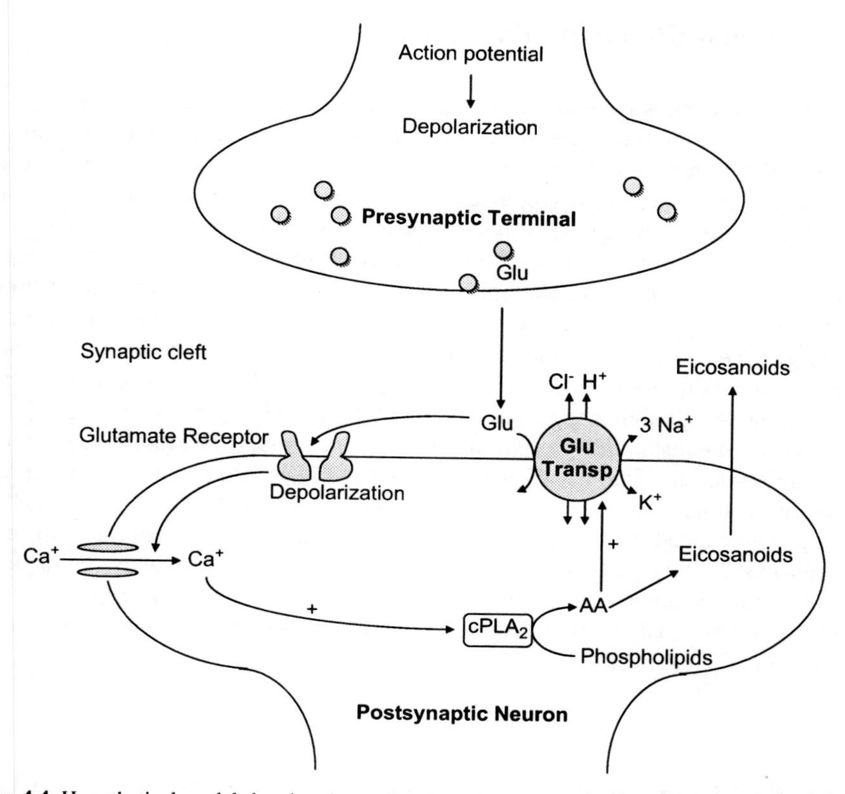

Fig. 4.4 Hypothetical model showing the modulation of glutamate transporter by arachidonic acid. Interactions of glutamate with its receptor result in depolarization and Ca^{2+} entry into the cell. Ca^{2+}-mediated stimulation of PLA_2 results in breakdown of neural membrane phospholipids and the release of arachidonic acid. Arachidonic acid not only modulates proton conductance associated with neuronal excitability, but also provides eicosanoids, which may control the glutamate transporter (modified from Fairman and Amara, 1999)

following ischemic insults (Kim et al., 2006). Collectively these studies suggest that alterations in glutamate transporters after ischemic injury contribute to the accumulation of extracellular glutamate and neuronal death. Neuronal damage is associated with excitotoxicity, a type of cell death triggered by the over-activation of glutamate receptors and the loss of Ca^{2+} homeostasis (Kosugi and Kawahara, 2006; Kim et al., 2006).

EAAT1/EAAT2 double knockout mice show multiple brain defects, including cortical and hippocampal defects, abnormal formation of the neocortex, and olfactory bulb disorganization with perinatal mortality (Matsugami et al., 2006). Furthermore, impairment in other essential aspects of neuronal development, such as stem cell proliferation, radial migration, and neuronal differentiation also occurs. These studies provide direct *in vivo* evidence that EAAT1 and EAAT2 are necessary for brain development through regulation of extracellular glutamate concentrations and show that an important mechanism is likely for the maintenance of glutamate-mediated synaptic transmission (Matsugami et al., 2006).

4.2 Neuronal Glutamate Transporters

Like the glial excitatory amino acid transporters (EEAT1 and EAAT2), neuronal EAAT play an important role in the termination of synaptic transmission and in extracellular glutamate homeostasis in the mammalian CNS. In hippocampus and cerebral cortex, glutamate uptake is attributed to neuronal transporters because in these regions many synapses are not surrounded by astrocytic processes (Bergles et al., 1999), so glutamate transmission is less dependent on the astrocytic contribution. A functional neuronal transporter is assembled as a trimer, of which each monomer appears to transport glutamate independently from the neighboring subunits. EAAT not only sustain a secondary-active glutamate transport, but also function as anion channels. EAAT-interacting proteins that modulate subcellular localization and glutamate transport activity of the EAAT also regulate, in part, the reuptake of glutamate. This process may have a direct impact on glutamatergic neurotransmission. Like glial EAAT, the neuronal EAAT mediate two distinct transport processes, a stoichiometrically coupled transport of glutamate, Na^+, K^+, and H^+, and a pore-mediated anion conductance. Neuronal EAAT may influence glutamatergic transmission by regulating the amount of glutamate available to activate pre- and post-synaptic metabotropic receptors and by altering neuronal excitability through a transporter-associated anion conductance that is activated by carrier substrates (Amara and Fontana, 2002).

4.2.1 EAAT3 in Brain

EAAT3 (EAAC1) has a molecular weight of 64 kDa. Its expression level is quite low in adult brain. It provides a presynaptic glutamate uptake mechanism to terminate the action of released glutamate at glutamatergic synapses. It is also expressed in γ-aminobutyric acid (GABA)-ergic cerebellar Purkinje cells, where it provides glutamate as a precursor for GABA synthesis. Its early expression during brain development, before the astrocytes are functional, suggests that this neuronal glutamate transporter is not only associated with the developmental effects of excitatory amino acids, but also with the biosynthesis of GABA, which is excitatory in nature in different brain regions during the earlier stages of brain development (Nieoullon et al., 2006). EAAT3 protein forms multimers. Homomultimers or heteromultimers can contribute to the maturation of the EAAT3 and its transporter pore. EAAT3 protein has approximately 8 to10 putative transmembrane membrane domains in α-helices. The amino acid sequence of EAAT3 has 50% identity at the carboxy-terminal with other glutamate transporters (Grunewald et al., 2002). Due to the presence of a large hydrophobic stretch near the C terminus (residues 357–444), several models with a different number of membrane spanning regions are conceivable. The N-terminal part of EAAT3 is hydrophilic and lacks a signal peptide. A motif-like cluster of serine residues is present at residues 331–334. Similar clusters have been identified in the ligand-binding sites of receptors for acetylcholine and biogenic amines.

The EAAT3 gene is moderately expressed in brain tissue compared to glial glutamate transporters. It is particularly expressed in the hippocampus, basal ganglia structures, and the olfactory bulb. EAAT3 is found in membranes and the cytosolic compartment (Conti et al., 1998). This is in contrast to EAAT1 and EAAT2, which are primarily located in membranes (Chaudhry et al., 1995).

Besides transporting glutamate, EAAT3 is also involved in the neuronal uptake of cysteine, which is required for the synthesis of glutathione, a major antioxidant. EAAT3 activity is regulated by intracellular signaling pathways involving protein kinase C-α (PKC-α) and phosphatidylinositol-3-kinase (Nieoullon et al., 2006). These regulatory processes may act either at the post-transductional or transcriptional level. Variations in neuronal Glu uptake are associated with rapid changes in the trafficking of the transporter protein altering the membrane location of the transporter. Astrocyte-secreted factors such as cholesterol can produce rapid changes in the location of EAAT3 between the plasma membrane and the cytoplasmic compartment. This process of EAAT3 activity/expression may have implications in the physiopathology of major diseases affecting signaling by excitatory amino acids in the brain (Nieoullon et al., 2006).

Cysteine uptake is the rate-limiting process in glutathione synthesis. Inhibitors of excitatory amino acid transporters (EAAT) significantly enhance glutamate toxicity via depletion of intracellular glutathione. The neuronal glutamate transporter EAAT3 is directly involved in cysteine uptake in cultured neurons. Neuronal cysteine uptake depends on extracellular Na^{+}, is suppressed by EAAT inhibitors, but not by substrates of cysteine transporters. Extracellular glutamate and aspartate, substrates of EAAT, suppress cysteine uptake (Himi et al., 2003; Chen and Swanson, 2003). Intracellular glutathione levels are down regulated by EAAT inhibitors, but not by inhibitors of cysteine transporters. Knock down of EAAT3 expression using the antisense oligonucleotide significantly down-regulates cysteine uptake, the intracellular glutathione level, and neuronal viability against oxidative stress. EAAT3 functions as a cysteine transporter (Himi et al., 2003), so neuronal EAAT3 activity may be a rate-limiting step for neuronal glutathione synthesis. The primary function of EAAT expressed by neurons *in vivo* may be to transport cysteine (Chen and Swanson, 2003). Genetically mutated EAAT3-null (Slc1a1(-/-)) mice also show reduced neuronal glutathione levels and, with aging, develop brain atrophy and behavioral changes. Neurons in the hippocampal slices of EAAT3(-/-) mice show reduced glutathione content, increased oxidant levels, and increased susceptibility to oxidant injury (Aoyama et al., 2006). These changes can be reversed by treating the EAAT3(-/-) mice with N-acetylcysteine, a membrane-permeable cysteine precursor. These findings suggest that EAAT3 may be the primary route for neuronal cysteine uptake and that EAAT3 deficiency thereby leads to impaired neuronal glutathione metabolism, oxidative stress, and age-dependent neurodegeneration (Aoyama et al., 2006).

Ischemic preconditioning up-regulates EAAT2 and EAAT3 but not EAAT1 expression (Pradillo et al., 2006). The TNF-α converting enzyme inhibitor BB1101 can down-regulate the increase in EAAT3 expression mediated by ischemic preconditioning. Intracerebral administration of either anti-TNF-α antibody or a TNFR1 antisense oligodeoxynucleotide also blocks up-regulation of EAAT3 induced by ischemic preconditioning. EAAT3 is expressed in both neuronal cytoplasm and plasma membrane, but immunohistochemical studies suggest that up-regulation of EAAT3 mediated by ischemic preconditioning is mainly localized at the plasma membrane level. Collectively the studies demonstrate that *in vivo* ischemic preconditioning increases the expression of EAAT2 and EAAT3 glutamate transporters. The up-regulation of the latter is at least partly mediated by the TNF-α converting enzyme/TNF-α/TNFR1 pathway (Pradillo et al., 2006). Changes in EAAT3 occur in experimental models of epilepsy (Crino et al., 2002), anoxia (Martin et al., 1997), and Alzheimer disease (Masliah et al., 2000). EAAT3 plays an active role in preventing neuronal damage following increases in extracellular glutamate concentration.

Detailed neurochemical investigations on EAAT3 indicate that a binding sequence for the regulatory factor X1 (RFX1) exists in the promoter region of the gene encoding for EAAT3, but not in the promoter regions of the genes encoding for glial cell glutamate transporters (Ma et al., 2006). RFX proteins are transcription factors binding to X-boxes of DNA sequences. Although RFX proteins are necessary for the normal function of sensory neurons in *Caenorhabditis elegans*, their roles in the mammalian brain are not known (Ma et al., 2006). RFX1 up-regulates the expression of EAAT3 in C6 glioma cells. RFX1 binding complexes are found in the nuclear extracts from C6 cells. RFX1 increases the activity of the EAAT3 promoter, as measured by luciferase reporter activity, in C6 cells and the neuron-like SH-SY5Y cells. However, RFX1 does not change the expression of EAAT2 proteins in NRK52E cells. RFX1 proteins are expressed in the neurons of rat brain. A high expression level of RFX1 proteins is found in the neurons of cerebral cortex and Purkinje cells. Antisense oligonucleotides of RFX1 down-regulate the expression of EAAT3 in rat cortical neurons. RFX1 may enhance the activity of the EAAT3 promoter to increase the expression of EAAT3 proteins (Ma et al., 2006).

4.2.2 EAAT4 in Brain

EAAT4 is a glutamate transporter that is expressed in the cerebellum and at lower levels in the brain stem, cortex and hippocampus. It has a mol mass of 50 kDa. cDNA encoding EAAT4 was obtained from human cerebellar messenger RNA (Rothstein et al., 1994). The amino acid sequence exhibits 65, 41, and 48% amino-acid identity to the human glutamate transporters EAAT1, EAAT2, and EAAT3, respectively. Like other glutamate transporters, the three subunits of EAAT4 transporter cooperate in activating anion channels and individually mediate coupled glutamate transport. The highest expression level is found in the cerebellum. Expression in the cortex is approximately 3.1% that of the cerebellum and the retina was found to have approximately 0.8% of the total cerebellar EAAT4 content (Ward et al., 2004). Human and rat cerebellar Purkinje cells have intense EAAT4 immunoreactivity with a somatodendritic localization. Other brain regions including neocortex, hippocampus, and striatum show faint neuropil staining of EAAT4. Immunogold localization studies identified EAAT4 protein at plasma membranes in Purkinje cell dendrites and spines. In the hippocampus and neocortex, EAAT4 immunoreactivity is found mainly in small caliber dendrites.

Human retina contains high levels of EAAT4 mRNA (Pignataro et al., 2005). EAAT4-immunoreactive proteins, corresponding to EAAT4 monomers and dimers, are present in human and mouse retinas. EAAT4 is also localized in rod and cone photoreceptor outer segments in the human retina, and in the outer and inner segments of mouse and ground squirrel retinas. In no case is EAAT4 found in the outer plexiform layer or in any other layer in the retina (Pignataro et al., 2005). Two proteins, namely GTRAP41 ($\sim$270 kDa) and GTRAP48 ($\sim$170 kDa) that specifically interact with the intracellular carboxy-terminal domain of EAAT4 and modulate its glutamate transport activity, are in brain and retinal tissues (Jackson et al., 2001; Pignataro et al., 2005). GTRAP41 has seventeen 16-amino acid spectrin repeats, 2 alpha-actinin domains, and a pleckstrin homology domain. GTRAP48 possesses a PDZ domain, a regulatory G protein-signaling sequence, tandem dbl homology and pleckstrin homology domains characteristic of guanine nucleotide exchange factors for the Rho family of G proteins, and 2 proline-rich sequences (Jackson et al., 2001; Pignataro et al., 2005).

Although earlier studies indicate that EAAT4 is a neuronal glutamate transporter, recent double-labeled fluorescent immunostaining and confocal image analysis indicates that EAAT4-like immunoreactivity colocalizes with an astrocytic marker, glial fibrillary acidic protein (GFAP) (Hu et al., 2003; Ward et al., 2004). The astrocytic localization of EAAT4 was further confirmed in astrocyte cultures by double-labeled fluorescent immunocytochemistry and Western blotting. Reverse transcriptase-polymerase chain reaction analysis also demonstrates mRNA expression of EAAT4 in astrocyte cultures. Sequencing confirmed the specificity of the amplified fragment. These results demonstrate that EAAT4 is expressed in astrocytes. This astrocytic localization of neuronal EAAT4 may reveal a new function of EAAT4 in the central nervous system (Hu et al., 2003). Collective evidence suggests that EAAT4 along with GTRAP41 and GTRAP48 functions as a transporter, reducing the amount of neurotransmitter available for activating postsynaptic receptors, and as a glutamate-gated chloride channel, modifying neuronal excitability by its capacity for enhancing chloride permeability.

4.2.3 EAAT5 in Brain

EAAT5 is a human glutamate transporter with its mRNA mostly expressed in the retina. EAAT5 has 46% identity with EAAT1, 43% with EAAT4, 37% with EAAT3 and 36% with EAAT2. In the EAAT5 C-terminus there is a sequence motif found in synaptic membrane proteins: E-S or T-X-V-COOH (Arriza et al., 1997). EAAT5 cDNA encodes a deduced 560-amino acid protein containing eight putative transmembrane domains and a single potential N-linked glycosylation site in the putative large extracellular loop. The C-termini interact with several PDZ (a modular protein-binding motif) domains in PSD-95 (postsynaptic density-95kDa protein). The presence of this PDZ domain suggests that the transporter is a component of the signal transduction pathway. Its channel-like properties may indicate a role in retinal physiology different from neurotransmitter clearance. Although EAAT5 shares the structural homologies of the EAAT gene family, a remarkable feature of the EAAT5 sequence is a carboxy-terminal motif identified previously in *N*-methyl-D-aspartate receptors (NMDA receptors) and potassium ion channels and shown to confer interactions with a family of synaptic proteins that promote ion channel clustering (Arriza et al., 1997). The functional properties of EAAT5 have been examined in the *Xenopus* oocyte expression system by measuring radiolabeled glutamate flux and two-electrode voltage clamp recording. EAAT5-mediated L-glutamate uptake is Na^+- and voltage-dependent and Cl^--independent. Transporter currents elicited by glutamate are also Na^+- and voltage-dependent, but ion substitution experiments suggest that this current is largely carried by chloride ions (Arriza et al., 1997).

These properties of EAAT5 are similar to the glutamate-elicited chloride conductances previously described in retinal neurons. This suggests that the EAAT5-associated chloride conductance may participate in visual processing. Like EAAT4, the chloride conductance of EAAT5 is of considerable interest because it suggests its association with ligand-gated chloride channels. Both pre- and postsynaptic glutamate-gated chloride conductances have been studied in the vertebrate retina. In perch retina, the cone-mediated light response in depolarizing bipolar cells is due to the closing of a postsynaptic chloride conductance that has the pharmacology and ionic dependence of a glutamate transporter (Arriza et al., 1997). Presynaptically, a glutamate-elicited chloride conductance with transporter-like properties in salamander cone photoreceptors responds to the release of glutamate from the same cell. The chloride equilibrium potential is more negative than the voltage-operating range of cones, so this EAAT5-like molecule may act as an inhibitory autoreceptor (Arriza et al., 1997). Collectively these studies indicate that EAAT5 acts as a major inhibitory presynaptic receptor at mammalian rod bipolar cell axon terminals. This feedback mechanism can control glutamate release at the ribbon synapses of a non-spiking neuron and increase the temporal contrast in the rod photoreceptor pathway (Wersinger et al., 2006).

In retinal amacrine and ganglion cells, EAAT5 colocalizes with the kinase SGK1, inducible with serum and glucocorticoid, a serine/threonine kinase known to regulate transport. It is proposed that SGK1, SGK3, and the closely related protein kinase B regulate EAAT5 (Boehmer et al., 2005). EAAT5 is coexpressed in *Xenopus laevis* oocytes with or without the respective kinases. Both currents mediated by EAAT5 and the abundance of EAAT5 protein at the cell surface are increased by a factor of 1.5–2 upon coexpression of SGK1 or SGK3, but not following coexpression of PKB. It is proposed that kinases SGK1 and SGK3 increase EAAT5 activity by increasing cell surface abundance of the carrier (Boehmer et al., 2005).

4.3 Vesicular Glutamate Transporters

In nerve terminals, glutamate is actively packaged into synaptic vesicles before its release by Ca^{2+}-dependent exocytosis. Glutamate transport into synaptic vesicles depends on the proton electrochemical gradient uptake system and is facilitated by vesicular glutamate transporter (VGLUT) proteins. The three VGLUT, 1, 2 and 3, are highly conserved proteins that display similar bioenergetic and pharmacological properties, but are expressed in different brain areas (Gras et al., 2005; Vinatier et al., 2006). VGLUT1 and VGLUT2 mRNAs are coexpressed in most of the sampled neurons from rat hippocampus, cortex, and cerebellum at postnatal day 14 (P14), but not at P60 (Danik et al., 2005). In accordance, changes in VGLUT1 and VGLUT2 mRNA concentrations are found to occur in these and other brain areas between P14 and P60. VGLUT1 and VGLUT2 coexpression in the hippocampal formation is supported further by hybridization data in situ showing that virtually all cells in the CA1-CA3 pyramidal and granule cell layers are highly positive for both transcripts until P14 (Danik et al., 2005).

Semiquantitative RT-PCR studies revealed that transcripts for VGLUT1 and VGLUT2 are also present in neurons of the cerebellum, striatum, and septum that expressed markers for gamma-aminobutyric acid (GABA)-ergic or cholinergic phenotypes, as well as in hippocampal cells containing transcripts for the glial fibrillary acidic protein. VGLUT1 and VGLUT2 proteins may often transport glutamate into vesicles within the same neuron, especially during early postnatal development. They are expressed widely in presumed glutamatergic, GABAergic, and cholinergic neurons, as well as in astrocytes (Danik et al., 2005). It should be noted that VGLUT represent the most specific marker for neurons that use glutamate as transmitter. In rat brain the expression of VGLUT1 and VGLUT2 increases linearly during postnatal development. In contrast, the VGLUT3 developmental pattern appears to have a more or less biphasic profile (Gras et al., 2005). A first peak of expression is centered on post-natal day 10 (P10). The second one is reached in the adult brain. Between P1 and P15, VGLUT3 is observed in the frontal brain, striatum, accumbens, and hippocampus, and in the caudal brain, colliculi, pons, and cerebellum. During a second phase extending from P15 to adulthood, the labeling of the caudal brain fades away. The adult pattern is reached at P21.

In the cerebellum, the expression of VGLUT3 corresponds to a temporary expression in Purkinje cells. At P10 VGLUT3 immunoreactivity is present both in the soma and terminals of Purkinje cells (PC), where it is colocalized with the vesicular inhibitory amino acid transporter (VIAAT). Recently the divergent C-terminus of VGLUT1 has been used as bait in a yeast two-hybrid screen to identify and map the interaction between a proline-rich domain of VGLUT1 and the Src homology domain 3 (SH3) domain of endophilin. Further interactions have been confirmed by using different glutathione-S-transferase-endophilin fusion proteins to pull down VGLUT1 from rat brain extracts. The expression profiles of the two genes and proteins have been compared in rat brain. Endophilin is most highly expressed in regions and cells expressing VGLUT1. Double immunofluorescence in the rat cerebellum shows that most VGLUT1-positive terminals co-express endophilin, whereas VGLUT2-expressing terminals are often devoid of endophilin. However, neither VGLUT1 transport activity, endophilin enzymatic activity, or VGLUT1 synaptic targeting are altered by this interaction. Overall, the discovery of endophilin as a partner for VGLUT1 in nerve terminals strongly suggests the existence of functional differences between VGLUT1 and VGLUT2 terminals in their abilities to replenish vesicle pools (Vinatier et al., 2006).

4.4 Glutamate Transporters in Neurological Disorders

Among the five known glutamate transporters, EAAT1 and EAAT2 are predominantly expressed by astrocytes, whilst EAAT3, EAAT4, and EAAT5 are localized primarily in neurons. EAAT4 and EAAT5 are restricted to cerebellar Purkinje cells and retina, respectively, but EAAT3 is widely expressed in neurons throughout the brain (Arriza et al., 1997). EAAT1 and EAAT2 are responsible for most glutamate uptake in brain and maintenance of physiological extracellular glutamate concentrations (Danbolt, 2001). The functions and regulation of EAAT3, EAAT4, and EAAT5 are not fully established. EAAT4 may act as a glutamate-gated chloride channel and modulate cell migration (Torres-Salazar and Fahlke, 2006; Beart and O'Shea, 2007). The mGluR5a receptor contributes to a dynamic control of EAAT2 function in activated astrocytes, acting as a glial sensor of the extracellular glutamate concentration in order to acutely regulate the excitatory transmission (Vermeiren et al., 2005).

Glutamate transporters are dependent on external Na^+, and thus on the activity of Na^+/K^+ ATPases, which maintain the Na^+ concentration gradient. During ischemic injury when the requirements of energy in brain are not fulfilled due to lack of blood supply (glucose and oxygen), the Na^+ gradient collapses leading to impaired glutamate removal, or even to the release of these amino acids through the reverse operation of their transporters. Such a scenario occurs during brain ischemia and hypoglycemia. In addition, down-regulation of glutamate transporters after the ischemic period, or the dysfunction induced by oxidation, contributes to the accumulation of extracellular glutamate and neuronal death through the overstimulation of glutamate receptors (Camacho and Massieu, 2006; Kim et al., 2002). The inhibition of synthesis of each glutamate transporter subtype by chronic antisense oligonucleotide administration, in vitro and *in vivo*, selectively and specifically blocks the protein expression and function of glutamate transporters (Rothstein et al., 1996; Perego et al., 2000). Inhibition of the glial glutamate transporters, EAAT1 and EAAT2, results in neurodegeneration, whereas the blockage of the neuronal glutamate transporter, EAAT3, does not elevate extracellular glutamate but does produce mild neurotoxicity (Perego et al., 2000). Glial glutamate transporters may provide the majority of functional glutamate transport and are essential for maintaining low extracellular glutamate and for preventing chronic glutamate neurotoxicity.

Neurochemical changes in acute neural trauma and neurodegenerative diseases result in cytokine overproduction, oxidative stress, and energy failure. Disturbance of ionic gradients and energy depletion decreases glutamate transporter activity (Hertz, 2006; Yi and Hazell, 2006). Similarly tumor necrosis factor-α (TNF-α) and interleukin-1β (lL-1β) potentiate glutamate-mediated oxidative stress and cause down-regulation of glutamate transporters. Cytokine overproduction, oxidative stress, and energy failure impair glutamate transport into astrocytes (Liao and Chen, 2001) and facilitate glutamate-mediated neuronal cell death. The collective evidence suggests that defects in glutamate transport subtypes are important component of glutamate-induced injury in acute neural trauma and neurodegenerative disorders (Table 4.1) and that excitatory transmission in the brain

Table 4.1 Involvement and status of glutamate transporters in neurological and neuropsychiatric disorders

Disorder	Glutamate transporter involved	Status	Reference
Ischemia	EAAT1	Increased	(Kim et al., 2006)
	EAAT2	Decreased	(Yeh et al., 2005)
	EAAT3	Increased	(Yamashita et al., 2006)
	EAAT4	Decreased	
Spinal cord trauma	EAAT1	Increased	(Vera-Portocarrero et al., 2002)
	EAAT2	Increased	
	EAAT3	Increased	
	EAAT4	Increased	
Head injury	EAAT1	Decreased	(Yi and Hazell, 2006)
	EAAT2	Decreased	(van Landeghem et al., 2006)
Hypoxia	EAAT2	Decreased	(Pow et al., 2004)
Epilepsy	EAAT3	Increased	(Voutsinos-Porche et al., 2006; Hoogland et al., 2004)
Alzheimer disease	EAAT1	Decreased	(Zoia et al., 2005)
	EAAT2	Decreased	(Tian et al., 2007; Maragakis et al., 2004)
Amyotrophic lateral sclerosis	EAAT2	Decreased	(Howland et al., 2002)(Guo et al., 2003)
Huntington disease	EAAT2	Decreased	(Lievens et al., 2001; Behrens et al., 2002)
Creutzfeldt-Jakob disease	EAAT1	No effect	(Chrétien et al., 2004)
AIDS dementia	EAAT1	Increased	(Vallat-Decouvelaere et al., 2003)
	EAAT2	Decreased	(Wang et al., 2003)

(continued)

Table 4.1 (continued)

Disorder	Glutamate	Status	Reference
Experimental	EAAT3	Increased	(Ohgoh et al., 2002)
Allergic	EAAT1	Decreased	(Hardin-Pouzet et al., 1997)
Encephalomyelitis	EAAT2	Decreased	(Hardin-Pouzet et al., 1997)
Multiple sclerosis	EAAT1	Increased	(Vallejo-Illarramendi et al., 2006)
	EAAT2	Increased	(Vallejo-Illarramendi et al., 2006)
Guam-type amyotrophic lateral sclerosis	EAAT2	Decreased	(Wilson et al., 2003)
Olivopontocerebellar atrophy	EAAT4	Decreased	(Dirson et al., 2002)
Schizophrenia	EAAT2	Increased	(Lauriat et al., 2006)
Pain	EAAT3	Decreased	(Wang et al., 2006)
Glaucoma	EAAT1	Decreased	(Wakabayashi et al., 2006)
Cancer	EAAT2	Decreased	(Ye et al., 1999)
Alcoholism	EAAT1	Increased	(Flatscher-Bader et al., 2005)

necessitates the existence of dynamic controls of the glutamate uptake achieved by astrocytes, both in physiological conditions and under pathological circumstances.

An animal model of Alzheimer disease, compared to wild-type, has decreased EAAT3 activity along with amyloid precursor protein overexpression (Masliah et al., 2000). Similarly in amyotrophic lateral sclerosis and Alzheimer disease, EAAT2 protein levels are significantly decreased in affected areas. The loss of functional EAAT2 and EAAT3 may lead to the accumulation of extracellular glutamate, resulting in cell death. The molecular mechanism associated with down-regulation of glutamate transporters in neurodegenerative diseases remains unknown. The occurrence of glutamate transporter splice variants in neurodegenerative diseases suggests that altered splicing or altered expression of glutamate transporters may play an important role in down-regulating the activity of EAAT2 (Lin et al., 1998). However, recent studies indicate that alterations in EAAT2 activity may not be the cause of ALS pathogenesis (Meyer et al., 1999; Flowers et al., 2001).

Another possibility is that differential distribution of various splice variants of EAAT2 in pre-symptomatic transgenic animals and ALS patients (Münch et al., 2002; Maragakis et al., 2004) may be responsible for changes in activities of glutamate transporters. This is supported by alterations in splice variants of EAAT2 in ALS/PD dementia complex, a disease caused by the ingestion of cycad flour in humans residing in Guam (Wilson et al., 2003). The specificity of glutamate splice variants in neurodegenerative diseases remains unknown because aberrant splicing of glutamate transporters was also reported in other neurological disorders including hypoxia (Pow et al., 2004), traumatic brain injury (Yi and Hazell, 2006), epilepsy (Hoogland et al., 2004), Alzheimer disease (Maragakis et al., 2004), Huntington disease (Hoogland et al., 2004), AIDS dementia complex (Wang et al., 2003), multiple sclerosis (Vallejo-Illarramendi et al., 2006), olivopontocerebellar atrophy (Dirson et al., 2002), schizophrenia (Lauriat et al., 2006), and glaucoma (Hossain et al., 2006). Altered glutamate transport and aberrant splicing of EAAT2 expression was also reported in glial cells from ALS patients infected with enterovirus (Legay et al., 2003). It still remains unknown whether changes in expression and distribution of splice variants of EAAT2 are the cause or consequence of EAAT2 activity. At present nothing is known about splice expression and distribution of EAAT1, EAAT3, EAAT4, and EAAT5 in brain tissue from other neurodegenerative diseases. More studies are required on expression of glutamate transporter subtypes in neurodegenerative diseases (Beart and O'Shea, 2007).

4.5 Conclusions

Glutamate is a component of neural proteins and plays an important role in signaling processes in brain tissue. Glutamate-mediated processes involve glutamate receptors for signal input, vesicular glutamate transporters for signal output, and glutamate transporters for signal termination. Glutamate transporters generate current by virtue of the net charge entry that occurs as a result of the ion movements coupled to glutamate movement. However, through neuronal glutamate transporters

EAAT4 and EAAT5, glutamate transport leads to the opening of chloride channels. For glial EAAT1 and EAAT2, the current flow through the anion conductance is too small to influence the astrocytic membrane potential. Thus under normal conditions the functional significance of chloride conductance in astrocytes is uncertain, but ischemic injury is known to decrease Na^+ pump activity due to the suppression of Na^+,K^+-ATPase. This causes reversal of the astrocytic glutamate transporters, EAAT1 and EAAT2. Under these conditions, the operation of EAAT2 is reversed, and three sodium ions are thus co-transported with a glutamate ion out of the cell (Jabaudon et al., 2000).

Astrocytes are less susceptible to ischemic injury and subsequent metabolic inhibition than neurons (Goldberg and Choi, 1993), because they have larger glycogen stores, lower density expression of voltage-sensitive sodium ion channels (Sontheimer and Ritchie, 1995), and less susceptibility to excitotoxic insult (Choi and Rothman, 1990). It is postulated that excessive influx of sodium ions into astroglial cells causes significant astroglial cell death (Takahashi et al., 2000) due to Ca^{2+} overload, probably via a reversed Na^+/Ca^{2+} exchanger. Alterations in glutamate uptake may represent an important adaptive and protective response of ischemic astrocytes by which many of them attempt to maintain homeostasis of ions, water, and glutamate under conditions of restricted energy and oxygen supply (Stanimirovic et al., 1997; Kosugi and Kawahara, 2006).

The situation is quite different in neuronal cells under ischemic conditions. Ionic changes in ischemic neurons automatically reverse the operation of glutamate transporters resulting in toxic levels that neuronal cell death. Thus unlike neurons, astrocytes undergo dramatic functional and morphological changes following ischemic injury to protect neurons and some of these changes may be related to glutamate transport mechanism involving glutamate transporters. Similar changes in glutamate transporter expression were also reported in neurodegenerative diseases. Trafficking and splicing of glutamate transporters may be responsible for the downregulation of glutamate transporters in neurodegenerative diseases (Nieoullon et al., 2006).

References

Amara S. G. and Fontana A. C. (2002). Excitatory amino acid transporters: keeping up with glutamate. Neurochem. Int. 41:313–318.

Aoyama K., Suh S. W., Hamby A. M., Liu J., Chan W. Y., Chen Y., and Swanson R. A. (2006). Neuronal glutathione deficiency and age-dependent neurodegeneration in the EAAC1 deficient mouse. Nat. Neurosci. 9:119–126.

Arriza J. L., Eliasof S., Kavanaugh M. P., and Amara S. G. (1997). Excitatory amino acid transporter 5, a retinal glutamate transporter coupled to a chloride conductance. Proc. Natl. Acad. Sci. USA 94:4155–4160.

Attwell D. (2000). Brain uptake of glutamate: food for thought. J. Nutr. 130:1023S–1025S.

Barbour B., Brew H., and Attwell D. (1988). Electrogenic glutamate uptake in glial cells is activated by intracellular potassium. Nature 335:433–435.

Beart P. M. and O'Shea R. D. (2007). Transporters for L-glutamate: An update on their molecular pharmacology and pathological involvement. Brit. J. Pharmacol. 150:5–17.

Behrens P. F., Franz P., Woodman B., Lindenberg K. S., and Landwehrmeyer G. B. (2002). Impaired glutamate transport and glutamate-glutamine cycling: downstream effects of the Huntington mutation. Brain 125:1908–1922.

Bergles D. E., Diamond J. S., and Jahr C. E. (1999). Clearance of glutamate inside the synapse and beyond. Curr. Opin. Neurobiol. 9:293–298.

Boehmer C., Rajamanickam J., Schniepp R., Kohler K., Wulff P., Kuhl D., Palmada M., and Lang F. (2005). Regulation of the excitatory amino acid transporter EAAT5 by the serum and glucocorticoid dependent kinases SGK1 and SGK3. Biochem. Biophys. Res. Commun. 329:738–742.

Camacho A. and Massieu L. (2006). Role of glutamate transporters in the clearance and release of glutamate during ischemia and its relation to neuronal death. Arch. Med. Res. 37:11–18.

Casado M., Bendahan A., Zafra F., Danbolt N. C., Aragón C., Giménez C., and Kanner B. I. (1993). Phosphorylation and modulation of brain glutamate transporters by protein kinase C. J. Biol. Chem. 268:27313–27317.

Chaudhry F. A., Lehre K. P., van Lookeren Campagne M., Ottersen O. P., Danbolt N. C., and Storm-Mathisen J. (1995). Glutamate transporters in glial plasma membranes: highly differentiated localizations revealed by quantitative ultrastructural immunocytochemistry. Neuron 15:711–720.

Chen Y. and Swanson R. A. (2003). The glutamate transporters EAAT2 and EAAT3 mediate cysteine uptake in cortical neuron cultures. J. Neurochem. 84:1332–1339.

Choi D. W. and Rothman S. M. (1990). The role of glutamate neurotoxicity in hypoxic-ischemic neuronal death. Annu. Rev. Neurosci. 13:171–182.

Chrétien F., Le Pavec G., Vallat-Decouvelaere A. V., Delisle M. B., Uro-Coste E., Ironside J. W., Gambetti P., Parchi P., Créminon C., Dormont D., Mikol J., Gray F., and Gras G. (2004). Expression of excitatory amino acid transporter-1 (EAAT-1) in brain macrophages and microglia of patients with prion diseases. J. Neuropathol. Exp. Neurol. 63: 1058–1071.

Conti F., DeBiasi S., Minelli A., Rothstein J. D., and Melone M. (1998). EAAC1, a high-affinity glutamate tranporter, is localized to astrocytes and gabaergic neurons besides pyramidal cells in the rat cerebral cortex. Cereb. Cortex 8:108–116.

Crino P. B., Jin H., Shumate M. D., Robinson M. B., Coulter D. A., and Brooks-Kayal A. R. (2002). Increased expression of the neuronal glutamate transporter (EAAT3/EAAC1) in hippocampal and neocortical epilepsy. Epilepsia 43:211–218.

Danbolt N. C. (2001). Glutamate uptake. Prog. Neurobiol. 65:1–105.

Danik M., Cassoly E., Manseau F., Sotty F., Mouginot D., and Williams S. (2005). Frequent coexpression of the vesicular glutamate transporter 1 and 2 genes, as well as coexpression with genes for choline acetyltransferase or glutamic acid decarboxylase in neurons of rat brain. J. Neurosci. Res. 81:506–521.

Dirson G., Desjardins P., Tannenberg T., Dodd P., and Butterworth R. F. (2002). Selective loss of expression of glutamate GluR2/R3 receptor subunits in cerebellar tissue from a patient with olivopontocerebellar atrophy. Metab Brain Dis. 17:77–82.

Dunlop J., Lou Z., Zhang Y., and McIlvain H. B. (1999). Inducible expression and pharmacology of the human excitatory amino acid transporter 2 subtype of L-glutamate transporter. Br. J. Pharmacol. 128:1485–1490.

Erecinska M. and Silver I. A. (1990). Metabolism and role of glutamate in mammalian brain. Prog. Neurobiol. 35:245–296.

Eriksson G., Peterson A., Iverfeldt K., and Walum E. (1995). Sodium-dependent glutamate uptake as an activator of oxidative metabolism in primary astrocyte cultures from newborn rat. Glia 15:152–156.

Fairman W. A. and Amara S. G. (1999). Functional diversity of excitatory amino acid transporters: ion channel and transport modes. Am. J. Physiol 277:F481–F486.

Farooqui, A. A., Anderson, D. K., and Horrocks, L. A. (1994). Potentiation of diacylglycerol and monoacylglycerol lipase activities by glutamate and its analogs. J. Neurochem. 62, S74B

Flatscher-Bader T., van der Brug M., Hwang J. W., Gochee P. A., Matsumoto I., Niwa S. I., and Wilce P. A. (2005). Alcohol-responsive genes in the frontal cortex and nucleus accumbens of human alcoholics. J. Neurochem. 93:359–370.

Flowers J. M., Powell J. F., Leigh P. N., Andersen P., and Shaw C. E. (2001). Intron 7 retention and exon 9 skipping EAAT2 mRNA variants are not associated with amyotrophic lateral sclerosis. Ann. Neurol. 49:643–649.

Furuta A., Rothstein J. D., and Martin L. J. (1997). Glutamate transporter protein subtypes are expressed differentially during rat CNS development. J. Neurosci. 17:8363–8375.

Gegelashvili G., Danbolt N. C., and Schousboe A. (1997). Neuronal soluble factors differentially regulate the expression of the GLT1 and GLAST glutamate transporters in cultured astroglia. J. Neurochem. 69:2612–2615.

Gendreau S., Voswinkel S., Torres-Salazar D., Lang N., Heidtmann H., Detro-Dassen S., Schmalzing G., Hidalgo P., and Fahlke C. (2004). A trimeric quaternary structure is conserved in bacterial and human glutamate transporters. J. Biol. Chem. 279:39505–39512.

Goldberg M. P. and Choi D. W. (1993). Combined oxygen and glucose deprivation in cortical cell culture: Calcium-dependent and calcium-independent mechanisms of neuronal injury. J. Neurosci. 13:3510–3524.

Gras C., Vinatier J., Amilhon B., Guerci A., Christov C., Ravassard P., Giros B., and El Mestikawy S. (2005). Developmentally regulated expression of VGLUT3 during early post-natal life. Neuropharmacology 49:901–911.

Grunewald M., Menaker D., and Kanner B. I. (2002). Cysteine-scanning mutagenesis reveals a conformationally sensitive reentrant pore-loop in the glutamate transporter GLT-1. J. Biol. Chem. 277:26074–26080.

Guo H., Lai L., Butchbach M. E., Stockinger M. P., Shan X., Bishop G. A., and Lin C. L. (2003). Increased expression of the glial glutamate transporter EAAT2 modulates excitotoxicity and delays the onset but not the outcome of ALS in mice. Hum. Mol. Genet. 12: 2519–2532.

Hardin-Pouzet H., Krakowski M., Bourbonnière L., Didier-Bazes M., Tran E., and Owens T. (1997). Glutamate metabolism is down-regulated in astrocytes during experimental allergic encephalomyelitis. Glia 20:79–85.

Hertz L. (2006). Glutamate, a neurotransmitter-And so much more. A synopsis of Wierzba III. Neurochem. Int. 48:416–425.

Hertz L., Dringen R., Schousboe A., and Robinson S. R. (1999). Astrocytes: glutamate producers for neurons. J. Neurosci. Res. 57:417–428.

Himi T., Ikeda M., Yasuhara T., Nishida M., and Morita I. (2003). Role of neuronal glutamate transporter in the cysteine uptake and intracellular glutathione levels in cultured cortical neurons. J. Neural Transm. 110:1337–1348.

Hoogland G., van Oort R. J., Proper E. A., Jansen G. H., van Rijen P. C., van Veelen C. W., van Nieuwenhuizen O., Troost D., and De Graan P. N. (2004). Alternative splicing of glutamate transporter EAAT2 RNA in neocortex and hippocampus of temporal lobe epilepsy patients. Epilepsy Res. 59:75–82.

Hossain M. A., Wakabayashi H., Izuishi K., Okano K., Yachida S., and Maeta H. (2006). The role of prostaglandins in liver ischemia-reperfusion injury. Curr. Pharmaceut. Design 12: 2935–2951.

Howland D. S., Liu J., She Y., Goad B., Maragakis N. J., Kim B., Erickson J., Kulik J., DeVito L., Psaltis G., DeGennaro L. J., Cleveland D. W., and Rothstein J. D. (2002). Focal loss of the glutamate transporter EAAT2 in a transgenic rat model of SOD1 mutant-mediated amyotrophic lateral sclerosis (ALS). Proc. Natl. Acad. Sci. USA 99:1604–1609.

Hu W. H., Walters W. M., Xia X. M., Karmally S. A., and Bethea J. R. (2003). Neuronal glutamate transporter EAAT4 is expressed in astrocytes. Glia 44:13–25.

Jabaudon D., Scanziani M., Gahwiler B. H., and Gerber U. (2000). Acute decrease in net glutamate uptake during energy deprivation. Proc. Natl. Acad. Sci. USA 97:5610–5615.

Jackson M., Song W., Liu M. Y., Jin L., Dykes-Hoberg M., Lin C. I., Bowers W. J., Federoff H. J., Sternweis P. C., and Rothstein J. D. (2001). Modulation of the neuronal glutamate transporter EAAT4 by two interacting proteins. Nature 410:89–93.

Kanai Y., Nussberger S., Romero M. F., Boron W. F., Hebert S. C., and Hediger M. A. (1995). Electrogenic properties of the epithelial and neuronal high affinity glutamate transporter. J. Biol. Chem. 270:16561–16568.

Kim A. H., Kerchner G. A., and Choi D. W. (2002). Blocking excitotoxicity. In: Marcoux F. W. and Choi D. W. (eds.), *CNS Neuroprotection*. Springer, New York, pp. 3–36.

Kim D. S., Kwak S. E., Kim J. E., Jung J. Y., Won M. H., Choi S. Y., Kwon O. S., and Kang T. C. (2006). Transient ischaemia affects plasma membrane glutamate transporter, not vesicular glutamate transporter, expressions in the gerbil hippocampus. Anat. Histol. Embryol. 35: 265–270.

Kosugi T. and Kawahara K. (2006). Reversed actrocytic GLT-1 during ischemia is crucial to excitotoxic death of neurons, but contributes to the survival of astrocytes themselves. Neurochem. Res. 31:933–943.

Lauriat T. L., Dracheva S., Chin B., Schmeidler J., McInnes L. A., and Haroutunian V. (2006). Quantitative analysis of glutamate transporter mRNA expression in prefrontal and primary visual cortex in normal and schizophrenic brain. Neuroscience 137:843–851.

Legay V., Deleage C., Beaulieux F., Giraudon P., Aymard M., and Lina B. (2003). Impaired glutamate uptake and EAAT2 downregulation in an enterovirus chronically infected human glial cell line. Eur. J. Neurosci. 17:1820–1828.

Lehre K. P., Levy L. M., Ottersen O. P., Storm-Mathisen J., and Danbolt N. C. (1995). Differential expression of two glial glutamate transporters in the rat brain: quantitative and immunocytochemical observations. J. Neurosci. 15:1835–1853.

Levy L. M., Lehre K. P., Walaas S. I., Storm-Mathisen J., and Danbolt N. C. (1995). Down-regulation of glial glutamate transporters after glutamatergic denervation in the rat brain. Eur. J. Neurosci. 7:2036–2041.

Li L. B., Toan S. V., Zelenaia O., Watson D. J., Wolfe J. H., Rothstein J. D., and Robinson M. B. (2006). Regulation of astrocytic glutamate transporter expression by Akt: evidence for a selective transcriptional effect on the GLT-1/EAAT2 subtype. J. Neurochem. 97: 759–771.

Liao S. L. and Chen C. J. (2001). Differential effects of cytokines and redox potential on glutamate uptake in rat cortical glial cultures. Neurosci. Lett. 299:113–116.

Lievens J. C., Woodman B., Mahal A., Spasic-Boscovic O., Samuel D., Kerkerian-Le Goff L., and Bates G. P. (2001). Impaired glutamate uptake in the R6 Huntington's disease transgenic mice. Neurobiol. Dis. 8:807–821.

Lin C. L., Bristol L. A., Jin L., Dykes-Hoberg M., Crawford T., Clawson L., and Rothstein J. D. (1998). Aberrant RNA processing in a neurodegenerative disease: the cause for absent EAAT2, a glutamate transporter, in amyotrophic lateral sclerosis. Neuron 20:589–602.

Ma K., Zheng S., and Zuo Z. (2006). The transcription factor regulatory factor X1 increases the expression of neuronal glutamate transporter type 3. J. Biol. Chem. 281: 21250–21255.

Maragakis N. J., Dykes-Hoberg M., and Rothstein J. D. (2004). Altered expression of the glutamate transporter EAAT2b in neurological disease. Ann. Neurol. 55:469–477.

Martin L. J., Brambrink A. M., Lehmann C., Portera-Cailliau C., Koehler R., Rothstein J., and Traystman R. J. (1997). Hypoxia-ischemia causes abnormalities in glutamate transporters and death of astroglia and neurons in newborn striatum. Ann. Neurol. 42:335–348.

Masliah E., Alford M., Mallory M., Rockenstein E., Moechars D., and van Leuven F. (2000). Abnormal glutamate transport function in mutant amyloid precursor protein transgenic mice. Exp Neurol. 163:381–387.

Matsugami T. R., Tanemura K., Mieda M., Nakatomi R., Yamada K., Kondo T., Ogawa M., Obata K., Watanabe M., Hashikawa T., and Tanaka K. (2006). Indispensability of the glutamate transporters GLAST and GLT1 to brain development. Proc. Natl. Acad. Sci. USA 103: 12161–12166.

Meyer T., Fromm A., Münch C., Schwalenstöcker B., Fray A. E., Ince P. G., Stamm S., Grön G., Ludolph A. C., and Shaw P. J. (1999). The RNA of the glutamate transporter EAAT2 is variably spliced in amyotrophic lateral sclerosis and normal individuals. J. Neurol. Sci. 170:45–50.

Münch C., Ebstein M., Seefried U., Zhu B., Stamm S., Landwehrmeyer G. B., Ludolph A. C., Schwalenstöcker B., and Meyer T. (2002). Alternative splicing of the 5'-sequences of the mouse EAAT2 glutamate transporter and expression in a transgenic model for amyotrophic lateral sclerosis. J. Neurochem. 82:594–603.

Nakajima K. and Kohsaka S. (2004). Microglia: neuroprotective and neurotrophic cells in the central nervous system. Curr. Drug Targets Cardiovasc. Haematol. Disord. 4:65–84.
Nieoullon A., Canolle B., Masmejean F., Guillet B., Pisano P., and Lortet S. (2006). The neuronal excitatory amino acid transporter EAAC1/EAAT3: does it represent a major actor at the brain excitatory synapse? J. Neurochem. 98:1007–1018.
Ohgoh M., Hanada T., Smith T., Hashimoto T., Ueno M., Yamanishi Y., Watanabe M., and Nishizawa Y. (2002). Altered expression of glutamate transporters in experimental autoimmune encephalomyelitis. J. Neuroimmunol. 125:170–178.
Pellerin L. and Magistretti P. J. (1994). Glutamate uptake into astrocytes stimulates aerobic glycolysis: a mechanism coupling neuronal activity to glucose utilization. Proc. Natl. Acad. Sci. USA 91:10625–10629.
Perego C., Vanoni C., Bossi M., Massari S., Basudev H., Longhi R., and Pietrini G. (2000). The GLT-1 and GLAST glutamate transporters are expressed on morphologically distinct astrocytes and regulated by neuronal activity in primary hippocampal cocultures. J. Neurochem. 75:1076–1084.
Pignataro L., Sitaramayya A., Finnemann S. C., and Sarthy V. P. (2005). Nonsynaptic localization of the excitatory amino acid transporter 4 in photoreceptors. Mol. Cell Neurosci. 28:440–451.
Pow D. V., Naidoo T., Lingwood B. E., Healy G. N., Williams S. M., Sullivan R. K., O'Driscoll S., and Colditz P. B. (2004). Loss of glial glutamate transporters and induction of neuronal expression of GLT-1B in the hypoxic neonatal pig brain. Brain Res. Dev. Brain Res. 153:1–11.
Pradillo J. M., Hurtado O., Romera C., Cárdenas A., Fernández-Tomé P., Alonso-Escolano D., Lorenzo P., Moro M. A., and Lizasoain I. (2006). TNFR1 mediates increased neuronal membrane EAAT3 expression after *in vivo* cerebral ischemic preconditioning. Neuroscience 138:1171–1178.
Rauen T., Wiessner M., Sullivan R., Lee A., and Pow D. V. (2004). A new GLT1 splice variant: cloning and immunolocalization of GLT1c in the mammalian retina and brain. Neurochem. Int. 45:1095–1106.
Robinson M. B. (1998). The family of sodium-dependent glutamate transporters: a focus on the GLT-1/EAAT2 subtype. Neurochem. Int. 33:479–491.
Rothstein J. D., Martin L., Levey A. I., Dykes-Hoberg M., Jin L., Wu D., Nash N., and Kuncl R. W. (1994). Localization of neuronal and glial glutamate transporters. Neuron 13:713–725.
Rothstein J. D., Dykes-Hoberg M., Pardo C. A., Bristol L. A., Jin L., Kuncl R. W., Kanai Y., Hediger M. A., Wang Y., Schielke J. P., and Welty D. F. (1996). Knockout of glutamate transporters reveals a major role for astroglial transport in excitotoxicity and clearance of glutamate. Neuron 16:675–686.
Saier M. H., Jr. (1999). Eukaryotic transmembrane solute transport systems. Int. Rev. Cytol. 190:61–136.
Seal R. P. and Amara S. G. (1998). A reentrant loop domain in the glutamate carrier EAAT1 participates in substrate binding and translocation. Neuron 21:1487–1498.
Sims K. D. and Robinson M. B. (1999). Expression patterns and regulation of glutamate transporters in the developing and adult nervous system. Crit. Rev. Neurobiol. 13:169–197.
Slotboom D. J., Konings W. N., and Lolkema J. S. (1999). Structural features of the glutamate transporter family. Microbiol. Mol. Biol. Rev. 63:293–307.
Sontheimer H. and Ritchie J. M. (1995). Voltage-gated sodium and calcium channels in neuroglia. In: Kettenmann H. and Ransom B. R. (eds.), *Neuroglia*. Oxford University Press, New York, pp. 202–220.
Stanimirovic D. B., Ball R., and Durkin J. P. (1997). Stimulation of glutamate uptake and Na,K-ATPase activity in rat astrocytes exposed to ischemia-like insults. Glia 19:123–134.
Storck T., Schulte S., Hofmann K., and Stoffel W. (1992). Structure, expression, and functional analysis of a Na^+-dependent glutamate/aspartate transporter from rat brain. Proc. Natl. Acad. Sci. USA 89:10955–10959.
Suárez I., Bodega G., and Fernández B. (2000). Modulation of glutamate transporters (GLAST, GLT-1 and EAAC1) in the rat cerebellum following portocaval anastomosis. Brain Res. 859:293–302.

Takahashi S., Shibata M., Gotoh J., and Fukuuchi Y. (2000). Astroglial cell death induced by excessive influx of sodium ions. Eur. J. Pharmacol. 408:127–135.

Tanaka K., Watase K., Manabe T., Yamada K., Watanabe M., Takahashi K., Iwama H., Nishikawa T., Ichihara N., Kikuchi T., Okuyama S., Kawashima N., Hori S., Takimoto M., and Wada K. (1997). Epilepsy and exacerbation of brain injury in mice lacking the glutamate transporter GLT-1. Science 276:1699–1702.

Tian G., Lai L., Guo H., Lin Y., Butchbach M. E., Chang Y., and Lin C. L. (2007). Translational control of glial glutamate transporter EAAT2 expression. J. Biol. Chem. 282:1727–1737.

Torp R., Danbolt N. C., Babaie E., Bjoras M., Seeberg E., Storm-Mathisen J., and Ottersen O. P. (1994). Differential expression of two glial glutamate transporters in the rat brain: an in situ hybridization study. Eur. J. Neurosci. 6:936–942.

Torres-Salazar D. and Fahlke C. (2006). Intersubunit interactions in EAAT4 glutamate transporters. J. Neurosci. 26:7513–7522.

Vallat-Decouvelaere A. V., Chrétien F., Gras G., Le Pavec G., Dormont D., and Gray F. (2003). Expression of excitatory amino acid transporter-1 in brain macrophages and microglia of HIV-infected patients. A neuroprotective role for activated microglia? J. Neuropathol. Exp. Neurol. 62:475–485.

Vallejo-Illarramendi A., Domercq M., Pérez-Cerdá F., Ravid R., and Matute C. (2006). Increased expression and function of glutamate transporters in multiple sclerosis. Neurobiol. Dis. 21:154–164.

van Landeghem F. K., Stover J. F., Bechmann I., Bruck W., Unterberg A., Buhrer C., and von Deimling A. (2001). Early expression of glutamate transporter proteins in ramified microglia after controlled cortical impact injury in the rat. Glia 35:167–179.

van Landeghem F. K. H., Weiss T., Oehmichen M., and von Deimling A. (2006). Decreased expression of glutamate transporters in astrocytes after human traumatic brain injury. J. Neurotrauma 23:1518–1528.

Vera-Portocarrero L. P., Mills C. D., Ye Z., Fullwood S. D., McAdoo D. J., Hulsebosch C. E., and Westlund K. N. (2002). Rapid changes in expression of glutamate transporters after spinal cord injury. Brain Res. 927:104–110.

Vermeiren C., Najimi M., Vanhoutte N., Tilleux S., de Hemptinne I., Maloteaux J. M., and Hermans E. (2005). Acute up-regulation of glutamate uptake mediated by mGluR5a in reactive astrocytes. J. Neurochem. 94:405–416.

Vinatier J., Herzog E., Plamont M. A., Wojcik S. M., Schmidt A., Brose N., Daviet L., El Mestikawy S., and Giros B. (2006). Interaction between the vesicular glutamate transporter type 1 and endophilin A1, a protein essential for endocytosis. J. Neurochem. 97: 1111–1125.

Voutsinos-Porche B., Koning E., Clément Y., Kaplan H., Ferrandon A., Motte J., and Nehlig A. (2006). EAAC1 glutamate transporter expression in the rat lithium-pilocarpine model of temporal lobe epilepsy. J. Cereb. Blood Flow Metab 26:1419–1430.

Wadiche J. I., Amara S. G., and Kavanaugh M. P. (1995). Ion fluxes associated with excitatory amino acid transport. Neuron 15:721–728.

Wakabayashi Y., Yagihashi T., Kezuka J., Muramatsu D., Usui M., and Iwasaki T. (2006). Glutamate levels in aqueous humor of patients with retinal artery occlusion. Retina 26:432–436.

Wang Z., Pekarskaya O., Bencheikh M., Chao W., Gelbard H. A., Ghorpade A., Rothstein J. D., and Volsky D. J. (2003). Reduced expression of glutamate transporter EAAT2 and impaired glutamate transport in human primary astrocytes exposed to HIV-1 or gp120. Virology 312: 60–73.

Wang S., Lim G., Yang L., Sung B., and Mao J. (2006). Downregulation of spinal glutamate transporter EAAC1 following nerve injury is regulated by central glucocorticoid receptors in rats. Pain 120:78–85.

Ward M. M., Jobling A. I., Puthussery T., Foster L. E., and Fletcher E. L. (2004). Localization and expression of the glutamate transporter, excitatory amino acid transporter 4, within astrocytes of the rat retina. Cell Tissue Res. 315:305–310.

Watase K., Hashimoto K., Kano M., Yamada K., Watanabe M., Inoue Y., Okuyama S., Sakagawa T., Ogawa S., Kawashima N., Hori S., Takimoto M., Wada K., and Tanaka K. (1998). Motor discoordination and increased susceptibility to cerebellar injury in GLAST mutant mice. Eur. J. Neurosci. 10:976–988.

Wersinger E., Schwab Y., Sahel J. A., Rendon A., Pow D. V., Picaud S., and Roux M. J. (2006). The glutamate transporter EAAT5 works as a presynaptic receptor in mouse rod bipolar cells. J. Physiol.

Wilson J. M., Khabazian I., Pow D. V., Craig U. K., and Shaw C. A. (2003). Decrease in glial glutamate transporter variants and excitatory amino acid receptor down-regulation in a murine model of ALS-PDC. NeuroMolecular Med. 3:105–118.

Yamashita A., Makita K., Kuroiwa T., and Tanaka K. (2006). Glutamate transporters GLAST and EAAT4 regulate postischemic Purkinje cell death: an *in vivo* study using a cardiac arrest model in mice lacking GLAST or EAAT4. Neurosci. Res. 55:264–270.

Ye Z. C., Rothstein J. D., and Sontheimer H. (1999). Compromised glutamate transport in human glioma cells: reduction-mislocalization of sodium-dependent glutamate transporters and enhanced activity of cystine-glutamate exchange. J. Neurosci. 19:10767–10777.

Yeh T. H., Hwang H. M., Chen J. J., Wu T., Li A. H., and Wang H. L. (2005). Glutamate transporter function of rat hippocampal astrocytes is impaired following the global ischemia. Neurobiol. Dis. 18:476–483.

Yi J. H. and Hazell A. S. (2006). Excitotoxic mechanisms and the role of astrocytic glutamate transporters in traumatic brain injury. Neurochem. Int. 48:394–403.

Zerangue N. and Kavanaugh M. P. (1996). Flux coupling in a neuronal glutamate transporter. Nature 383:634–637.

Zoia C. P., Tagliabue E., Isella V., Begni B., Fumagalli L., Brighina L., Appollonio I., Racchi M., and Ferrarese C. (2005). Fibroblast glutamate transport in aging and in AD: correlations with disease severity. Neurobiol. Aging 26:825–832.

Chapter 5
Excitatory Amino Acid Receptors and Their Association with Neural Membrane Glycerophospholipid Metabolism

Glutamate mediates synaptic transmission by interacting with both ionotropic and metabotropic types of glutamate receptors (Farooqui and Horrocks, 1994, 1997). Brain tissue contains huge amounts of glutamate (about 10 mmol/kg, wet weight). Under normal conditions, due to the presence of glutamate transporters in both glial cells and neurons, the concentration of glutamate is maintained at a low level (1 μM) in the extracellular compartment (Danbolt, 1994). At low concentrations, intracellular glutamate has a trophic effect on neural cells (Balazs, 2006; Hetman and Kharebava, 2006). Following an ischemic episode or mechanical neural trauma (head injury and spinal cord injury), the extracellular concentration of glutamate increases rapidly and reaches as high as 2–3 mM (Farooqui and Horrocks, 1991, 1994). High levels of glutamate cause excessive excitation of glutamate receptors, alter activities of enzymes associated with the synthesis and degradation of neural membrane glycerophospholipids, and a number of Ca^{2+}-dependent enzymes such as protein kinase C (Suzuki et al., 1997).

High levels of extracellular glutamate also interfere with cystine uptake through the cystine/glutamate antiporter, which normally transports cystine into cells at the expense of the outflow of glutamate (Davis and Maher, 1994). This reduction in cystine uptake decreases intracellular levels of glutathione, a cysteine-containing tripeptide whose deficiency causes oxidative stress-mediated cell death. This mechanism of cell death is oxidative glutamate toxicity.

Exposure of astroglial, oligodendroglial, and microglial cell cultures to glutamate produces glial cell demise by the oxidative glutamate toxicity mechanism involving the inhibition of cystine uptake and depletion of glutathione leaving glial cells vulnerable to toxic free radicals (Oka et al., 1993; Matute et al., 2006). Similarly those neurons that either do not contain glutamate receptors or are not enriched in glutamate receptors also degenerate via oxidative glutamate toxicity (Fig. 5.1). In acute glutamate toxicity, an interaction between excitotoxicity and oxidative glutamate toxicity may play an important role in neuronal and glial cell death in brain tissue. Thus glutamate-mediated neural cell damage is accompanied by excessive Ca^{2+} influx, stimulation of Ca^{2+}-dependent enzymes (Table 5.1), generation of free radicals and reactive oxygen species (ROS), and reduced levels of glutathione (Farooqui and Horrocks, 1991, 1994). These processes, along with mitochondrial dysfunction and reduction in ATP levels, may be responsible for apoptotic and necrotic

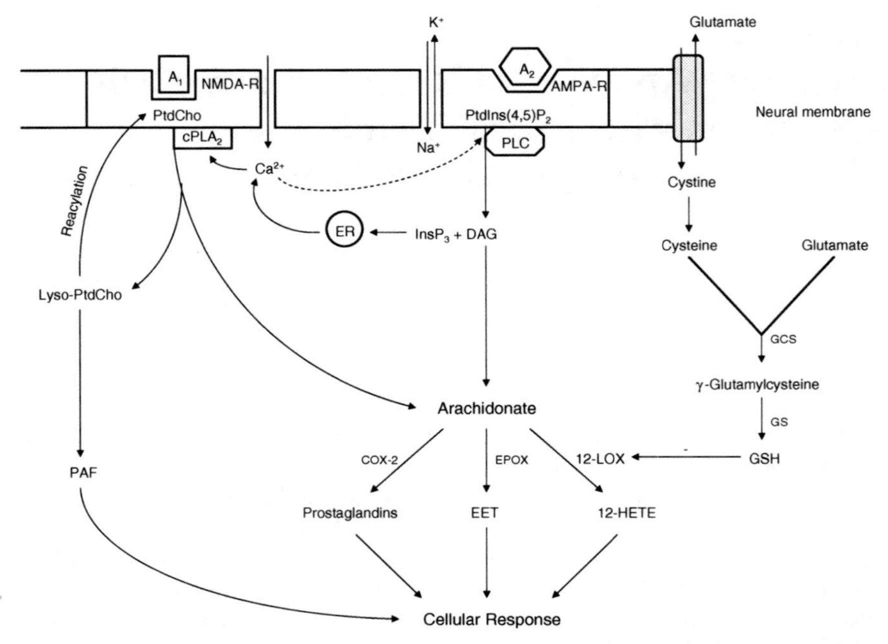

Fig. 5.1 Hypothetical model showing the interaction between excitotoxicity and oxidative glutamate toxicity. NMDA-R, *N*-methyl-D-aspartate receptor; AMPA-R, α-amino-3-hydroxy-5-methyl-4-isoxazole propionate receptor; PtdCho, phosphatidylcholine; lyso-PtdCho, lyso-phosphatidylcholine; PtdIns(4,5)P_2, inositol 4,5-bisphosphate; InsP_3, inositol 1,4,5-trisphosphate; DAG, diacylglycerol; ER, endoplasmic reticulum; PAF, platelet-activating factor; cPLA$_2$, cytosolic phospholipase A$_2$; PLC, phospholipase C; COX-2, cyclooxygenase-2; 12-LOX, 12-lipoxygenase; EPOX, epoxygenase; EET, 14,15 *cis*-epoxyeicosatrienoic acids; Cys-Glu-A. cystine/glutamate antiporter; GCS, γ-glutamylcysteine synthase; and GS, glutathione synthetase

Table 5.1 Stimulatory effect of glutamate-mediated Ca^{2+} influx on enzymic activities involved in excitotoxicity

Enzyme	Reference
Phospholipase A$_2$	(Kim et al., 1995)
Phospholipase C	(Li et al., 1998)
Diacylglycerol lipase	(Farooqui et al., 1989)
Calpain	(Siman and Noszek, 1988; Ong et al., 1997)
Calcineurin	(Halpain et al., 1990)
Protein kinase C	(Nishizuka, 1986)
Calmodulin-dependent protein kinase	(Chin et al., 1985)
Guanylate cyclase	(Novelli et al., 1987)
Nitric oxide synthase	(Gally et al., 1990; Milatovic et al., 2002)
Endonuclease	(Siesjö, 1990)

neural cell death in glutamate and its analog-mediated neurotoxicity (Farooqui and Horrocks, 1994; Wang et al., 2005; Nicholls, 2004).

5.1 Effects of Glutamate on Glycerophospholipid Synthesis

Exposure of cerebellar granule cells to glutamate, to glucose deprivation, and to the combination of both, inhibits phosphatidylcholine (PtdCho) synthesis by 71, 92 and 91%, respectively. The inhibition of PtdCho synthesis is accompanied by a decrease in the incorporation of [^{3}H]choline ([^{3}H]Cho) into phosphocholine and by an increase of the intracellular content of free [^{3}H]Cho, indicating that these treatments inhibit the synthesis of PtdCho by inhibiting choline kinase activity (Gasull et al., 2002). However, only the combined treatment with glutamate and glucose deprivation produces a prolonged inhibition of PtdCho synthesis that extends after the end of treatment. Excitotoxic death is associated with the sustained inhibition of PtdCho synthesis and suggests that this effect of the combined treatment with glutamate and glucose deprivation on PtdCho synthesis is caused by an action on an enzymic step downstream of choline kinase activity (Gasull et al., 2002). The stimulation of NMDA receptors not only increases extracellular choline in cortical neuronal cultures (Zapata et al., 1998; Gasull et al., 2000), but also reduces levels of PtdCho *in vivo*, as studied with microdialysis. Both processes are Ca^{2+}-dependent and are blocked by MK-801, an irreversible NMDA receptor antagonist. NMDA receptor-mediated choline release precedes and is directly related to excitotoxic necrotic neuronal cell death (Gasull et al., 2000).

In cortical cultures, AMPA receptor activation is also linked to choline release and inhibition of PtdCho synthesis (Gasull et al., 2001). In contrast, the selective stimulation of KA receptors has no affect on the release of choline and does not reduce PtdCho synthesis. Sensitization of AMPA receptors plays an important role during this process. Thus in the presence of cyclothiazide (CTZ), an inhibitor of AMPA receptor desensitization (Bertolino et al., 1993), AMPA receptor over stimulation becomes neurotoxic after 24 h of treatment. However, significant neuronal death does not occur until the first hour of treatment. In contrast, stimulation of non-sensitized AMPA receptors causes a marked inhibition of PtdCho synthesis within 5 min of treatment. These observations suggest that AMPA-mediated inhibition of PtdCho synthesis precedes membrane permeability changes that may cause excitotoxic necrotic neuronal cell death (Gasull et al., 2001).

It is postulated that inhibition of PtdCho synthesis and the release of choline are key steps associated with excitotoxicity and are common to NMDA and AMPA receptor stimulation. The mechanism of inhibition of PtdCho is not fully understood. Metabolic labeling experiments in cortical cultures demonstrate that NMDA receptor over activation does not modify the activity of phosphocholine or phosphoethanolamine cytidylyltransferases but strongly inhibits choline and ethanolamine phosphotransferase activities. This effect is observed well before any significant membrane damage and cell death. Moreover, cholinephosphotransferase activity is lower in microsomes from NMDA-treated cells. These results show that membrane damage by NMDA is preceded by inhibition of glycerophospholipid synthesis and not by glycerophospholipid degradation in the early stages of the excitotoxic process, and that NMDA receptor over activation decreases phosphatidylcholine and phosphatidylethanolamine synthesis by inhibiting choline and ethanolamine phosphotransferase activities (Gasull et al., 2003).

In cerebellar slices, the lowering of oxygen availability, achieved by bubbling N_2 in the medium, inhibits the incorporation of radioactive serine into phosphatidylserine (PtdSer). 7-Hydroxyiminocyclopropan[b]chromen-1a-carboxylic acid ethyl ester (CPCCOEt), an antagonist of metabotropic glutamate receptors type 1 (mGluR1), blocks this effect, whereas antagonists of NMDA or AMPA receptors have no effect (Buratta et al., 2004). In oxygenated slices, agonists of Group I mGluRs, which include mGluR1, inhibit PtdSer synthesis. CPCCOEt also retards this effect. These observations suggest that glutamate inhibits PtdSer synthesis by acting on mGluR1. This may be important in the release of glutamate under hypoxia-ischemia conditions. In cerebellar Purkinje cells, mGluR1 are involved in the generation of mGluR-EPSP evoked by parallel fiber stimulation. The administration of l-serine to cerebellar slices reduces in a dose-dependent manner the mGluR-EPSP evoked by parallel fiber stimulation. The effect is mostly due to the increased synthesis of PtdSer. Thus inhibition of PtdSer synthesis, induced by mGluR1, may participate in the generation of mGluR-EPSP (Buratta et al., 2004).

In rat hippocampal slices, L-AP3 (L-amino-3-phosphonopropionate), a putative antagonist of metabotropic glutamate receptors (Ikeda, 1993), inhibits the synthesis of [^{3}H]inositol phosphates that is mediated by (1S,3R)1-aminocyclopentane-1,3-dicarboxylic acid (t-ACPD). This inhibition is accompanied by decreased in the levels of [^{3}H]PtdIns. Preincubation of slices with L-AP3 also inhibits the incorporation of *myo*-[^{3}H]inositol into PtdIns fractions. The effect of L-AP3 contrasts with that of a typical receptor antagonist, atropine; atropine inhibited carbachol-induced phosphoinositide hydrolysis, but the levels of [^{3}H]PtdIns were not affected. These observations suggest that the inhibition of phosphoinositide hydrolysis by L-AP3 is not due to receptor antagonism, but may be caused by the inhibition of PtdIns synthesis. Collectively the above studies indicate that glutamate and its analogs modulate several signal transduction pathways that may or may not involve glutamate receptor-mediated phosphoinositide hydrolysis (Gasull et al., 2002, 2003; Buratta et al., 2004; Ikeda, 1993).

5.2 Effects of Glutamate on Glycerophospholipid Degradation

Injections of high levels of glutamate and its analogs produce toxic effects through the over-stimulation of NMDA and non-NMDA types of glutamate receptors (Lipton and Rosenberg, 1994; Farooqui and Horrocks, 1994; Matute et al., 2006) (Fig. 5.1). This process is excitotoxicity. Morphologically, excitotoxicity results in neuronal swelling, vacuolization, and eventual neurodegeneration (Farooqui and Horrocks, 1994). Biochemically, the over-stimulation of NMDA and non-NMDA receptors results in increased degradation of neuronal membrane glycerophospholipids with production of free radicals and ROS. ROS generation induces mitochondrial dysfunction. These processes, along with changes in neural membrane fluidity and permeability and with alterations in energy status and redox status of neuronal cells, are closely associated with apoptotic and necrotic cell death in mechanical and metabolic trauma and neurodegenerative diseases (Farooqui and Horrocks, 1994; Farooqui et al., 2001; Wang et al., 2005). In neurotoxicity mediated by glutamate and its analogs, the activation of astrocytes and microglial cells (astrogliosis and microgliosis) also occurs with neuronal injury. These processes contribute to the increase in pro-inflammatory cytokines. These cytokines stimulate transcription factors and cause induction of a number of genes, including inducible nitric oxide synthase (iNOS), cyclooxygenase-2 (COX-2), and $sPLA_2$ (Farooqui et al., 2007). Collectively, glutamate neurotoxicity is accompanied by glial cell activation and neuronal cell death.

Two major enzymic mechanisms mediated by glutamate are responsible for increased degradation of neural membrane phospholipids. A direct mechanism for arachidonic acid (AA) release involves phospholipase A_2 (PLA_2). The indirect mechanism utilizes the phospholipase C (PLC)/ diacylglycerol lipase pathway to release AA (Farooqui et al., 1989). The release of AA within brain tissue is of considerable physiological importance because this fatty acid is a potent modulator of glutamatergic neurotransmission and a potential mediator of both long-term potentiation and long-term depression (Williams et al., 1989). Glutamate, N-methyl-D-aspartate (NMDA), and kainate (KA) act on cortical, striatal, hippocampal, and hypothalamic neurons and cerebellar granule cells to produce a dose-dependent increase in PLA_2 activity and arachidonic acid release (Dumuis et al., 1988; Lazarewicz et al., 1990; Farooqui et al., 2003b) (Fig. 5.2-A and B). The NMDA antagonist, 2-amino-5-phosphonoverate (APV) and the AMPA receptor antagonist,

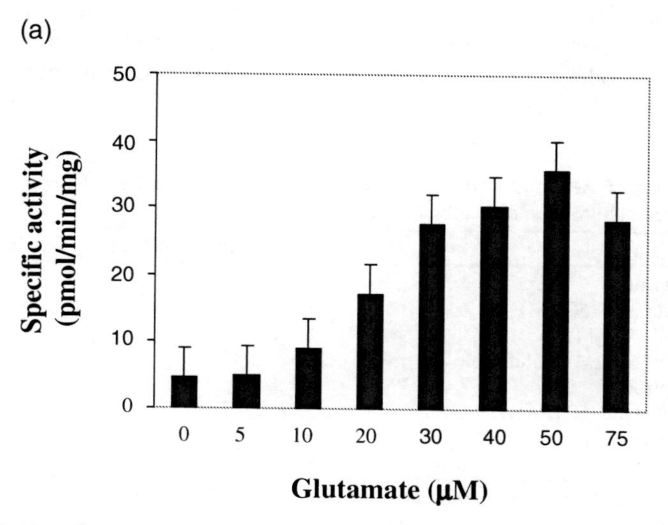

Fig. 5.2 (Continued)

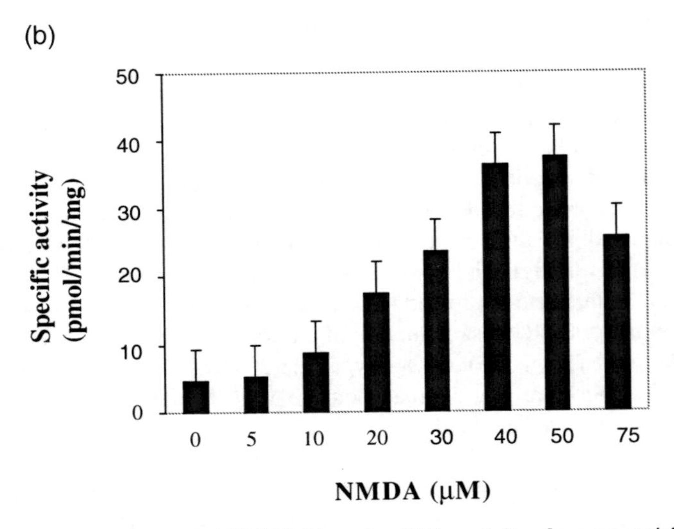

Fig. 5.2 Effect of glutamate (a) and NMDA (b) on the $cPLA_2$ activity of neuron-enriched cultures from rat cerebral cortex. $cPLA_2$ specific activity is expressed as pmol/min/mg protein

6-cyano-7-nitroquinoxaline (CNQX) block this release in a dose-dependent manner (Table 5.2). As pH shifts from 7.2 to 7.8, the glutamate-mediated release of arachidonic acid is enhanced 3-fold (Stella et al., 1995). Quinacrine, a $cPLA_2$ non-specific inhibitor (Sanfeliu et al., 1990; Kim et al., 1995; Farooqui et al., 2003b), as well as arachidonoyl trifluoromethylketone, a potent inhibitor of $cPLA_2$, also inhibit this increase in $cPLA_2$ activity mediated by glutamate (Table 5.3).

Systemic administration of KA into adult rats markedly increases $cPLA_2$ immunoreactivity in neurons at 1 and 3 days after injection (Sandhya et al., 1998). KA injection increases the $cPLA_2$ immunoreactivity in astrocytes after 1, 2, 4, 11 weeks. Increased $cPLA_2$ activity in neurons in KA-mediated toxicity may be involved in neurodegeneration, whereas the elevation of $cPLA_2$ immunoreactivity in astrocytes is associated with gliosis (Sandhya et al., 1998; Farooqui et al., 1997c)

Table 5.2 Effect of APV and CNQX on glutamate-mediated stimulation of $cPLA_2$ activity of neuron-enriched culture from rat cerebral cortex

Treatment	Sp. Activity (pmol/min/mg protein)
Control	5.2 ± 1.4
Glutamate (50 μM)	33.3 ± 6.2
APV (50 μM)	4.9 ± 1.7
APV (50 μM) +Glutamate (50 μM)	8.6 ± 2.3
Control	5.4 ± 1.5
Glutamate (50 μM)	33.7 ± 5.7
CNQX (20 μM)	5.1 ± 1.2
CNQX (20 μM) + Glutamate (50 μM)	8.7 ± 1.5

$cPLA_2$ activity was determined by procedure described earlier. Results are the means ± SEM for three different cultures. APV, 2-amino-5- phosphonovalerate; and CNQX, 6-cyano-7-nitroquinox aline.

Table 5.3 Effect of PLA_2 inhibitors on glutamate-mediated stimulation of $cPLA_2$ activity of neuron-enriched cultures of rat cerebral cortex

Treatment	Sp. Activity (pmol/min/mg protein)
Control	5.4 ± 1.5
Glutamate (50 μM)	35.7 ± 5.8
Quinacrine (50 μM)	5.2 ± 1.2
Quinacrine (50 μM) + Glutamate (50 μM)	7.5 ± 1.6
Control	5.2 ± 1.3
Glutamate (50 μM)	33.8 ± 5.7
$AACOCF_3$ (10 μM)	4.9 ± 1.8
$AACOCF_3$ (10 μM) + Glutamate (50 μM)	7.3 ± 1.9

$cPLA_2$ activity was determined by procedure described earlier. Results are the means $\pm$ SEM for three different cultures. $AACOCF_3$, arachidonoyl trifluoromethylketone.

(Fig. 5.3). Injections of glutamate and its analogs induce a marked increase in $cPLA_2$ mRNA and protein levels (Kim et al., 1995; Sandhya et al., 1998; Ong et al., 2003). Glutamate receptor antagonists (Sanfeliu et al., 1990; Kim et al., 1995), and also PLA_2 inhibitors (Ong et al., 2003), can prevent this increase in PLA_2, indicating that generation of arachidonic acid is a receptor-mediated process. These observations strongly suggest that the stimulation of $cPLA_2$ activity in neuronal cultures is a process mediated by glutamate receptors and PLA_2 is the main effector responsible for the release of AA in response to glutamate receptors. Unlike the α-adrenergic receptors, in which PLA_2 is coupled to the receptor complex through a G protein (Axelrod et al., 1988; Axelrod, 1990), the coupling of NMDA receptor to PLA_2 in cerebellar granule cells is not mediated by a G protein (Lazarewicz et al., 1990). This suggests that elevated intracellular Ca^{2+} ions triggered by the opening of NMDA receptor channels may serve as the second messenger function in translocation and activation of PLA_2. Increased arachidonic acid release is accompanied with an increase in lipid peroxidation as evidenced by accumulation of lipid peroxides, eicosanoids, 4-hydroxynonenal (4-HNE) and 4-HNE-modified protein (Farooqui and Horrocks, 2006).

Similarly, the glutamate analog kainic acid also stimulates plasmalogen selective-PLA_2 (PlsEtn-PLA_2) in a dose- and time-dependent manner. Bromoenol lactone (BEL), a PlsEtn-PLA_2 inhibitor, and CNQX, an KA/AMPA antagonist, block this stimulation (Farooqui et al., 2003a), suggesting that the stimulation of PlsEtn-PLA_2 is also a receptor-mediated process. Kainate also stimulates the $sPLA_2$ activity of neural cell cultures, like $cPLA_2$ and PlsEtn-PLA_2 (Thwin et al., 2003). Collective evidence suggests that several isoforms of PLA_2 are stimulated by glutamate and its analogs in neural cell culture systems. This stimulation of these isoforms releases arachidonic acid, decreases levels of PtdCho and PlsEtn, and increases degradation products of glycerophospholipids such as phosphomonoesters (PME) and phosphodiesters (PDE) (Farooqui et al., 2006).

PLC is linked to metabotropic glutamate receptors in neural cell cultures (Nakamura et al., 2004; Miyata et al., 2003; Hannan et al., 2001; Nicolle et al., 1999). A 2-h exposure to NMDA or KA produces a marked reduction, about 75%, in quisqualate (QA)-mediated hydrolysis of PtdIns in neural cell cultures. The reported

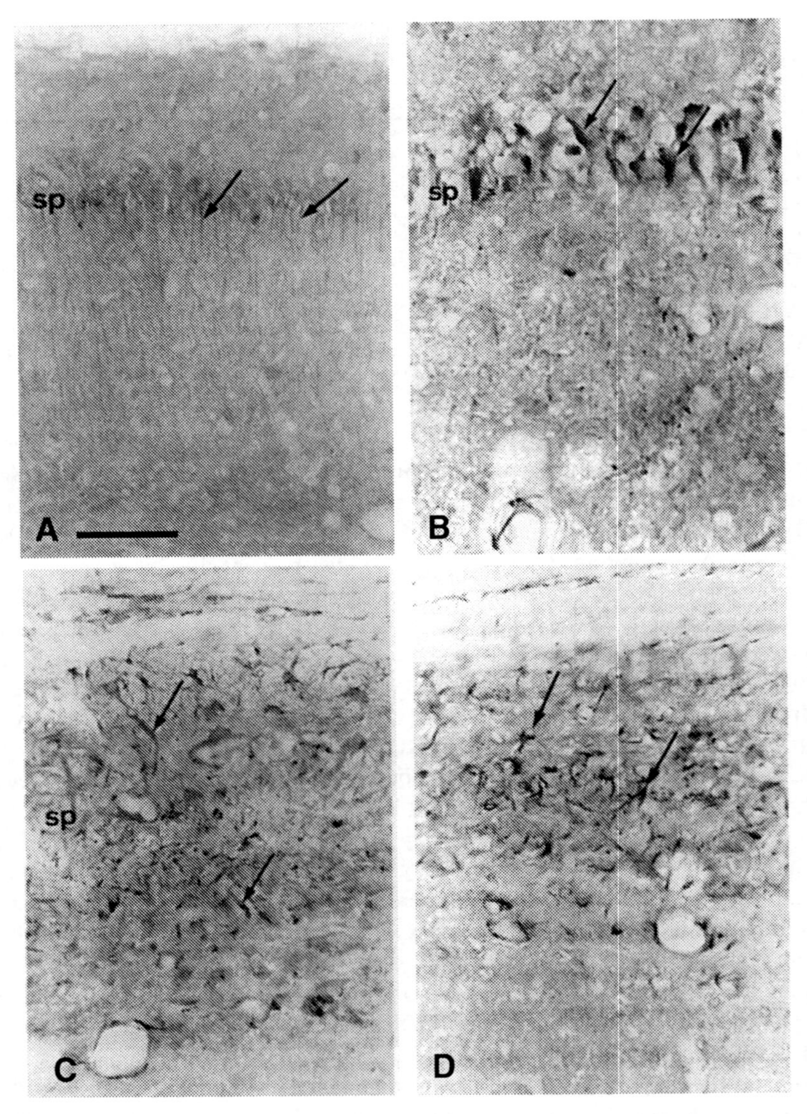

Fig. 5.3 $cPLA_2$ immunoreactivity in field CA1 of the hippocampus. A: section from a normal rat, showing light $cPLA_2$ immunoreactivity in pyramidal neurons (arrows). B: section from a rat that had been injected with kainate 3 days earlier, showing increased labeling of pyramidal neurons (arrows). C, D: sections from rats that have been injected with kainate 2 weeks (C) and 11 weeks earlier (D), showing increased labeling in astrocytes (arrows). Abbreviation: sp: stratum pyramidale. Scale: A = 200 μm, B = 80 μm, C,D = 100 μm. Reproduced with kind permission from Sandhya et al., 1998, Brain Research 788:223–231, Elsevier

efficacy of these agonists is similar, but the potencies are quite different with IC50 for NMDA about 35 μM and for KA about 70 μM. NMDA-mediated depression of PtdIns hydrolysis evoked by QA is relatively long lasting but reversible. In nominally Ca^{2+}-free medium neither NMDA nor KA attenuate PtdIns hydrolysis evoked by QA (Facchinetti et al., 1998). The effect of NMDA can be retarded by the NMDA receptor antagonist MK801, but not by the wide spectrum protein kinase inhibitor staurosporin nor by the nitric oxide synthase inhibitor *N*-ω-nitro-L-arginine. The effect of KA is blocked by the selective non-NMDA receptor antagonist 2,3-dihydroxy-6-nitro-7-sulfamoyl-benzo(F)quinoxaline (NBQX) and CNQX. Voltage-sensitive Ca^{2+} channel antagonists together with MK801 do not counteract the inhibition by KA of the QA-mediated response. The muscarinic receptor agonist carbachol stimulates, about 30%, PtdIns hydrolysis attenuated by NMDA or KA. This suggests that the activation of iGluR exerts a relatively general inhibitory effect on the function of different PLC-coupled metabotropic receptors. These observations indicate that ionotropic glutamate receptor stimulation induces a long lasting suppression of QA-mediated PtdIns hydrolysis through a Ca^{2+}-dependent mechanism that seems to involve receptor-coupled transduction systems downstream from mGluR. Such a Ca^{2+}-dependent cross-talk involving ionotropic and metabotropic receptors may play a role in certain events of synaptic plasticity (Facchinetti et al., 1998).

PLC isoforms are strongly expressed distinctly and not overlapping in the mouse cerebellum in subsets of Purkinje cells (Webber and Hajra, 1993). Based on their distribution pattern and electrophysiological studies, PLC isoforms may be involved in long-term depression. PtdIns turnover in young and aged Long-Evans rats indicates that maximal PtdIns turnover occurs in the hippocampus (Nicolle et al., 1999). These rats are behaviorally characterized for spatial learning in the Morris water maze. Stimulation of PtdIns turnover with 1S,3R ACPD indicates that metabotropic glutamate receptors mediate PtdIns turnover in the aged rats. The magnitude of the decrease in PtdIns turnover significantly correlates with age-related spatial memory decline. The decrease in mGluR-mediated PtdIns turnover occurs without changes in the protein level of either the mGluR or the G-protein coupled to those receptors. A significant decrease in the immunoreactivity of PLCβ-1 has observed in the hippocampus of aged rats and this PLCβ-1 immunoreactivity correlates with spatial learning only when the young and aged rats are considered together (Nicolle et al., 1999). The decrease in mGluR-mediated signal transduction in the hippocampus that is related to cognitive impairment in aging may be attributable, at least in part, to a deficiency in the enzyme PLCβ-1 (Nicolle et al., 1999). Collective evidence suggests that age-related alteration in this signal transduction system associated with PLCβ-1 may provide a functional basis for cognitive decline independent of any loss of neurons in the hippocampus.

The involvement of PLC in ionotropic and metabotropic glutamate receptor-mediated neurotoxicity has also been reported in primary cultures of cerebellar neurons (Miyata et al., 2003; Facchinetti et al., 1998; Nicolle et al., 1999). It is shown that 1-[6-[[(17b)-3-methoxyestra-1,3, 5(10)-trien-17-yl]amino] hexyl]-1H-pyrrole-2,5-dione (U-73122) and 1-*O*-octadecyl-2-*O*-methyl-*rac*-glycero-3-phosphocholine (Et-18-OCH_3), two agents that inhibit PLC, block glutamate and

NMDA-mediated neurotoxicity (Llansola et al., 2000). Both compounds prevent glutamate neurotoxicity at concentrations lower than those required to inhibit carbachol-induced hydrolysis of inositol phospholipids. Furthermore, there is a good correlation between the concentrations of U-73122 and Et-18-OCH_3 required to prevent NMDA-induced hydrolysis of glycerophospholipids and those required to prevent glutamate and NMDA neurotoxicity. NMDA-induced hydrolysis of glycerophospholipids is also inhibited by nitroarginine, an inhibitor of nitric-oxide synthase, and is mimicked by the nitric oxide-generating agent *S*-nitroso-*N*-acetyl-penicillamine. Glutamate neurotoxicity mediated by activation of NMDA receptors leads to activation of nitric-oxide synthase and increase in the generation of nitric oxide. This results in increased PLC activity. Inhibition of PLC by U-73122 or Et-18-OCH_3 prevents glutamate-mediated neuronal death (Llansola et al., 2000).

The treatment of neuron-enriched cultures with glutamate or NMDA results in a dose- and time-dependent stimulation of diacylglycerol and monoacylglycerol lipase activities (Figs. 5.4, 5.5, and 5.6) (Farooqui et al., 1993). The NMDA antagonist, dextrorphan (Table 5.4) and the diacylglycerol lipase inhibitor, RHC 80267 (Farooqui et al., 1993, 2003b) can block this increase (Table 5.5). Like the glutamate-mediated stimulation in vitro of diacylglycerol lipase and monoacyl-glycerol lipase activities, ibotenate also causes a 2- to 4-fold increase in enzymic activities in plasma membrane (PM) and synaptosomal plasma membrane (SPM) fractions at 10 days after injection (Figs. 5.7 and 5.8) (Farooqui et al., 1991). PM and SPM fractions prepared from brains of animals killed at 1 day and 3 day after ibotenate injection do not show any increase in diacylglycerol lipase and monoa-cylglycerol lipase activities. However, membrane preparations from rats killed 6 months after injection still show 3- to 4-fold increases in the specific activities of diacylglycerol and monoacylglycerol lipases (Fig. 5.9 and 5.10).

Unlike glutamate and NMDA, the treatment of neuron-enriched cultures with KA (100 μM) has no effect on diacylglycerol and monoacylglycerol lipase activities (Farooqui et al., 1993; Farooqui and Horrocks, 1994). This suggests that

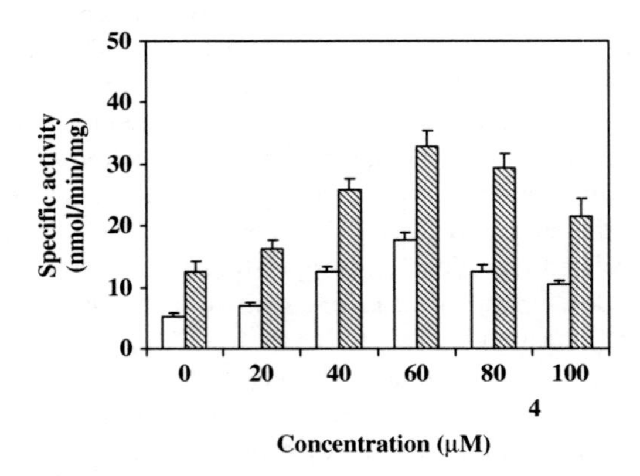

Fig. 5.4 Effects of glutamate on diacylglycerol lipase and monoacylglycerol lipase activities of neuron-enriched cultures from fetal mouse spinal cord. Diacylglycerol lipase (clear bars) and monoacylglycerol glycerol lipase (hatched bars). Data modified from Farooqui et al., 1993

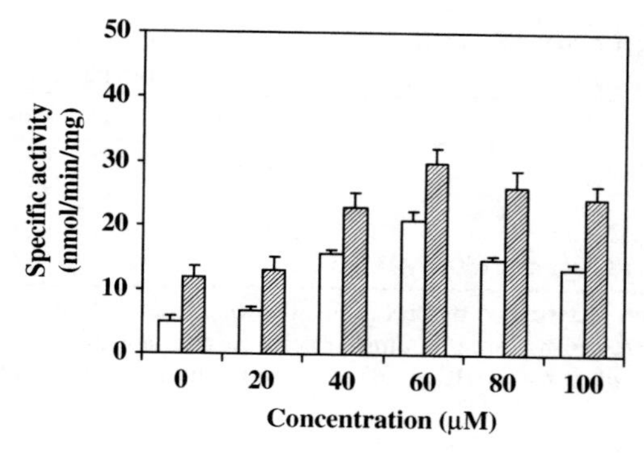

Fig. 5.5 Effects of NMDA on diacylglycerol lipase and monoacylglycerol lipase activities of neuron-enriched cultures from fetal mouse spinal cord. Diacylglycerol lipase (clear bars) and monoacylglycerol glycerol lipase (hatched bars). Data modified from Farooqui et al., 1993

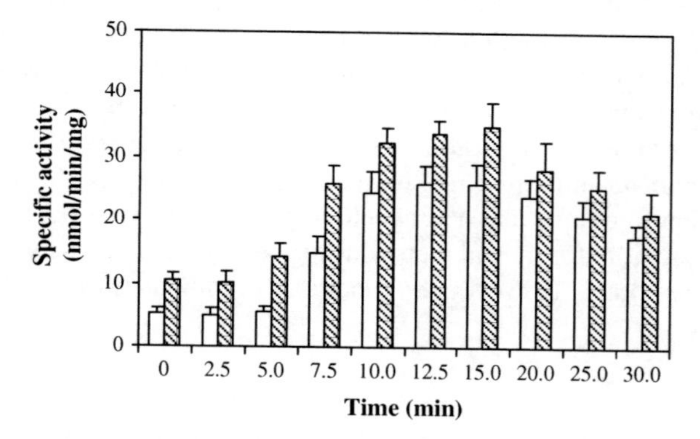

Fig. 5.6 Time-dependence of NMDA effect on diacylglycerol lipase and monoacylglycerol lipase activities of neuron-enriched cultures from fetal mouse spinal cord . Diacylglycerol lipase (clear bars) and monoacylglycerol glycerol lipase (hatched bars). Data modified from Farooqui et al., 1993

either KA receptors are not linked to diacylglycerol and monoacylglycerol lipases or these cultures do not have fully developed KA receptors. No information is available on diacylglycerol and monoacylglycerol lipase activities in KA-injected rat brain. Collectively, the studies above indicate that both PLA_2 isoforms and

Table 5.4 Effect of dextrorphan on glutamate-mediated stimulation of diacylglycerol lipase and monoacylglycerol lipase activities of neuron-enriched cultures from rat cerebral cortex

Treatment	Diacylglycerol lipase	Monoacylglycerol lipase
Control	5.0 ± 0.7	11.7 ± 1.2
Glu (50 μM)	20.7 ± 1.6	25.7 ± 2.9
Dextror (50 μM)	4.9 ± 0.5	10.7 ± 1.1
Dextror (50 μM) + Glu (50 μM)	6.9 ± 0.8	14.7 ± 1.8

Enzymic activities were determined by procedures described earlier (Farooqui et al., 1993). Specific activity was expressed as nmol/min/mg protein. Results are the means ± SEM for three different cultures. Glu, glutamate; and Dextror, Dextrorphan.

Table 5.5 Effect of RHC 80267 on glutamate-mediated stimulation of diacylglycerol lipase and monoacylglycerol lipase activities of neuron-enriched cultures from rat cerebral cortex

Treatment	Diacylglycerol lipase	Monoacylglycerol lipase
Control	5.3 ± 0.5	12.3 ± 1.8
Glu (50 μM)	19.7 ± 1.8	26.7 ± 2.7
RHC 80267 (50 μM)	5.6 ± 0.5	12.8 ± 1.5
RHC 80267 (50 μM) + Glutamate (50 μM)	5.7 ± 0.8	13.6 ± 1.7

Enzymic activities were determined by procedures described earlier (Farooqui et al., 1993). Specific activity was expressed as nmol/min/mg protein. Results are the means ± SEM for three different cultures. Glu, glutamate. RHC 80267 is a potent inhibitor of diacylglycerol lipase.

PLC/ diacylglycerol lipase pathway participate in the release of AA in neural cells. However, the relative contributions of these pathways for AA release are still obscure. Furthermore, in glutamate neurotoxicity, high levels of extracellular glutamate inhibit the cellular uptake of cystine. PtdCho-specific PLC, an enzyme responsible for the degradation of PtdCho and generation of diacylglycerol, modulates this process. As stated above, this second messenger is a substrate for diacylglycerol lipase in brain tissue (Li et al., 1998). Glutamate mediates both PtdCho-PLC/diacylglycerol lipase-induced arachidonic acid release and the uncoupling of the cystine/glutamate antiporter in brain tissue. These processes may be responsible for glutamate neurotoxicity and neural cell death in brain tissue (Li et al., 1998).

In brain tissue, ionotropic and metabotropic receptors are closely located on the postsynaptic membrane. However, the functional significance of the close location

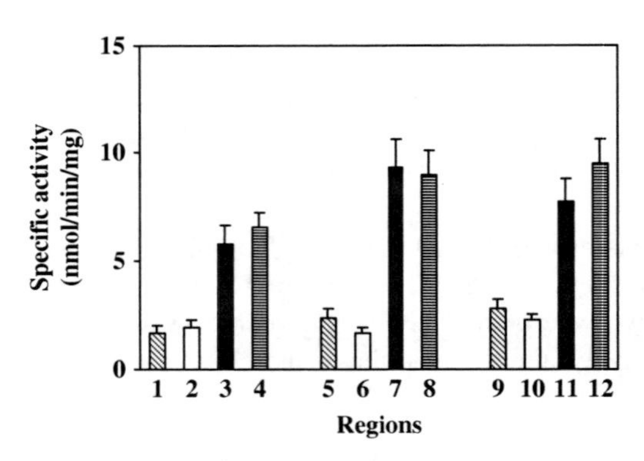

Fig. 5.7 Effect of ibotenate injections on diacylglycerol lipase activity of plasma and synaptosomal plasma membrane fractions prepared from different regions of rat brain. Ibotenate was injected in the right hippocampal region. Controls were injected with saline on the left side. Coordinates for injections were 0.9 mm posterior and 2.6 mm lateral from bregma and 6.8 mm ventral from the surface of the brain (tooth bar set at 2 mm below the interural line. Brains were removed 10 days after ibotenate injections, dissected into frontal cortex (fractions 1–4), hippocampal region (fractions 5–8), and midbrain (fractions 9–12) and used for the preparation of PM and SPM. Saline injected PM (1); saline injected SPM (2); ibotenate injected PM (3); ibotenate injected SPM (4); saline injected PM (5); saline injected SPM (6); ibotenate injected PM (7); ibotenate injected SPM (8); saline injected PM (9); saline injected SPM (10); ibotenate injected PM (11); ibotenate injected SPM (12). Data modified from Farooqui et al., 1991

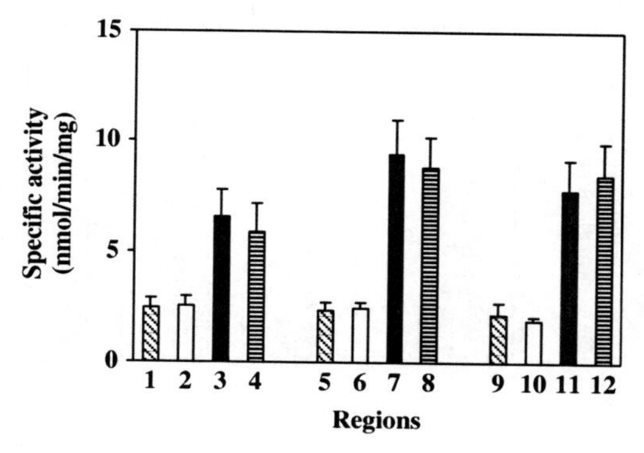

Fig. 5.8 Effect of ibotenate injections on monoacylglycerol lipase activity of plasma and synaptosomal plasma membrane fractions prepared from different regions of rat brain. Ibotenate was injected in the right hippocampal region. Controls were injected with saline on the left side. Coordinates for injections were 0.9 mm posterior and 2.6 mm lateral from bregma and 6.8 mm ventral from the surface of the brain (tooth bar set at 2 mm below the interural line. Brains were removed 10 days after ibotenate injections, dissected into frontal cortex (fractions 1–4), hippocampal region (fractions 5–8), and midbrain (fractions 9–12) and used for the preparation of PM and SPM. Saline injected PM (1); saline injected SPM (2); ibotenate injected PM (3); ibotenate injected SPM (4); saline injected PM (5); saline injected SPM (6); ibotenate injected PM (7); ibotenate injected SPM (8); saline injected PM (9); saline injected SPM (10); ibotenate injected PM (11); ibotenate injected SPM (12). Data modified from Farooqui et al., 1991

of the two types of glutamate receptors remains unknown. Based on pharmacological, electrophysiological, and two photon laser scanning microscopic studies, functional interaction and a robust cross talk between ionotropic and metabotropic glutamate receptors may occur in inositol 1,4,5-trisphosphate (InsP_3) synthesis induced in parallel fibers in cerebellar Purkinje cells (Okubo et al., 2004). Collective evidence suggests an extensive cross talk between second messengers generated by signal transduction processes mediated by ionotropic and metabotropic glu-

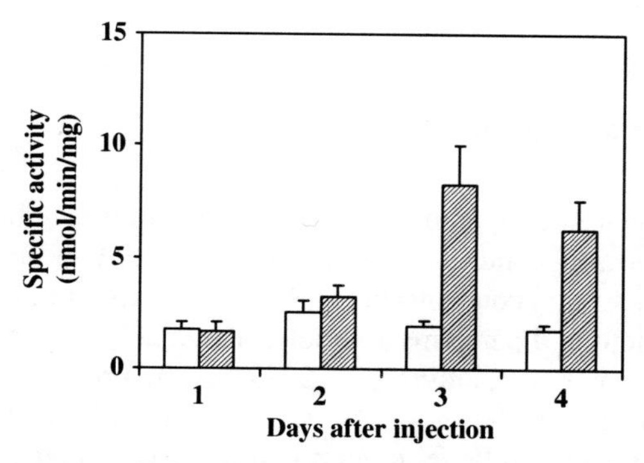

Fig. 5.9 Temporal changes in the diacylglycerol lipase activity after ibotenate injection. Sodium chloride injected (clear bars) and ibotenate injected (hatched bar). Brains were removed after 1 day (1), 3 days (2), 10 days (3), and 6 months (4) after ibotenate injections. Diacylglycerol lipase activity was determined in 10,000 rpm supernatant (having PM and SPM) fraction from hippocampal region. Data modified from Farooqui et al., 1991

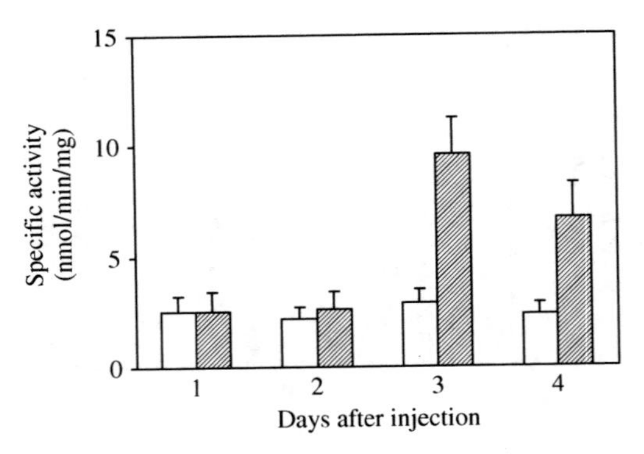

Fig. 5.10 Temporal changes in the monoacylglycerol lipase activity after ibotenate injection. Sodium chloride injected (clear bars) and ibotenate injected (hatched bar). Brains were removed after 1 day (1), 3 days (2), 10 days (3), and 6 months (4) after ibotenate injections. Monoacylglycerol lipase activity was determined in 10,000 rpm supernatant (having PM and SPM) fraction from hippocampal region. Data modified from Farooqui et al., 1991

tamate receptors in response to prolonged stimulation by glutamate. These processes involve neural membrane glycerophospholipid degradation through PLA_2 and PLC. Detailed investigations are crucial for understanding the molecular mechanisms associated with cross talk among second messengers derived from the action of PLA_2 and PLC on synaptic membrane glycerophospholipids. The cross talk between two types of glutamate receptors has been implicated in neuronal plasticity associated with long-term potentiation and long-term depression, neurotoxicity, actions of abused drugs, and neurodegenerative diseases.

5.3 Physiological and Pathophysiological Effects of Released AA in Brain

5.3.1 Physiological Effects of AA

AA is a major polyunsaturated fatty acid present in neural membrane glycerophospholipids. Under normal conditions, the levels of free AA in brain tissue are very low, <10 μmol/kg. At this concentration AA acts as a second messenger molecule in the nervous system and performs a variety of functions. It regulates the activity of many enzyme proteins including protein kinase A, protein kinase C, NADPH oxidase, choline acetyltransferase, and caspase-3. In addition, AA modulates ion channels, neurotransmitter release, induction of long-term potentiation, and neural cell differentiation (Farooqui et al., 2002). AA may act as a facilitatory retrograde neuromodulator in glutamatergic synapses (Katsuki and Okuda, 1995), because it is released upon activation of glutamate receptors. In the nucleus, AA interacts with elements of gene structure, such as promoters, enhancers, suppressors, etc., to modulate gene expression in a specific manner that is not shared by eicosanoids or other fatty acids (Farooqui et al., 1997a).

AA has trophic effects on PC12 cells, hippocampal neurons, and cerebellar neurons in vitro. At 1 μM this fatty acid significantly potentiates the elongation of neurites in hippocampal cultures (Katsuki and Okuda, 1995). AA activates PKC, an enzyme involved in neuronal plasticity and the differentiation and elongation of neurites (Murakami and Routtenberg, 2003). The addition of AA to rat dentate gyrus produces a slow onset and persistent increase in synaptic activity, which is accompanied by a marked elevation in the release of glutamate (Lynch and Voss, 1990). AA generated by NMDA receptor stimulation at the postsynaptic level may cross the synaptic cleft to act at the presynaptic level and thus may act as a retrograde messenger for long-term potentiation. The age-dependent suppression of LTP may be a consequence of the AA decline in the hippocampal dentate gyrus due to long-term exposure of neurons to oxidative stress (McGahon et al., 1997) or suppression by oxygen free radicals. Supplementation of AA in neural membranes influences both the pre-and postsynaptic membranes through glycerophospholipid metabolic pathways to facilitate the mobility of functional proteins in the plasma membrane due to increased fluidity (McGahon et al., 1997). Collectively these results suggest that low levels of AA are involved in maintaining the structural integrity of neural membranes, determining the neural membrane fluidity, and thereby regulating neuronal transmission and long-term potentiation.

5.3.2 Pathophysiological Effects of AA

During ischemic injury, glutamate-mediated activation of PLA_2 isozymes results in a general increase in the release and accumulation of all fatty acids. However, the highest increase is observed for AA (Zhang and Sun, 1995). Under pathological conditions such as ischemia, AA is released from membrane glycerophospholipids and accumulates, 0.5 mmol/kg. At high concentrations, AA produces a variety of detrimental effects on neural cell structure and function. AA causes intracellular acidosis and uncouples oxidative phosphorylation, which results in mitochondrial dysfunction (Schapira, 1996). AA produces mitochondrial swelling in neurons and induces changes in membrane permeability by regulating ion channels (Farooqui et al., 1997a,b). AA has a profound adverse effect on the ATP-producing capacity of mitochondria. It also affects the activity of membrane-bound enzymes, together with neurotransmitter release and uptake in neuronal preparations (Schapira, 1996). Arachidonic acid activates nuclear factor-κB (NF-κB) and decreases neuronal viability (Toborek et al., 1999).

The accumulation of free fatty acids can trigger an uncontrolled "arachidonic acid cascade". This sets the stage for increased production of reactive oxygen species (ROS). ROS include oxygen free radicals (superoxide radicals, hydroxyl and alkoxyl radicals, lipid peroxy radicals), and peroxides (hydrogen peroxide and lipid hydroperoxide). At low levels, ROS can function as signaling intermediates in the regulation of fundamental cell activities such as growth and adaptation responses.

At higher concentrations, ROS contribute to neural membrane damage when the balance between reducing and oxidizing (redox) forces shifts toward oxidative stress. The elimination of ROS is controlled, in part by, free radical scavengers including superoxide dismutase, catalase, and glutathione. Thus an uncontrolled sustained increase in calcium ion influx through increased phospholipid degradation can lead to increased membrane permeability and stimulation of many enzymes associated with lipolysis, proteolysis, and disaggregation of microtubules with a disruption of cytoskeleton and membrane structure (Farooqui and Horrocks, 1994).

AA is metabolized to 4-hydroxynonenal (4-HNE). 4-HNE is an α, β unsaturated aldehyde which reacts with nucleophilic sites of proteins on lysine, cysteine, and histidine residues. 4-HNE is capable of altering the function of key membrane proteins including glucose transporter, glutamate transporter, and Na^+,K^+-ATPase (Jamme et al., 1995; Friguet et al., 1994; Kooy et al., 1994). Such modification can compromise their function and promote protein aggregation. Inhibition of Na^+,K^+-ATPase by 4-HNE can result in the depolarization of neuronal membranes leading to the opening of NMDA receptor channels and influx of additional Ca^{2+} into the cell. Electrophysiological studies indicate that 4-HNE exerts a biphasic effect on the NMDA current. Thus, in hippocampal neurons an early enhancement of NMDA current within the first 30–120 min is followed by a delayed decrease in current that is significantly less than the basal current at 6 h after 4-HNE exposure (Lu et al., 2001). The early enhancement of NMDA current may involve increased phosphorylation of NR1, a NMDA receptor subunit, and the delayed suppression of NMDA current by 4-HNE may be due to the depletion of ATP resulting in impairment of the NMDA receptor channel function (Lu et al., 2001). In cortical neurons, 4-HNE disrupts G-protein-linked muscarinic cholinergic receptors (mAChR) and metabotropic glutamate receptors (mGluRs). This may alter the activity of phospholipase C, indicating that 4-HNE modulates signal transduction processes in brain tissue.

5.4 Physiological and Pathophysiological Effects of Lyso-Glycerophospholipids

5.4.1 Physiological Effects of Lyso-Glycerophospholipids

The stimulation of PLA_2 isoforms mediated by glutamate generates lyso-glycerophospholipids. These metabolites are either hydrolyzed to fatty acid and glycerophosphobase by lyso-phospholipases or reacylated to the native phospholipid by CoA-dependent or CoA-independent acyltransferase reactions (Farooqui et al., 2000). Lysophospholipid: lysophospholipid transacylase also regulates lyso-glycerophospholipid levels in neural membranes. These reactions keep the levels of lyso-glycerophospholipids very low in brain tissue (Farooqui et al., 2000). In endothelial cells lyso-PtdCho modulates Ca^{2+} signals and inhibits the phosphorylation of nitric oxide synthase and $cPLA_2$ (Millanvoye-Van Brussel et al., 2004).

Lyso-PtdCho is not simply a lipid metabolite producing neurotrophic and neurotoxic effects. It participates in many signal transduction processes. Lyso-PtdCho activates protein kinases such as protein kinase C, protein kinase A, and c-jun terminal kinase. Lyso-PtdCho stimulates phospholipase D and inhibits CTP-phosphocholine cytidylyltransferase (Gómez-Muñoz et al., 1999; Boggs et al., 1995). Lyso-PtdCho acts as an agonist for certain G-protein coupled receptors and can be converted to another bioactive lipid, lyso-phosphatidic acid.

Lyso-PtdCho promotes the activation of microglia and other immune cells and induces the de-ramification of murine microglia (Schilling et al., 2004). The de-ramification of microglial cells can be prevented by inhibition of non-selective cation channels and K^+-Cl^- co-transporters. It is proposed that de-ramification results in complete retraction of cell extensions and increased size of cell bodies with amoeboid morphology. Lyso-PtdCho also stimulates cell motility and releases proinflammatory cytokines. Lyso-PtdCho modulates ion channel permeability in various brain preparations through at least three different mechanisms. By incorporation into neural membranes, it can perturb the orderly packing of phospholipid bilayers inducing alterations of the normal conformation of integral membrane proteins such as ion channels. Secondly, lyso-PtdCho can interact directly with ion channel proteins, and finally, lyso-PtdCho can modulate ion channels by modulating signal transduction processes. Thus in neurons, lyso-PtdCho produces prolonged hyperpolarization of K^+channels (Maingret et al., 2000). Under certain conditions, lyso-PtdCho also causes cell fusion. Thus, lyso-PtdCho may be involved in cell-cell and membrane-membrane interactions, and neurotransmitter release.

5.4.2 Pathophysiological Effects of Lyso-Glycerophospholipids

In pathological situations such as ischemia and epileptic conditions, overstimulation of PLA_2 isoforms mediated by glutamate results in PtdCho hydrolysis and subsequent accumulation of lyso-PtdCho and free fatty acids in brain tissue (Sun and Foudin, 1984; Yegin et al., 2002). Stimulation of phospholipid methylation mediated by methyl-4-phenyl-1,2,3,6-tetrahydropyridine (MPTP) suggests the involvement of lyso-PtdCho in the pathogenesis of Parkinson's disease (PD) (Lee et al., 2004, 2005b,c). The addition of S-adenosylmethionine (AdoMet) to the cell culture medium increases MPP^+-induced cytotoxicity. Interestingly, the addition of lyso-PtdCho to PC12 cell cultures also has a similar effect, but with at least 10 times higher potency. Thus lyso-PtdCho induces dopamine release and inhibits dopamine uptake in PC12 cells. Lyso-PtdCho also causes a decreased mitochondrial potential and increases the formation of reactive oxygen species in PC12 cells. These results indicate that the phospholipid methylation pathway may be involved in MPP^+ neurotoxicity and lyso-PtdCho may play an important role in MPP^+-induced neurotoxicity. Also, injections of lyso-PtdCho into the lateral ventricle of rat brain cause alterations in locomotor activity and induce changes in biogenic amine levels. Quinacrine, a non-specific PLA_2 inhibitor, protects against changes induced by lyso-PtdCho in this PD model. The hypokinesia observed following the administration of lyso-PtdCho might be related to the decline in DA turnover in the striatum in response to lyso-PtdCho exposure (Lee et al., 2005b,c). Collective evidence thus suggests that inhibitory effect of lyso-PtdCho on dopaminergic neurotransmission is one of the contributing factors in AdoMet and MPP^+-induced PD-like changes and may be associated with the pathogenesis of PD (Lee et al., 2004, 2005b,c).

Focal injections of lyso-PtdCho produce demyelination (Woodruff and Franklin, 1999; Birgbauer et al., 2004). At high concentrations, lyso-PtdCho acts as a detergent and solubilizes neural membranes (Woodruff and Franklin, 1999; Birgbauer et al., 2004). The collective evidence suggests that the accumulation of high levels of lyso-glycerophospholipids in pathological situations induces neural cell demyelination and injury.

5.5 Physiological and Pathophysiological Effects of PAF

5.5.1 Physiological Effects of PAF

PLA_2 stimulation mediated by glutamate generates 1-alkyl-2-lyso-*sn*-glycero-3-phosphocholine (lyso-platelet activating factor, lyso-PAF). Lyso-PAF is then acetylated by acetyltransferase at the *sn*-2 position yielding platelet-activating factor (PAF). PAF is a short-lived biologically active ether lipid with diverse physiological and pathophysiological activities (Snyder, 1995). PAF increases mitogen-activated protein (MAP) kinases in primary hippocampal neurons in vitro (Mukherjee et al., 1999; DeCoster et al., 1998). Extracellular signal-regulated kinases, c-Jun N-terminal kinase, and p38 kinases are activated by PAF in hippocampal neurons. The receptor antagonist BN 50730 can prevent this activation. The PAF receptor antagonist BN 50730 also retards activation of kainate receptors (DeCoster et al., 1998). In contrast, CNQX has no effect on PAF activation of the kinases, indicating that PAF is downstream to kainate activation. Co-application of submaximal concentrations of PAF and kainate produces a less than additive activation, suggesting similar routes of activation by the two agonists. Both CNQX and BN 50730 block kainate-mediated neurotoxicity. These results indicate that PAF and kainate activate similar kinase pathways (Mukherjee et al., 1999; DeCoster et al., 1998).

Long-term potentiation (LTP) is a long-lasting enhancement of synaptic efficacy due to repeated stimulation of postsynaptic NMDA receptors. An influx of Ca^{2+}, stimulation of $cPLA_2$, and the release of arachidonic acid accompany LTP. It depends on gene expression, protein synthesis, and the establishment of new neuronal connections. PAF antagonists block the development of LTP (del Cerro et al., 1990; Kato et al., 1994; Kato and Zorumski, 1996; Kornecki et al., 1996), indicating that PAF modulates LTP. In rats, the administration of PAF antagonists impairs spatial learning and inhibitory avoidance tests, whilst treatment with a synthetic non-hydrolysable analog of PAF, 1-*O*-hexadecyl-2-methylcarbamoyl-*sn*-glycerol-3-phosphocholine, enhances memory (Packard et al., 1996). PAF may also be involved in plasticity responses because PAF leads to the expression of early response genes and subsequent gene cascades. The PAF antagonist, BN 50730, blocks PAF-mediated induction of gene expression. Besides this, PAF is involved in stimulation and modulation of PLA_2, PLC, PLD, and COX activities (Farooqui and Horrocks, 2004). Collectively these studies suggest that PAF plays an important role not only in learning and memory processes but also in gene expression.

5.5.2 Pathophysiological Effects of PAF

PAF is involved in inflammation, allergic reactions, and immune responses. It is a potent inducer of gene expression in CNS. A wide variety of cells including macrophages, platelets, endothelial cells, mast cells, neutrophils, and neural cells release PAF. It exerts its biological effects by activating PAF receptors that consequently activate leukocytes, stimulate platelet aggregation, and induce the release of cytokines and expression of cell adhesion molecules (Snyder, 1995; Maclennan et al., 1996; Ishii et al., 2002; Honda et al., 2002). The NMDA receptor/ nitric oxide (NO) signaling pathway is associated with PAF-induced neurotoxicity. The degree of neuronal death in cultures increased in a dose-dependent manner during exposure to PAF for 24 h (Xu and Tao, 2004). The PAF antagonist (BN52021), and also MK-801, an NMDA antagonist, and L-NAME, a nitric oxide synthase (NOS) inhibitor retarded the neurodegenerative effect of PAF significantly. Moreover, BN52021 and MK-801 dramatically block the increases in NOS activity and neuronal NOS expression induced by chronic exposure of the cultured neurons to PAF. Collectively these studies suggest that the NMDA receptor/ NO signaling pathway may contribute to the pathological mechanism of cell death triggered via PAF receptor activation (Xu and Tao, 2004). Furthermore, PSD, a molecular adaptive protein binds to and clusters the NMDA receptor and assembles a specific set of signaling proteins, such as neuronal nitric oxide synthase, around the NMDA receptor at synapses in the brain tissue. Targeted disruption of the PSD-93 gene significantly retards the neurotoxicity triggered by PAF receptor activation dependent on NMDA receptor-nitric oxide signaling (Xu et al., 2004). In addition, the deficiency of PSD-93 markedly attenuates the PAF-induced increase in cGMP and blocks PAF-promoted formation of the NMDA receptor-neuronal nitric oxide synthase complex. Thus, PSD-93 is involved in the NMDA receptor-nitric oxide-mediated pathological processing of neuronal damage triggered via platelet-activating factor receptor activation (Xu et al., 2004). Since platelet-activating factor is a potent neuronal injury mediator during the development of brain trauma, seizures, and ischemia, PSD-93 may contribute to the molecular mechanisms of neuronal damage. The addition of PAF to isolated rat brain mitochondrial preparation induces mitochondrial swelling, membrane permeability transition (MPT), and release of cytochrome c (Parker et al., 2002). The PAF antagonist BN50730 can block this process. This is further evidence for the involvement of PAF in neural cell injury.

5.6 Physiological and Pathophysiological Effects of Eicosanoids

5.6.1 Neurotrophic Effects

Glutamate and its analogs modulate the generation of prostaglandins, leukotrienes, and thromboxanes from AA. These metabolites are known collectively as eicosanoids (Phillis et al., 2006). They differ from other intracellular messengers in one important way. They are not stored in neural cells, but are synthesized rapidly in response to receptor-mediated stimulation as modulators of neural cell response. Eicosanoids can leave the cell in which they are generated, crossing the membrane, to act on neighboring cells because of their amphiphilic nature. This process is observed in all mammalian tissues including brain and may be involved in the cross talk and interplay among various neural cells. In brain tissue, both neurons and glial cells produce prostaglandins, whereas cerebral microvessels and the choroid plexus mainly synthesize thromboxanes. Eicosanoids act through specific superficial or intracellular receptors, modulating signal transduction pathways and gene transcription. Thus PGD_2 activates the DP receptors, PGE_2 activates the EP receptors, and $PGF_{2\alpha}$, PGI_2, and TXA_2, respectively, stimulate the FP, IP, and TP receptors (Coleman et al., 1994). Eicosanoid receptors typically have seven transmembrane segments coupled with G proteins. They have an extracellular amino terminus and intracellular carboxyl terminus. These receptors are involved in the generation of cyclic AMP, diacylglycerol, and phosphatidylinositol 1,4,5-trisphosphate and modulation of Ca^{2+} influx. By interacting with eicosanoid receptors on astrocytes, the prostaglandins regulate glutamate release into the synaptic cleft (Bazan, 2003). Released glutamate modulates neuronal excitability and synaptic transmission at the presynaptic level. In contrast, the uptake of glutamate by astrocytes prevents its neurotoxic accumulation in the synaptic cleft. Eicosanoids modulate glutamate receptors in the hippocampus (Chabot et al., 1998). Collectively these studies suggest cross talk among glutamate, prostaglandin, leukotriene, and thromboxane receptors. Under normal conditions, this cross talk refines their communication, but under pathological situations, it promotes neuronal injury that depends on the magnitude of PLA_2 expression and generation of arachidonic acid metabolites. Our emphasis on the interaction between glutamate and eicosanoid receptors does not rule out the participation of other mechanisms involved in neural cell injury. However, it is timely to suggest that high levels of glutamate and eicosanoids in brain tissue and interactions between glutamate and eicosanoid receptors may modulate the intensity of oxidative stress, neuroinflammation, and neural cell injury in neural trauma and various neurodegenerative diseases (Farooqui and Horrocks, 2006).

NMDA receptor-mediated generation of the eicosanoids PGI_2 and PGE_2 is associated with nociceptive processing. In the facial carrageenan injection model of pain, significantly greater responses have been observed in mice injected with PGE_1 or PGE_3 plus carrageenan at 8 and 72 h after injection, compared with controls injected with DMSO plus carrageenan (Vahidy et al., 2006). Injection of PGE_2 also produces greater responses compared to the above controls at 8 h, but reduced responses at

24 h after injection. These studies suggest that prostaglandins play an important role in nociceptive processing, although the molecular mechanism associated with these processes is not fully understood. However, the stimulation of spinal NMDA receptors initiates activation of the p38 mitogen-activated protein kinase (p38 MAPK) pathway, leading to spinal release of prostaglandins during nociceptive processing (Svensson et al., 2003a,b). Intrathecal injections of SD-282 (a selective p38 MAPK inhibitor) modulate the NMDA-mediated release of PGE_2 and thermal hyperalgesia. Inhibition of spinal p38 MAPK attenuates both NMDA-evoked release of PGE_2 and thermal hyperalgesia. NMDA injections result in increased phospho-p38 MAPK immunoreactivity in superficial (I-II) dorsal laminae. Activated p38 MAPK is predominantly in microglia but also in a small subpopulation of neurons. Taken together these studies suggest a role for p38 MAPK in NMDA-induced PGE_2 release and hyperalgesia, and that microglia are involved in spinal nociceptive processing (Svensson et al., 2003a,b).

At low concentrations, eicosanoids stimulate PKC and other enzymes and protect cortical neurons against glutamate toxicity (Cazevieille et al., 1994) with the order of protection potency $PGF_{2\alpha}$, = PGE_2 > PGE_1 >PGD_2 > PGI_2 > 6-keto-$PGF_{1\alpha}$. Prostaglandins may interact with EP_2 receptors (a PGE_2 receptor subtype) and suppress the generation of nitric oxide triggered by Ca^{2+} influx through NMDA receptors (Martínez-Cayuela, 1995). Eicosanoids may also be involved in synaptic plasticity related to long-term potentiation and long-term depression. At low concentrations, eicosanoids induce trophic effects related to normal brain function (Wolfe and Horrocks, 1994).

5.6.2 Pathophysiological Effects of Eicosanoids

Increased arachidonic acid levels following injections of glutamate and its analogs result in accumulation of eicosanoids in brain tissue (Phillis et al., 2006; Yoshikawa et al., 2006). High levels of eicosanoids have degenerative affects on differentiated murine neuroblastoma cells in cultures (Prasad et al., 1998; Kwon et al., 2005). *In vivo*, prostaglandins are involved in the regulation of cytokine levels and maintenance of the inflammatory cascade. For example, when released from activated microglial cells, PGE_1 and PGE_2 stimulate the expression of interleukin-6 in astrocytes (Fiebich et al., 1997). This process in turn, initiates the synthesis of additional prostaglandins. Leukotriene B_4 is produced not only by activated neutrophils and macrophages, but also by astrocytes and oligodendrocytes. It induces its neurochemical effects by interacting with specific G protein-coupled receptors. Eicosanoids contribute to the development of cytotoxicity and vasogenic brain edema, as well as neuronal damage through the processes mentioned above (Phillis et al., 2006).

In glial cell cultures, high levels of extracellular glutamate block the cellular uptake of cystine thereby depleting glutathione (Li et al., 1998). This antioxidant depletion activates soluble guanylate cyclase, causing an elevation in intracellular cGMP. cGMP is known to activate Ca^{2+} channels resulting in Ca^{2+} influx and its accumulation (Li et al., 1997b). Thus the inhibition of the cystine/glutamate antiporter by glutamate is central to oxidative glutamate toxicity (Li et al., 1998). The glutathione depletion may also be enhanced by ROS-dependent activation of the extracellular signal-regulated kinases, ERK-1/2 signaling pathway (de Bernardo et al., 2004).

Glutamate induces arginine release from astrocytes and makes it available to neurons. Neurons use this amino acid for the generation of high levels of nitric oxide that are neurotoxic (Gensert and Ratan, 2006). In neural cells, glutamate-mediated modulation of arachidonic acid generation and metabolism and glutamate transport, depletion of glutathione, and decreased ATP levels are involved in initiation, maintenance, and modulation of oxidative stress and also in inflammatory processes, apoptosis, and synaptic activity (Gilroy et al., 2004; Maccarrone et al., 2001; Li et al., 1997a; Uz et al., 1998; Chabot et al., 1998; Manev et al., 2000).

5.7 Neuroprotective Effects of NMDA Receptors

Besides excitotoxicity, the NMDA receptor promotes activity-dependent cell survival (Lee et al., 2005a; Balazs, 2006). The protective effect of glutamate or NMDA is time- and concentration-dependent, suggesting that sufficient agonist and time are required to establish an intracellular neuroprotective state. The mechanism of this process is not fully understood. However, the activation of kinases, Ras-MAPK/ERK and PI3-K-Akt, transcription factors, CREB, SRF, MEF-2, NF-κB, and growth factor gene expression, brain-derived neurotrophic factor, may play an important role during activity-dependent cell survival. Activity-dependent cell survival comprises two mechanistically distinct phases that differ in their spatial requirements for Ca^{2+} and in their reliance on the CREB family (Papadia et al., 2005). The first phase involves nuclear Ca^{2+} signaling and cAMP response element (CRE)-mediated gene expression. The second phase is associated with the activation of the phosphatidylinositol 3-kinase/Akt pathway, serine/threonine protein kinase B (PKB)/Akt (Papadia et al., 2005). The phosphatidylinositol 3-kinase/Akt pathway transduces signals to Ras-MARK cascades that convey information between the plasma membrane and the nucleus. Ras-MARK cascades, through activating their specific nuclear transcription factor, regulate gene expression (Wang et al., 2004). Comparison of cellular and molecular signaling events that couple excitotoxic insult and activity-dependent cell survival indicates that NMDA receptor-mediated synaptic activity triggers a sustained CREB phosphorylation (pCREB) at serine 133. In contrast, brief stimulation with an excitotoxic concentration of NMDA triggers transient pCREB. The duration of pCREB depends on calcineurin activity (Lee et al., 2005a). Excitotoxic levels of NMDA stimulate calcineurin activity, whereas synaptic activity has no effect. Calcineurin inhibition reduces NMDA toxicity and converts the transient increase in pCREB into a sustained increase. Thus sustained pCREB does not require persistent kinase activity. The sequence of stimulation with excitotoxic levels of NMDA and neuroprotective synaptic activity determines which stimulus exerts control over pCREB duration. Collective evidence implicates CREB in synaptic activity-dependent neuroprotection against NMDA-induced excitotoxicity. Together, these findings provide a framework for understanding how the neuroprotective and excitotoxic effects of NMDA receptor activity function in an antagonistic manner at the level of the CREB/CRE transcriptional pathway (Lee et al., 2005a; Papadia et al., 2005). The NMDA receptor may also support neuronal survival by regulating pro-survival trophic factor signaling. Thus the physiological stimulation of NMDA receptors can mediate trophic effects through gene expression. This process promotes neuronal plasticity (Wang et al., 2004). Collective evidence suggests that NMDA receptors play fundamental roles in the survival, migration, differentiation, and activity-dependent maturation of neural cells (Okabe et al., 1998; Hetman and Kharebava, 2006).

References

Axelrod J. (1990). Receptor-mediated activation of phospholipase A_2 and arachidonic acid release in signal transduction. Biochem. Soc. Trans. 18:503–507.

Axelrod J., Burch R. M., and Jelsema C. L. (1988). Receptor-mediated activation of phospholipase A2 via GTP-binding proteins: arachidonic acid and its metabolites as second messengers. Trends Neurosci. 11:117–123.

Balazs R. (2006). Trophic effect of glutamate. Curr. Top. Medicinal Chem. 6:961–968.

Bazan N. G. (2003). Synaptic lipid signaling: significance of polyunsaturated fatty acids and platelet-activating factor. J. Lipid Res. 44:2221–2233.

Bertolino M., Baraldi M., Parenti C., Braghiroli D., DiBella M., Vicini S., and Costa E. (1993). Modulation of AMPA/kainate receptors by analogues of diazoxide and cyclothiazide in thin slices of rat hippocampus. Receptors & Channels 1:267–278.

Birgbauer E., Rao T. S., and Webb M. (2004). Lysolecithin induces demyelination in vitro in a cerebellar slice culture system. J. Neurosci. Res. 78:157–166.

Boggs K. P., Rock C. O., and Jackowski S. (1995). Lysophosphatidylcholine and 1-*O*-octadecyl-2-*O*- methyl-*rac*-glycero-3-phosphocholine inhibit the CDP-choline pathway of phosphatidylcholine synthesis at the CTP: phosphocholine cytidylyltransferase step. J. Biol. Chem. 270:7757–7764.

Buratta S., Mambrini R., Miniaci M. C., Tempia F., and Mozzi R. (2004). Group I metabotropic glutamate receptors mediate the inhibition of phosphatidylserine synthesis in rat cerebellar slices: a possible role in physiology and pathology. J. Neurochem. 89:730–738.

Cazevieille C., Muller A., Meynier F., Dutrait N., and Bonne C. (1994). Protection by prostaglandins from glutamate toxicity in cortical neurons. Neurochem. Int. 24:395–398.

Chabot C., Gagné J., Giguère C., Bernard J., Baudry M., and Massicotte G. (1998). Bidirectional modulation of AMPA receptor properties by exogenous phospholipase A_2 in the hippocampus. Hippocampus 8:299–309.

Chin J. H., Buckholz T. M., and DeLorenzo R. J. (1985). Calmodulin and protein phosphorylation: implications in brain ischemia. Prog. Brain Res. 63:169–184.

Coleman R. A., Smith W. L., and Narumiya S. (1994). International Union of Pharmacology classification of prostanoid receptors: properties, distribution, and structure of the receptors and their subtypes. Pharmacol. Rev. 46:205–229.

Danbolt N. C. (1994). The high affinity uptake system for excitatory amino acid in brain. Prog. Neurobiol. 44:377–396.

Davis J. B. and Maher P. (1994). Protein kinase C activation inhibits glutamate-induced cytotoxicity in a neuronal cell line. Brain Res. 652:169–173.

de Bernardo S., Canals S., Casarejos M. J., Solano R. M., Menendez J., and Mena M. A. (2004). Role of extracellular signal-regulated protein kinase in neuronal cell death induced by glutathione depletion in neuron/glia mesencephalic cultures. J. Neurochem. 91:667–682.

DeCoster M. A., Mukherjee P. K., Davis R. J., and Bazan N. G. (1998). Platelet-activating factor is a downstream messenger of kainate-induced activation of mitogen-activated protein kinases in primary hippocampal neurons. J. Neurosci. Res. 53:297–303.

del Cerro S., Arai A., and Lynch G. (1990). Inhibition of long-term potentiation by an antagonist of platelet-activating factor receptors. Behav. Neural Biol. 54:213–217.

Dumuis A., Sebben M., Haynes L., Pin J.-P., and Bockaert J. (1988). NMDA receptors activate the arachidonic acid cascade system in striatal neurons. Nature 336:68–70.

Facchinetti F., Hack N. J., and Balazs R. (1998). Calcium influx via ionotropic glutamate receptors causes long lasting inhibition of metabotropic glutamate receptor-coupled phosphoinositide hydrolysis. Neurochem. Int. 33:263–270.

Farooqui A. A. and Horrocks L. A. (1991). Excitatory amino acid receptors, neural membrane phospholipid metabolism and neurological disorders. Brain Res. Rev. 16:171–191.

Farooqui A. A. and Horrocks L. A. (1994). Excitotoxicity and neurological disorders: involvement of membrane phospholipids. Int. Rev. Neurobiol. 36:267–323.

Farooqui A. A. and Horrocks L. A. (1997). Nitric oxide synthase inhibitors do not attenuate diacylglycerol or monoacylglycerol lipase activities in synaptoneurosomes. Neurochem. Res. 22:1265–1269.

Farooqui A. A. and Horrocks L. A. (2004). Plasmalogens, platelet activating factor, and other ether lipids. In: Nicolaou A. and Kokotos G. (eds.), *Bioactive Lipids*. Oily Press, Bridgwater, England, pp. 107–134.

Farooqui A. A. and Horrocks L. A. (2006). Phospholipase A_2-generated lipid mediators in the brain: the good, the bad, and the ugly. Neuroscientist 12:245–260.

Farooqui A. A., Rammohan K. W., and Horrocks L. A. (1989). Isolation, characterization and regulation of diacylglycerol lipases from bovine brain. Ann. N. Y. Acad. Sci. 559:25–36.

Farooqui A. A., Wallace L. J., and Horrocks L. A. (1991). Stimulation of mono- and diacylglycerol lipase activities in ibotenate-induced lesions of nucleus basalis magnocellularis. Neurosci. Lett. 131:97–99.

Farooqui A. A., Anderson D. K., and Horrocks L. A. (1993). Effect of glutamate and its analogs on diacylglycerol and monoacylglycerol lipase activities of neuron-enriched cultures. Brain Res. 604:180–184.

Farooqui A. A., Rosenberger T. A., and Horrocks L. A. (1997a). Arachidonic acid, neurotrauma, and neurodegenerative diseases. In: Yehuda S. and Mostofsky D. I. (eds.), *Handbook of Essential Fatty Acid Biology*. Humana Press, Totowa, NJ, pp. 277–295.

Farooqui A. A., Yang H. C., Rosenberger T. A., and Horrocks L. A. (1997b). Phospholipase A_2 and its role in brain tissue. J. Neurochem. 69:889–901.

Farooqui A. A., Yang H.-C., and Horrocks L. A. (1997c). Involvement of phospholipase A_2 in neurodegeneration. Neurochem. Int. 30:517–522.

Farooqui A. A., Horrocks L. A., and Farooqui T. (2000). Deacylation and reacylation of neural membrane glycerophospholipids. J. Mol. Neurosci. 14:123–135.

Farooqui A. A., Ong W. Y., Lu X. R., Halliwell B., and Horrocks L. A. (2001). Neurochemical consequences of kainate-induced toxicity in brain: involvement of arachidonic acid release and prevention of toxicity by phospholipase A_2 inhibitors. Brain Res. Rev. 38:61–78.

Farooqui A. A., Ong W. Y., Lu X. R., and Horrocks L. A. (2002). Cytosolic phospholipase A_2 inhibitors as therapeutic agents for neural cell injury. Curr. Med. Chem. - Anti-Inflammatory & Anti-Allergy Agents 1:193–204.

Farooqui A. A., Ong W. Y., and Horrocks L. A. (2003a). Plasmalogens, docosahexaenoic acid, and neurological disorders. In: Roels F., Baes M., and de Bies S. (eds.), *Peroxisomal Disorders and Regulation of Genes*. Kluwer Academic/Plenum Publishers, London, pp. 335–354.

Farooqui A. A., Ong W. Y., and Horrocks L. A. (2003b). Stimulation of lipases and phospholipases in Alzheimer disease. In: Szuhaj B. and van Nieuwenhuyzen W. (eds.), *Nutrition and Biochemistry of Phospholipids*. AOCS Press, Champaign, pp. 14–29.

Farooqui A. A., Ong W. Y., and Horrocks L. A. (2006). Inhibitors of brain phospholipase A_2 activity: Their neuropharmacologic effects and therapeutic importance for the treatment of neurologic disorders. Pharmacol. Rev. 58:591–620.

Farooqui A. A., Horrocks L. A., and Farooqui T. (2007). Modulation of inflammation in brain: a matter of fat. J. Neurochem. 101:577–599.

Fiebich B. L., Hüll M., Lieb K., Gyufko K., Berger M., and Bauer J. (1997). Prostaglandin E_2 induces interleukin-6 synthesis in human astrocytoma cells. J. Neurochem. 68:704–709.

Friguet B., Stadtman E. R., and Szweda L. I. (1994). Modification of glucose-6-phosphate dehydrogenase by 4-hydroxy-2-nonenal. J. Biol. Chem. 269:21639–21643.

Gally J. A., Montague P. R., Reeke G. N., Jr., and Edelman G. M. (1990). The NO hypothesis: Possible effects of a short-lived, rapidly diffusible signal in the development and function of the nervous system. Proc. Natl. Acad. Sci. USA 87:3547–3551.

Gasull T., DeGregorio-Rocasolano N., Zapata A., and Trullas R. (2000). Choline release and inhibition of phosphatidylcholine synthesis precede excitotoxic neuronal death but not neurotoxicity induced by serum deprivation. J. Biol. Chem. 275:18350–18357.

Gasull T., DeGregorio-Rocasolano N., and Trullas R. (2001). Overactivation of α-amino-3-hydroxy-5-methylisoxazole-4-propionate and *N*-methyl-D-aspartate but not kainate receptors inhibits phosphatidylcholine synthesis before excitotoxic neuronal death. J. Neurochem. 77:13–22.

Gasull T., DeGregorio-Rocasolano N., Enguita M., Hurtan J. M., and Trullas R. (2002). Inhibition of phosphatidylcholine synthesis is associated with excitotoxic cell death in cerebellar granule cell cultures. Amino Acids 23:19–25.

Gasull T., Sarri E., DeGregorio-Rocasolano N., and Trullas R. (2003). NMDA receptor overactivation inhibits phospholipid synthesis by decreasing choline-ethanolamine phosphotransferase activity. J. Neurosci. 23:4100–4107.

Gensert J. M. and Ratan R. R. (2006). The metabolic coupling of arginine metabolism to nitric oxide generation by astrocytes. Antioxidants & Redox Signaling 8:919–928.

Gilroy D. W., Newson J., Sawmynaden P. A., Willoughby D. A., and Croxtall J. D. (2004). A novel role for phospholipase A_2 isoforms in the checkpoint control of acute inflammation. FASEB J. 18:489–498.

Gómez-Muñoz A., O'Brien L., Hundal R., and Steinbrecher U. P. (1999). Lysophosphatidylcholine stimulates phospholipase D activity in mouse peritoneal macrophages. J. Lipid Res. 40: 988–993.

Halpain S., Girault J.-A., and Greengard P. (1990). Activation of NMDA receptors induces dephosphorylation of DARPP-32 in rat striatal slices. Nature 343:369–372.

Hannan A. J., Blakemore C., Katsnelson A., Vitalis T., Huber K. M., Bear M., Roder J., Kim D., Shin H. S., and Kind P. C. (2001). PLC-β1, activated via mGluRs, mediates activity-dependent differentiation in cerebral cortex. Nat. Neurosci. 4:282–288.

Hetman M. and Kharebava G. (2006). Survival signaling pathways activated by NMDA receptors. Curr. Top. Medicinal Chem. 6:787–799.

Honda Z., Ishii S., and Shimizu T. (2002). Platelet-activating factor receptor. J. Biochem. 131:773–779.

Ikeda M. (1993). Reduction of phosphoinositide hydrolysis by L-amino-3-phosphonopropionate may be caused by the inhibition of synthesis of phosphatidylinositols. Neurosci. Lett. 157:87–90.

Ishii S., Nagase T., and Shimizu T. (2002). Platelet-activating factor receptor. Prostaglandins Other Lipid Mediat. 68-69:599–609.

Jamme I., Petit E., Divoux D., Gerbi A., Maxient J. M., and Nouvelot A. (1995). Modulation of mouse cerebral Na^+,K^+-ATPase activity by oxygen free radicals. NeuroReport 7:333–337.

Kato K. and Zorumski C. F. (1996). Platelet-activating factor as a potential retrograde messenger. J. Lipid Mediat. Cell Signal. 14:341–348.

Kato K., Clark G. D., Bazan N. G., and Zorumski C. F. (1994). Platelet-activating factor as a potential retrograde messenger in CA1 hippocampal long-term potentiation. Nature 367:175–179.

Katsuki H. and Okuda S. (1995). Arachidonic acid as a neurotoxic and neurotrophic substance. Prog. Neurobiol. 46:607–636.

Kim D. K., Rordorf G., Nemenoff R. A., Koroshetz W. J., and Bonventre J. V. (1995). Glutamate stably enhances the activity of two cytosolic forms of phospholipase A_2 in brain cortical cultures. Biochem. J. 310:83–90.

Kooy N., Royall J., Ischoropoulos H., and Beckman J. (1994). Peroxynitrite-mediated oxidation of dihydrorhodamine 123. Free Radic. Biol. Med. 16:149–156.
Kornecki E., Wieraszko A., Chan J. C., and Ehrlich Y. H. (1996). Platelet activating factor (PAF) in memory formation: Role as a retrograde messenger in long-term potentiation. J. Lipid Mediat. Cell Signal. 14:115–126.
Kwon K. J., Jung Y. S., Lee S. H., Moon C. H., and Baik E. J. (2005). Arachidonic acid induces neuronal death through lipoxygenase and cytochrome P450 rather than cyclooxygenase. J. Neurosci. Res. 81:73–84.
Lazarewicz J. W., Wroblewski J. T., and Costa E. (1990). N-methyl-D-aspartate-sensitive glutamate receptors induce calcium-mediated arachidonic acid release in primary cultures of cerebellar granule cells. J. Neurochem. 55:1875–1881.
Lee E. S. Y., Chen H. T., Shepherd K. R., Lamango N. S., Soliman K. F. A., and Charlton C. G. (2004). Inhibitory effects of lysophosphatidylcholine on the dopaminergic system. Neurochem. Res. 29:1333–1342.
Lee B., Butcher G. Q., Hoyt K. R., Impey S., and Obrietan K. (2005a). Activity-dependent neuroprotection and cAMP response element-binding protein (CREB): kinase coupling, stimulus intensity, and temporal regulation of CREB phosphorylation at serine 133. J. Neurosci. 25:1137–1148.
Lee E. S. Y., Chen H., Charlton C. G., and Soliman K. F. A. (2005b). The role of phospholipid methylation in 1-methyl-4-phenyl-pyridinium ion (MPP^+)-induced neurotoxicity in PC12 cells. NeuroToxicology 26:945–957.
Lee E. S. Y., Soliman K. F. A., and Charlton C. G. (2005c). Lysophosphatidylcholine decreases locomotor activities and dopamine turnover rate in rats. NeuroToxicology 26:27–38.
Li Y., Maher P., and Schubert D. (1997a). A role for 12-lipoxygenase in nerve cell death caused by glutathione depletion. Neuron 19:453–463.
Li Y., Maher P., and Schubert D. (1997b). Requirement for cGMP in nerve cell death caused by glutathione depletion. J. Cell Biol. 139:1317–1324.
Li Y., Maher P., and Schubert D. (1998). Phosphatidylcholine-specific phospholipase C regulates glutamate-induced nerve cell death. Proc. Natl. Acad. Sci. USA 95:7748–7753.
Lipton S. A. and Rosenberg P. A. (1994). Mechanisms of disease: Excitatory amino acids as a final common pathway for neurologic disorders. New Eng. J. Med. 330:613–622.
Llansola M., Monfort P., and Felipo V. (2000). Inhibitors of phospholipase C prevent glutamate neurotoxicity in primary cultures of cerebellar neurons. J. Pharmacol. Exp. Ther. 292: 870–876.
Lu C., Chan S. L., Haughey N., Lee W. T., and Mattson M. P. (2001). Selective and biphasic effect of the membrane lipid peroxidation product 4-hydroxy-2,3-nonenal on *N*-methyl-D-aspartate channels. J. Neurochem. 78:577–589.
Lynch M. A. and Voss K. L. (1990). Arachidonic acid increases inositol phospholipid metabolism and glutamate release in synaptosomes prepared from hippocampal tissue. J. Neurochem. 55:215–221.
Maccarrone M., Melino G., and Finazzi-Agro A. (2001). Lipoxygenases and their involvement in programmed cell death. Cell Death. Differ. 8:776–784.
Maclennan K. M., Smith P. F., and Darlington C. L. (1996). Platelet-activating factor in the CNS. Prog. Neurobiol. 50:585–596.
Maingret F., Patel A. J., Lesage F., Lazdunski M., and Honoré E. (2000). Lysophospholipids open the two-pore domain mechano-gated K^+ channels TREK-1 and TRAAK. J. Biol. Chem. 275:10128–10133.
Manev H., Uz T., Sugaya K., and Qu T. Y. (2000). Putative role of neuronal 5-lipoxygenase in an aging brain. FASEB J. 14:1464–1469.
Martínez-Cayuela M. (1995). Oxygen free radicals and human disease. Biochimie 77:147–161.
Matute C., Domercq M., and Sánchez-Gómez M. V. (2006). Glutamate-mediated glial injury: Mechanisms and clinical importance. Glia 53:212–224.
McGahon B., Clements M. P., and Lynch M. A. (1997). The ability of aged rats to sustain long-term potentiation is restored when the age-related decrease in membrane arachidonic acid concentration is reversed. Neuroscience 81:9–16.

Milatovic D., Gupta R. C., and Dettbarn W. D. (2002). Involvement of nitric oxide in kainic acid-induced excitotoxicity in rat brain. Brain Res. 957:330–337.

Millanvoye-Van Brussel E., Topal G., Brunet A., Do Phaw T., Deckert V., Rendu F., and David-Dufilho M. (2004). Lysophosphatidylcholine and 7-oxocholesterol modulate Ca^{2+} signals and inhibit the phosphorylation of endothelial NO synthase and cytosolic phospholipase A_2. Biochem. J. 380:533–539.

Miyata M., Kashiwadani H., Fukaya M., Hayashi T., Wu D. Q., Suzuki T., Watanabe M., and Kawakami Y. (2003). Role of thalamic phospholipase Cβ4 mediated by metabotropic glutamate receptor type 1 in inflammatory pain. J. Neurosci. 23:8098–8108.

Mukherjee P. K., DeCoster M. A., Campbell F. Z., Davis R. J., and Bazan N. G. (1999). Glutamate receptor signaling interplay modulates stress-sensitive mitogen-activated protein kinases and neuronal cell death. J. Biol. Chem. 274:6493–6498.

Murakami K. and Routtenberg A. (2003). The role of fatty acids in synaptic growth and plasticity. In: Peet M., Glen L., and Horrobin D. F. (eds.), *Phospholipid Spectrum Disorders in Psychiatry and Neurology*. Marius Press, Carnforth, Lancashire, pp. 77–92.

Nakamura M., Sato K., Fukaya M., Araishi K., Aiba A., Kano M., and Watanabe M. (2004). Signaling complex formation of phospholipase Cβ4 with metabotropic glutamate receptor type 1α and 1,4,5-trisphosphate receptor at the perisynapse and endoplasmic reticulum in the mouse brain. Eur. J. Neurosci. 20:2929–2944.

Nicholls D. G. (2004). Mitochondrial dysfunction and glutamate excitotoxicity studied in primary neuronal cultures. Curr. Mol. Med. 4:149–177.

Nicolle M. M., Colombo P. J., Gallagher M., and McKinney M. (1999). Metabotropic glutamate receptor-mediated hippocampal phosphoinositide turnover is blunted in spatial learning-impaired aged rats. J. Neurosci. 19:9604–9610.

Nishizuka Y. (1986). Studies and perspectives of protein kinase C. Science 233:305–312.

Novelli A., Nicoletti F., Wroblewski J. T., Alho H., Costa A. E., and Guidotti A. (1987). Excitatory amino acid receptors coupled with guanylate cyclase in primary cultures of cerebellar granule cells. J. Neurosci. 7:40–47.

Oka A., Belliveau M. J., Rosenberg P. A., and Volpe J. J. (1993). Vulnerability of oligodendroglia to glutamate: pharmacology, mechanisms, and prevention. J. Neurosci. 13:1441–1453.

Okabe S., Vicario-Abejón C., Segal M., and McKay R. D. (1998). Survival and synaptogenesis of hippocampal neurons without NMDA receptor function in culture. Eur. J. Neurosci. 10:2192–2198.

Okubo Y., Kakizawa S., Hirose K., and Iino M. (2004). Cross talk between metabotropic and ionotropic glutamate receptor- mediated signaling in parallel fiber-induced inositol 1,4,5-trisphosphate production in cerebellar Purkinje cells. J. Neurosci. 24:9513–9520.

Ong W. Y., He Y., Suresh S., and Patel S. C. (1997). Differential expression of apolipoprotein D and apolipoprotein E in the kainic acid-lesioned rat hippocampus. Neuroscience 79:359–367.

Ong W. Y., Lu X. R., Ong B. K. C., Horrocks L. A., Farooqui A. A., and Lim S. K. (2003). Quinacrine abolishes increases in cytoplasmic phospholipase A_2 mRNA levels in the rat hippocampus after kainate-induced neuronal injury. Exp. Brain Res. 148: 521–524.

Packard M. G., Teather L. A., and Bazan N. G. (1996). Effects of intrastriatal injections of platelet-activating factor and the PAF antagonist BN 52021 on memory. Neurobiol. Learn. Mem. 66:176–182.

Papadia S., Stevenson P., Hardingham N. R., Bading H., and Hardingham G. E. (2005). Nuclear Ca^{2+} and the cAMP response element-binding protein family mediate a late phase of activity-dependent neuroprotection. J. Neurosci. 25:4279–4287.

Parker M. A., Bazan H. E. P., Marcheselli V., Rodriguez de Turco E. B., and Bazan N. G. (2002). Platelet-activating factor induces permeability transition and cytochrome c release in isolated brain mitochondria. J. Neurosci. Res. 69:39–50.

Phillis J. W., Horrocks L. A., and Farooqui A. A. (2006). Cyclooxygenases, lipoxygenases, and epoxygenases in CNS: their role and involvement in neurological disorders. Brain Res. Rev. 52:201–243.

Prasad K. N., La Rosa F. G., and Prasad J. E. (1998). Prostaglandins act as neurotoxin for differentiated neuroblastoma cells in culture and increase levels of ubiquitin and beta-amyloid. In Vitro Cell Dev. Biol. Anim 34:265–274.
Sandhya T. L., Ong W. Y., Horrocks L. A., and Farooqui A. A. (1998). A light and electron microscopic study of cytoplasmic phospholipase A_2 and cyclooxygenase-2 in the hippocampus after kainate lesions. Brain Res. 788:223–231.
Sanfeliu C., Hunt A., and Patel A. J. (1990). Exposure to N-methyl-D-aspartate increases release of arachidonic acid in primary cultures of rat hippocampal neurons and not in astrocytes. Brain Res. 526:241–248.
Schapira A. H. (1996). Oxidative stress and mitochondrial dysfunction in neurodegeneration. Curr. Opin. Neurol. 9:260–264.
Schilling T., Lehmann F., Ruckert B., and Eder C. (2004). Physiological mechanisms of lysophosphatidylcholine-induced de-ramification of murine microglia. J. Physiol. (London) 557:105–120.
Siesjö B. K. (1990). Calcium in the brain under physiological and pathological conditions. Eur. Neurol. 30:3–9.
Siman R. and Noszek J. C. (1988). Excitatory amino acids activate calpain I and induce structural protein breakdown *in vivo*. Neuron 1:279–287.
Snyder F. (1995). Platelet-activating factor: the biosynthetic and catabolic enzymes. Biochem. J. 305:689–705.
Stella N., Pellerin L., and Magistretti P. J. (1995). Modulation of the glutamate-evoked release of arachidonic acid from mouse cortical neurons: Involvement of a pH-sensitive membrane phospholipase A_2. J. Neurosci. 15:3307–3317.
Sun G. Y. and Foudin L. L. (1984). On the status of lysolecithin in rat cerebral cortex during ischemia. J. Neurochem. 43:1081–1086.
Suzuki Y. J., Forman H. J., and Sevanian A. (1997). Oxidants as stimulators of signal transduction. Free Radical Biology & Medicine 22:269–285.
Svensson C. I., Hua X. Y., Protter A. A., Powell H. C., and Yaksh T. L. (2003a). Spinal p38 MAP kinase is necessary for NMDA-induced spinal PGE_2 release and thermal hyperalgesia. NeuroReport 14:1153–1157.
Svensson C. I., Marsala M., Westerlund A., Calcutt N. A., Campana W. M., Freshwater J. D., Catalano R., Feng Y., Protter A. A., Scott B., and Yaksh T. L. (2003b). Activation of p38 mitogen-activated protein kinase in spinal microglia is a critical link in inflammation-induced spinal pain processing. J. Neurochem. 86:1534–1544.
Thwin M. M., Ong W. Y., Fong C. W., Sato K., Kodama K., Farooqui A. A., and Gopalakrishnakone P. (2003). Secretory phospholipase A_2 activity in the normal and kainate injected rat brain, and inhibition by a peptide derived from python serum. Exp. Brain Res. 150: 427–433.
Toborek M., Malecki A., Garrido R., Mattson M. P., Hennig B., and Young B. (1999). Arachidonic acid-induced oxidative injury to cultured spinal cord neurons. J. Neurochem. 73:684–692.
Uz T., Pesold C., Longone P., and Manev H. (1998). Aging-associated up-regulation of neuronal 5-lipoxygenase expression: putative role in neuronal vulnerability. FASEB J. 12:439–449.
Vahidy W. H., Ong W. Y., Farooqui A. A., and Yeo J. F. (2006). Effects of intracerebroventricular injections of free fatty acids, lysophospholipids, or platelet activating factor in a mouse model of orofacial pain. Exp. Brain Res. 174:781–785.
Wang J. Q., Tang Q. S., Parelkar N. K., Liu Z. G., Samdani S., Choe E. S., Yang L., and Mao L. M. (2004). Glutamate signaling to Ras-MAPK in striatal neurons - Mechanisms for inducible gene expression and plasticity. Mol. Neurobiol. 29:1–14.
Wang Q., Yu S., Simonyi A., Sun G. Y., and Sun A. Y. (2005). Kainic acid-mediated excitotoxicity as a model for neurodegeneration. Mol. Neurobiol. 31:3–16.
Webber K. O. and Hajra A. K. (1993). Purification of dihydroxyacetone phosphate acyltransferase from guinea pig liver peroxisomes. Arch. Biochem. Biophys. 300:88–97.
Williams J. H., Errington M. L., Lynch M. A., and Bliss T. V. P. (1989). Arachidonic acid induces a long-term activity dependent enhancement of synaptic transmission in the hippocampus. Nature 341:739–742.

Wolfe L. S. and Horrocks L. A. (1994). Eicosanoids. In: Siegel G. J., Agranoff B. W., Albers R. W., and Molinoff P. B. (eds.), *Basic Neurochemistry*. Raven Press, New York, pp. 475–490.

Woodruff R. H. and Franklin R. J. M. (1999). The expression of myelin protein mRNAs during remyelination of lysolecithin-induced demyelination. Neuropathol. Appl. Neurobiol. 25:226–235.

Xu Y. and Tao Y. X. (2004). Involvement of the NMDA receptor/nitric oxide signal pathway in platelet-activating factor-induced neurotoxicity. NeuroReport 15:263–266.

Xu Y., Zhang B. S., Hua Z. C., Johns R. A., Bredt D. S., and Tao Y. X. (2004). Targeted disruption of PSD-93 gene reduces platelet-activating factor-induced neurotoxicity in cultured cortical neurons. Exp. Neurol. 189:16–24.

Yegin A., Akbas S. H., Ozben T., and Korgun D. K. (2002). Secretory phospholipase A_2 and phospholipids in neural membranes in an experimental epilepsy model. Acta Neurol. Scand. 106:258–262.

Yoshikawa K., Kita Y., Kishimoto K., and Shimizu T. (2006). Profiling of eicosanoid production in the rat hippocampus during kainic acid-induced seizure - Dual phase regulation and differential involvement of COX-1 and COX-2. J. Biol. Chem. 281:14663–14669.

Zapata A., Capdevila J. L., and Trullas R. (1998). Region-specific and calcium-dependent increase in dialysate choline levels by NMDA. J. Neurosci. 18:3597–3605.

Zhang J. P. and Sun G. Y. (1995). Free fatty acids, neutral glycerides, and phosphoglycerides in transient focal cerebral ischemia. J. Neurochem. 64:1688–1695.

Chapter 6
Glutamate Receptors and Their Association with Other Neurochemical Parameters in Excitotoxicity

Intracerebroventricular or systemic injections of glutamate and its analogs result in collapse of mitochondrial potential, deregulation of calcium homeostasis, and generation of high levels of ROS and 4-hydroxynonenal (4-HNE). Opening of NMDA channels allows Ca^{2+} influx whereas AMPA receptor activation allows Na^{+} influx. Changes in Ca^{2+} and Na^{+} influx occur in parallel resulting in depolarization, amplification, and continuation of glutamate toxicity by releasing more glutamate. In neural cell cultures, Na^{+} influx is followed by passive influx of chloride ions together with influx of water. This increased intracellular water is probably responsible for the swelling of intracellular organelles and cellular lysis (Al Noori and Swann, 2000). The increase in intracellular Na^{+} also causes the reversal of normal Ca^{2+} extrusion through the 3 Na^{+}/Ca^{2+} antiporter, thereby causing a higher Ca^{2+} influx that may lead to mitochondrial dysfunction and depletion of ATP. Decreased ATP levels result in the failure of the Na^{+}/K^{+} pump that is responsible for maintaining the resting neural membrane potential.

Besides affecting neural membrane glycerophospholipid metabolism (Farooqui and Horrocks, 1994), the stimulation of neurotoxicity mediated through glutamate receptors also involves alterations in cholesterol synthesis, ceramide synthesis, energy metabolism, and protein phosphorylation, and in the expression of immediate early genes, enzymes, growth factors, heat shock proteins, and divalent metal ion transporters. In addition to phospholipases, an increase in cytosolic Ca^{2+} also causes the activation of other Ca^{2+}-sensitive enzymes such as proteases, protein kinases/phosphatases, calmodulin-dependent nitric oxide synthase, and endonucleases (Farooqui and Horrocks, 1994). Activation of these enzymes along with decreased energy metabolism, the generation of ROS including superoxide, nitric oxide, hydrogen peroxide, hydroxyl, and alkoxyl radicals, the alteration in ion homeostasis, decreased cellular redox, the alterations in the organization of the cytoskeleton, and the activation of cell death associated genetic signals are closely associated with glutamate, its neurotoxicity mediated by analogs, and neural injury to both neurons and glial cells (Matute et al., 2006).

6.1 Glutamate Toxicity and Production of Free Radicals and Lipid Peroxides

During neurotoxicity induced by glutamate, an uncontrolled arachidonic acid cascade generates prostaglandins, leukotrienes, thromboxanes, and isoprostanes (Phillis et al., 2006). This process sets the stage for the increased production of free radicals and hence for lipid peroxidation and oxidative damage to neural membrane lipids, proteins, and nucleic acids (Wang et al., 2005b). Increased immunoreactivities of PLA_2, cyclooxygenase-2, and 5-lipoxygenase are observed in brain tissue during neurotoxicity induced by KA (Sandhya et al., 1998; Manev et al., 2000; Phillis et al., 2006). The time-course of free radical generation and prostaglandin synthesis correlates well with neuronal damage (Sandhya et al., 1998; Yoshikawa et al., 2006). PLA_2 inhibitors, quinacrine and $AACOCF_3$, cyclooxygenase inhibitors, indomethacin and NS-398, and the lipoxygenase inhibitor BW755C largely prevent neuronal damage induced by KA (Pepicelli et al., 2002; Lu et al., 2001b; Manev et al., 2000).

Peroxidative membrane injury results from the reaction of free radical species with proteins and unsaturated lipids in plasma membrane, leading to chemical cross-linking. This depletion of unsaturated lipids is associated with an alteration in membrane fluidity and changes in activity of membrane-bound enzymes, ion channels, and receptors (Farooqui and Horrocks, 2006). ROS also attack DNA bases causing damage through hydroxylation, ring opening, and fragmentation. This reaction generates 8-hydroxy-2'-deoxyguanosine (8-OHdGua) and 2,6-diamino-4-hydroxy-5-formamidopyrimidine (FapyGua) (Jenkinson et al., 1999) (Table 6.1).

The presence of peroxidized glycerophospholipids due to neurotoxicity may also produce a membrane-packing defect, making the *sn*-2 ester bond at glycerol moiety more accessible to the action of calcium-independent PLA_2. In fact, glycerophospholipid hydroperoxides are a better substrate for PLA_2 than are the native glycerophospholipids. Phospholipid hydroperoxides inhibit the reacylation of lyso-glycerophospholipids in neuronal membranes (Zaleska and Wilson, 1989;

Table 6.1 Neurochemical consequences of toxicity mediated by KA

Parameter	Effect on content and immunoreactivity	Reference
ATP	Decreased	(Retz and Coyle, 1982)
Intracellular calcium	Increased	(Brorson et al., 1994)
Polyamines	Decreased	(de Vera et al., 1991)
Glutathione	Decreased	(Ong et al., 2000a)
Free radicals & lipid peroxides	Increased	(Sun et al., 1992)
Prostaglandins & leukotrienes	Increased	(Yoshikawa et al., 2006)
4-Hydroxynonenal	Increased	(Ong et al., 2000b)
8-OHdGua	Increased	(Liang et al., 2000)
Cholesterol	Increased	(Ong et al., 2003)

ATP, Adenosine triphosphate; GFAP, glial fibrillary acidic protein; β-APP, β-amyloid precursor protein; 8-OHdGua, 8-hydroxy-2-deoxyguanosine; and p53, a tumor-suppressor gene.

The elevated expression of murine GSHPx-1 in transgenic mice surprisingly results in increased rather than decreased neurotoxicity mediated by KA in hippocampus and increased seizure activity (Boonplueang et al., 2005). Isolated primary culture neurons from the transgenic hippocampus also display increased susceptibility to KA treatment compared with those from wild type animals. This may be due to alterations in the redox state of the glutathione system resulting in elevated glutathione disulfide (GSSG) levels that, in turn, may directly activate NMDA receptors or enhance the response of the NMDA receptor (Boonplueang et al., 2005; Leslie et al., 1992). Collective evidence thus suggests that the glutathione cycle in neural cells not only works as a peroxide scavenger, but also regulates the redox state of the neural cells under pathological situations. Thus the maintenance of appropriate intracellular glutathione (GSH) levels is crucial for cellular defense against excitotoxicity and oxidative damage.

6.3 4-Hydroxynonenal Generation in Neurotoxicity Mediated by Glutamate

Oxidative damage to the polyunsaturated fatty acid moieties of neural membrane glycerophospholipids, i.e. lipid peroxidation, produces several reactive aldehydes capable of modifying proteins. These aldehydes include malondialdehyde and 4-hydroxynonenal (4-HNE). 4-HNE is the most prevalent. It's an electrophilic and cytotoxic α, β-unsaturated aldehyde that reacts with proteins nucleophiles to form Michael adducts (Nadkarni and Sayre, 1995). Intracerebroventricular injections of KA produce dense 4-HNE staining at 1 day post injection (Fig. 6.2), before there is any histological evidence of neurodegeneration (Ong et al., 2000b). Degenerating neurons in CA1 and CA3 regions of hippocampus are seen at 3 days and 1 week after KA injection. The increased immunoreactivity remains confined to a cluster of neurons at the edge of the degenerating CA1 and CA3 regions at 2 and 3 weeks after KA injections (Ong et al., 2000b). No 4-HNE immunoreactivity is observed in reactive astrocytes.

The molecular mechanism by which 4-HNE kills neurons in the CA1 and CA3 regions of the rat hippocampus is not fully understood. However, 4-HNE impairs the activities of key metabolic enzymes, including Na^+, K^+-ATPase, glucose 6-phosphate dehydrogenase, and several kinases. It stimulates stress-activated protein kinases (Camandola et al., 2000; Tamagno et al., 2003) such as c-jun amino-terminal kinase (JNK) and p38 mitogen-activated protein kinase. 4-HNE also disrupts transmembrane signaling and the glucose and glutamate transporters in astrocytes (Mark et al., 1997). The inhibition of Na^+, K^+-ATPase depolarizes neuronal membranes leading to the opening of NMDA receptor channels and influx of additional Ca^{2+} into neurons. In hippocampal cultures, 4-HNE exerts biphasic effects on the NMDA current. After 4-HNE exposure, an early enhancement of NMDA current within the first 30–120 min is followed at 6 h by a delayed decrease in current that is significantly less than the basal current

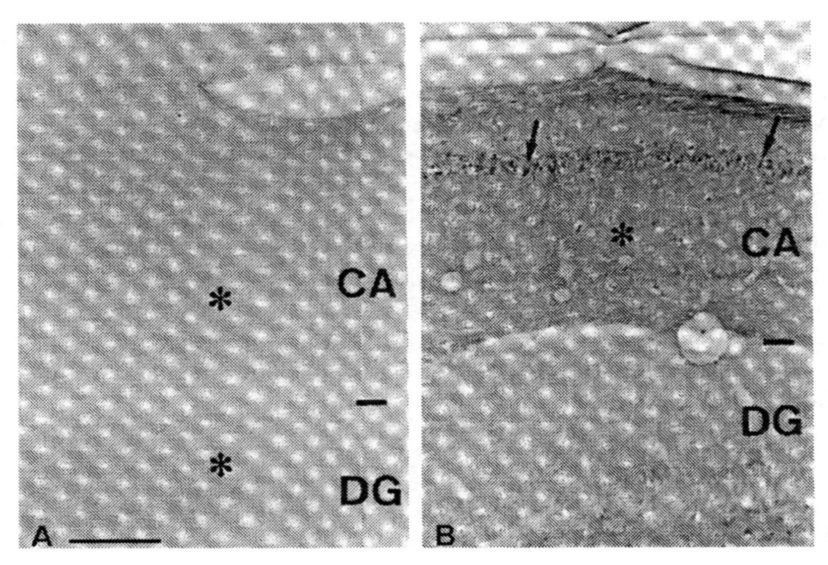

Fig. 6.2 HNE immunoreactivity in the hippocampus. A: section from a normal rat showing no staining in field CA1 and the dentate gyrus (asterisks). B: section from a 1 day post-kainate injected rat, showing dense staining in pyramidal cell bodies (arrows), and the neuropil (asterisk), in an affected area. Abbreviations: CA, field CA1. DG, dentate gyrus. Scale = 250 μm. Reproduced with kind permission from Ong et al., 2000b, Free Radical Biology and Medicine 28:1214–1221. Elsevier

(Lu et al., 2001a). The molecular mechanism of this biphasic action of 4-HNE on NMDA current is unknown. However, the early enhancement of NMDA current may involve increased phosphorylation of NR1, a NMDA receptor subunit, and the delayed suppression of NMDA current by 4-HNE may be due to the depletion of ATP resulting in impairment of NMDA receptor channel function (Lu et al., 2001a). These processes may result in a sustained increase and overloading of Ca^{2+} that is very cytotoxic to neural cells. In addition the inhibition of the glutamate transporter impairs glutamate uptake inducing a more pronounced glutamate-mediated excitotoxicity by inhibiting the removal of glutamate from extracellular space (Mark et al., 1997; Lauderback et al., 2001).

In cell cultures, 4-HNE induces glutathione depletion and inactivation of glutathione peroxidase (Table 6.2). It disrupts the coupling of muscarinic cholinergic and glutamate receptors to phospholipase C-linked GTP-binding proteins in cultured rat cerebrocortical neurons (Blanc et al., 1997). 4-HNE also produces inhibition of DNA and RNA synthesis, calcium homeostasis disturbances, and inhibition of mitochondrial respiration. Collective evidence thus suggests that abnormalities induced by KA in neural membrane glycerophospholipid metabolism lead to the generation of 4-HNE that produces neuronal cell death by interacting with many enzyme systems in brain tissue.

Table 6.2 Alterations in enzymic activities of brain tissue induced by glutamate and its analogs

Enzyme	Effects on enzymic activity and immunoreactivity	Reference
Na^+, K^+-ATPase	Decreased	(Sun et al., 1992)
Phospholipase A_2	Increased	(Sandhya et al., 1998)
Cyclooxygenase-2	Increased	(Sandhya et al., 1998)
DAG/ MAG-lipase	No effect	(Farooqui et al., 1993)
PLD	Increased	(Kim et al., 2004)
Lipoprotein lipase	Increased	(Paradis et al., 2004)
Acetylcholine esterase	Decreased	(Cohen et al., 1987)
Calpain II	Increased	(Ong et al., 1997a)
Caspase-3	Increased	(Faherty et al., 1999; Djebaili et al., 2002)
Nitric oxide synthase	Increased	(Milatovic et al., 2002)
Endonuclease	Increased	(Rothstein and Kuncl, 1995)
Protein kinase C isozymes	Increased	(McNamara et al., 1999)
MAP-kinase	Increased	(Liu et al., 1999)
Ca^{2+}/Calmodulin kinase II	Increased	(Lee et al., 2001a)
Ornithine decarboxylase	Increased	(de Vera et al., 1991)
Mn Superoxide dismutase	Decreased	(Kim et al., 2000b)
Glutathione oxidase and catalase	Decreased	(Cohen et al., 1987)
Aconitase	Decreased	(Liang et al., 2000)
Xanthine oxidase	Increased	(Atlante et al., 1997)

6.4 NF-κB in Glutamate Neurotoxicity

NF-κB is a nuclear transcription factor involved in the control of fundamental cellular functions including regulation of cell survival. It is also implicated in neurotoxicity mediated by glutamate (Grilli et al., 1996; Furukawa and Mattson, 1998; Qin et al., 1999; Djebaili et al., 2000; Cruise et al., 2000). NF-κB has several subunits. They include p65 (RelA), p50, p52, c-Rel, and relB. This transcription factor exists in the cytoplasm of neural cells as a heteromeric protein consisting of a dimer of two of these subunits complexed with an inhibitory subunit, IκB. Phosphorylation of the IκB subunit through an extracellular stimulus results in dissociation from NF-κB dimer. This allows the dimer to translocate to the nucleus where it binds to target sequences in the genome. The mechanism by which NF-κB mediates glutamate neurotoxicity remains unknown. Glutamate activates only the p50 and p65 proteins subunits of NF-κB. Furthermore, NF-κB also interacts with ROS (McInnis et al., 2002). This suggestion is based on the observation that M40403, an antioxidant, not only blocks NMDA-mediated neurotoxicity but also prevents NF-κB translocation to the nucleus (McInnis et al., 2002). In addition, the interactions between the inhibitory protein IκB and NF-κB proteins (p50/RelA and p65) are modulated by protein kinases, such as NF-κB-inducing kinase (NIK) and members of the MAP kinase family, that contain several redox-sensitive cysteine residues in critical kinase domains (Cruise et al., 2000).

NF-κB responds to changes in the redox state of the cytoplasm and translocates to the nucleus within 5 min in response to NMDA (Ko et al., 1998). NMDA treatment also enhances the DNA binding activity of nuclear NF-κB in a Ca^{2+} -dependent manner. The ability of NF-κB to up-regulate p53 expression through phosphorylation may also contribute to NF-κB participation in glutamate neurotoxicity (McInnis et al., 2002). Collective evidence suggests that glutamate and its analogs activate NF-κB through synergistic interplay among ROS generation, Ca^{2+} influx, and Ca^{2+}-dependent activation of Ras followed by a cascade involving raf, MEK, and MAP kinase (p44/p42) (Rusanescu et al., 1995) or, alternatively through Ca^{2+}/calmodulin dependent protein kinase IV (Enslen et al., 1996). ROS-mediated translocation of NF-κB to the nucleus may intensify that activation of NF-κB.

6.5 Protein Kinase C in Neurotoxicity Mediated by Glutamate

Temporal and regional alterations in protein kinase C isoform expression and subcellular distribution occur following NMDA and KA treatment (Guglielmetti et al., 1997; McNamara and Lenox, 2000). This is accompanied by alterations in expression of mRNA for primary protein kinase C (PKC) substrates such GAP-43, RC3, MARCKS, and MLP at early and late phases of NMDA and KA-induced limbic motor seizures and during the early and late phases of hippocampal neuronal degeneration (McNamara and Lenox, 2000). As stated in Chapter 3, protein phosphorylation is one of the important mechanisms associated with the post-translational modulation of ionotropic and metabotropic glutamate receptors. Constitutive and regulatory phosphorylation occurs at distinct sites on serine, threonine, or tyrosine in the intracellular C-terminal domain of almost all subunits capable of assembling a functional channel. Several key protein kinases, such as protein kinase A, protein kinase C, Ca^{2+}/calmodulin-dependent protein kinases, and tyrosine kinases are involved in the site-specific catalysis and regulation of NMDA and AMPA receptor phosphorylation (Wang et al., 2005a,b; Moriguchi et al., 2006; Dell'Acqua et al., 2006).

Through the phosphorylation mechanism, these protein kinases as well as protein phosphatases control biochemical properties (biosynthesis, delivery, and subunit assembling), subcellular distribution, and interactions of these receptors with various synaptic proteins, which ultimately modify the efficacy and strength of excitatory synapses containing NMDARs and AMPA receptors and many forms of synaptic plasticity (Wang et al., 2006). Thus AMPA receptors are constitutively internalized through both basal PKC and MEK-ERK1/2 (extracellular signal-regulated kinase 1/2) activities at parallel fiber Purkinje cell (PF-PC) synapses (Tatsukawa et al., 2006). Collective evidence suggests that ionotropic and metabotropic glutamate receptors are directly phosphorylated and functionally modulated by protein kinases. The phosphorylation may be involved in synergistic interplay between ionotropic and metabotropic receptors that occurs during LTP induction (Kotecha and MacDonald, 2003).

In addition to phosphorylating glutamate receptors, PKCα selectively mediates redistribution of the neuronal glutamate transporter, EAAT3 (Davis et al., 1998; González et al., 2005). PKCα co-immunoprecipitates with EAAT3. When the glial glutamate transporter EAAT2 is transfected into C6 glioma cells, this transporter is internalized in response to activation of PKCα. Phorbol ester-dependent internalization of EAAT2 is inhibited by a general inhibitor of PKC, bisindolylmaleimide II, and by concentrations of Go6976 that selectively block classical PKCs, but not by rottlerin, an inhibitor of PKCδ. PKCα immunoreactivity is detected in EAAT2 immunoprecipitates obtained from transfected C6 cells and from crude rat brain synaptosomes, a milieu that better mimics *in vivo* conditions. The amount of PKCα in both types of immunoprecipitate is modestly increased by phorbol ester, and a PKC antagonist blocks this increase. These observations support the view that PKCα may be involved in the regulated redistribution of EAAT2 (Davis et al., 1998; González et al., 2005).

In neurotoxicity related to glutamate and its analogs, the activation of ionotropic, NMDA, KA, and metabotropic glutamate receptors produces Ca^{2+} influx, diacylglycerol (DAG) generation, and the subsequent activation of the neuron-specific PKCγ isoform in hippocampal neurons (Codazzi et al., 2006). PKCγ phosphorylates a broad range of effectors such as cell surface receptors, ion channels, transcription factors, as well as cytoskeletal proteins. An elevation of both Ca^{2+} and DAG is necessary for sustained translocation and activation of EGFP (enhanced green fluorescent protein)-PKCγ, as shown with a combination of Ca^{2+} imaging with total internal reflection microscopy analysis of specific biosensors. Both DAG generation and PKCγ translocation are localized processes, typically observed within discrete microdomains along the dendritic branches (Codazzi et al., 2006). Moderate activation of NMDA receptors or moderate electrical stimulation produces Ca^{2+} influx but no DAG signals, whereas mGluR activation generates DAG but no Ca^{2+} signals. Both processes are needed for PKCγ activation. Based on the co-requirement of Ca^{2+} and DAG, an interplay or cross talk may occur between NMDA and metabotropic glutamate receptors during PKCγ activation. However, the requirement for co-stimulation with metabotropic receptors may be overcome by maximal stimulation of NMDA receptors associated with direct production of DAG via activation of the Ca^{2+}-sensitive PLCδ isoform (Codazzi et al., 2006). Furthermore, the stimulation of metabotropic glutamate receptors is sufficient for PKCγ activation in neurons in which Ca^{2+} stores were loaded by previous electrical activity. Collectively these studies suggest that a dual activation requirement for PKCγ provides a plausible molecular interpretation for different synergistic contributions of metabotropic glutamate receptors to long-term potentiation and other synaptic plasticity processes (Codazzi et al., 2006).

Collectively these studies suggest that protein kinases modulate Na^{+} and Ca^{2+} channels, neurotransmitter release, synaptogenesis, and neurite outgrowth. These processes may be related to the synaptic reorganization and neuroplasticity associated with long-term potentiation, long-term depression, and neurodegeneration (Guglielmetti et al., 1997; McNamara et al., 1999; Xia et al., 2000; Kondo et al., 2005; Moriguchi et al., 2006; Wang et al., 2007; Dell'Acqua et al., 2006).

6.6 Ornithine Decarboxylase and Polyamines in Neurotoxicity Mediated by Glutamate

Intracerebellar or systemic injections of the glutamate analog, KA, significantly upregulate ornithine decarboxylase activity (Table 6.2), a rate limiting enzyme responsible for the conversion of ornithine to putrescine (Ciani and Contestabile, 1993). Ornithine decarboxylase has a short half-life in mammalian tissues, 10 min. Traumatic and neurodegenerative situations significantly increase its activity (Lombardi et al., 1993). Products of the ornithine decarboxylase reaction, ornithine and putrescine, are metabolized to spermidine and spermine (de Vera et al., 1991). Treatment of rats with MK-801 can prevent this upregulation in ornithine decarboxylase activity (Reed and De Belleroche, 1990), suggesting that upregulation of ornithine decarboxylase is a process mediated by glutamate receptors. In fact, NMDA receptors contain a modulatory site sensitive to the polyamines spermine and spermidine. At present little is known about the toxicity of the input to the polyamine site during normal levels of NMDA receptor activation. However, only micromolar concentrations of polyamines are required for NMDA receptor activation in vitro (Fage et al., 1992).

Injections of KA also alter spermidine content during the acute convulsant phase in frontal cortex, but not in hippocampus. Putrescine contents are markedly increased in frontal cortex and hippocampus after KA treatment. This increase in putrescine content may be involved in producing behavioral changes and histological lesions. The ornithine decarboxylase inhibitor, α-difluoromethylornithine, antagonizes behavioral alterations and lowers putrescine levels in the frontal cortex of convulsant animals, but does not block neuronal injury. In primary cultures of rat cerebellar granule neurons, exogenous application of the polyamines spermidine and spermine, but not putrescine, potentiates the delayed neurotoxicity mediated by NMDA receptor stimulation with glutamate (Lombardi et al., 1993). Furthermore, both toxic and nontoxic concentrations of glutamate and NMDA stimulate ornithine decarboxylase activity, and also increase the concentration of ornithine decarboxylase mRNA in cerebellar granule neurons but not in glial cells.

α-Difluoromethylornithine prevents glutamate-mediated ornithine decarboxylase activation but has no effect on neurotoxicity mediated by glutamate (Lombardi et al., 1993). These studies suggest that high levels of extracellular polyamines potentiate glutamate-triggered neuronal death. The glutamate-induced increase in neuronal ornithine decarboxylase activity may play an important role in intracellular events associated with neural cell injury during neurotoxicity mediated by glutamate. However, the increase in ornithine decarboxylase activity is only transient after systemic KA injection (Najm et al., 1992). Blockade of ornithine decarboxylase activity does not alter the levels of putrescine and extent of spectrin degradation suggesting that ornithine decarboxylase may not be involved in KA-mediated neural cell injury. Thus the role of ornithine decarboxylase in neural cell injury mediated by glutamate and its analogs remains controversial. The treatment of organotypic hippocampal slice cultures with N1,N(2)-bis(2,3-butadienyl)-1,4-butanediamine (MDL 72527), an irreversible inhibitor of polyamine oxidase, causes a partial but significant neuronal protection (Liu et al., 2001). In addition, pretreatment with MDL 72527 attenuates the KA-mediated increase in lipid peroxidation, cytosolic cytochrome C release, and glial cell activation (Liu et al., 2001). Furthermore, pre-treatment with a combination of cyclosporin A, an inhibitor of the mitochondrial permeability transition pore, and MDL 72527 results in an additive and almost total neuronal protection against KA toxicity, whereas the combination of MDL 72527 and EUK-134, a synthetic catalase/superoxide dismutase mimetic, does not provide additive protection. The polyamine interconversion pathway partially contributes to KA-induced neural injury via the production of reactive oxygen species (Liu et al., 2001).

6.7 MAP Kinases in Neurotoxicity Mediated by Glutamate

In oligodendrocyte progenitor cells, KA produces a time- and dose-dependent increase in MAP-kinase activity. CNQX, a competitive AMPA/KA receptor antagonist, can block this effect (Liu et al., 1999) (Table 6.2). Furthermore, a specific inhibitor of MAPK-kinase, PD098059, also inhibits MAP-kinase activation and reduces KA-induced c-fos gene expression. It selectively inactivates mitochondrial aconitase, but not fumarase activity, in the hippocampus of KA treated rat (Liang et al., 2000). This suggests involvement of the citric acid cycle in KA-induced toxicity. Systemic KA administration specifically increases mitochondrial superoxide production. Transgenic mice over expressing mitochondrial superoxide dismutase-2 are resistant to KA-induced neuronal loss. The alteration in activities of cellular antioxidant enzymes is a good index of oxidative stress (Cohen et al., 1987; Kim et al., 2000a). An activation of caspase-3 activity is also observed in rat hippocampus following KA administration (Faherty et al., 1999), suggesting that KA-induced apoptotic cell death may contribute to neuronal damage.

6.8 Nitric Oxide Synthase in Neurotoxicity Mediated by Glutamate

Nitric oxide (NO) is formed from L-arginine by Ca^{2+}/calmodulin-dependent nitric oxide synthase (NOS). In brain, the highest amounts of NOS are found in the cerebellum in which NO generation and cyclic GMP synthesis are increased by glutamate and its analogs (Bolanos et al., 1997). NO, at low levels, is an important signaling molecule. Under certain circumstances, excessive NO production produces neurotoxicity. Either the nitric oxide synthase (NOS) inhibitor *N*-omega-nitro-L-arginine methyl ester (NAME) or the NMDA receptor antagonist 2-amino-5-phosphonopentanoate (APV) (Almeida et al., 1998) can prevent the neurotoxic effects of NO. Excessive glutamate-receptor stimulation leads to neuronal death through a mechanism involving NO and superoxide production and the generation of peroxynitrite, $ONOO^-$. The reactivity of $ONOO^-$ is roughly the same as that of hydroxyl radical, •OH.

Astroglial cells respond to high levels of glutamate and cytokines by generating large amounts of NO, which may be deleterious to the neighboring neurons and oligodendrocytes. The exact molecular mechanism involved in NO-mediated neuronal death is not known. However, $ONOO^-$ interacts with nitrate and sulfhydryl groups, and can hydroxylate the aromatic rings of amino acid residues (Beckman et al., 1992). Thus, $ONOO^-$ reacts with lipid, proteins, and DNA (Radi et al., 1991a,b; Moreno and Pryor, 1992; Qi et al., 2000). In addition, $ONOO^-$ reduces mitochondrial respiration, inhibits membrane pumps, depletes cellular glutathione, and damages DNA, thus activating poly (ADP-ribose) synthase, an enzyme that leads to cellular energy depletion (Pryor and Squadrito, 1995). Collectively these studies suggest that ONOO- interferes with key enzymes of the tricarboxylic acid cycle, the mitochondrial respiratory chain, and mitochondrial Ca^{2+} metabolism (Bolanos et al., 1997). All these processes may be involved in neuronal energy deficiency and neurotoxicity caused by glutamate and its analogs.

6.9 Expression of Apolipoproteins D and E in Neurotoxicity Mediated by Glutamate

A marked increase in apolipoprotein D immunoreactivity is observed in hippocampus following neurotoxicity induced by KA (Table 6.3) (Ong et al., 1997). This increase in apolipoprotein D immunoreactivity occurs on the first day after KA injection, mainly in the cell bodies and dendrites of neurons (Fig. 6.3A, 3B). Unlike the gray matter, no increase is observed in white matter tracts (Ong et al., 1997a). The significance of this upregulation of apolipoprotein D remains unknown. However, apolipoprotein D binds to arachidonic acid (Morais Cabral et al., 1995), which is released by PLA_2 stimulation. This could be a protective mechanism to limit the oxidative products from arachidonic acid, e.g. 4-HNE, after kainate injection.

Similarly, apolipoprotein E expression increases in neurotoxicity mediated by KA (Table 6.3) (Boschert et al., 1999). Apolipoprotein E is a major lipoprotein in the brain. It is involved in the transport, distribution, and other aspects of cholesterol homeostasis. Apolipoprotein E also plays a dominant role in the mobilization and redistribution of brain lipids in repair, growth, and maintenance of nerve cells (Mahley, 1988). The secretion of apolipoproteins E and D may be differentially regulated in cultured astrocytes. In cell culture systems this depends upon the extracellular lipid milieu (Patel et al., 1995). During neurotoxicity mediated by KA, apolipoprotein E levels increase moderately in astrocytes and apolipoprotein E mRNA increases very strongly in clusters of CA1 and CA3 pyramidal neurons. Based on hybridization in situ and immunohistochemical studies, expression of apolipoprotein E in neurons may be a part of a rescue program to counteract neurodegeneration mediated by KA (Boschert et al., 1999).

Table 6.3 Other neurochemical parameters associated with neurotoxicity mediated by glutamate and its analogs

Parameter	Effect	Reference
NF-κB	Activation and translocation	(Cruise et al., 2000)
Cyclin D and Cdk4	Activation	(Park et al., 2000; Verdaguer et al., 2002)
Divalent metal ion transporter	Increased	(Ong et al., 2006)
GABA transporter BGT-1	Up-regulation	(Zhu and Ong, 2004)
GAT-1 and GAT-3	Down-regulated	(Zhu and Ong, 2004)
GFAP	Increased	(Sohl et al., 2000)
Calmodulin	Up-regulation	(Sola et al., 1997)
P53	Increased	(Sakhi et al., 1994)
TGF-β3	Up-regulation	(Kim et al., 2002)
Other growth factors (NGF, BDNF, IGF, GDNF)	Increased	(Sperk, 1994; Zafra et al., 1991)
β-APP	Increased	(Shoham and Ebstein, 1997)
Apolipoprotein E and D	Increased	(Ong et al., 1997; Boschert et al., 1999)
14-3-3 ζ	Up-regulation	(van der Brug et al., 2002)
c-fos and c-jun	Up-regulation	(Lerea et al., 1995; Lee et al., 2001b)

NF-κB, Nuclear factor-kappaB; GFAP, glial fibrillary acidic protein; β-APP, β-amyloid precursor protein; TGF-β3, transforming growth factor-β3; Cdk4, cyclin-dependent kinase; and p53, a tumor-suppressor gene.

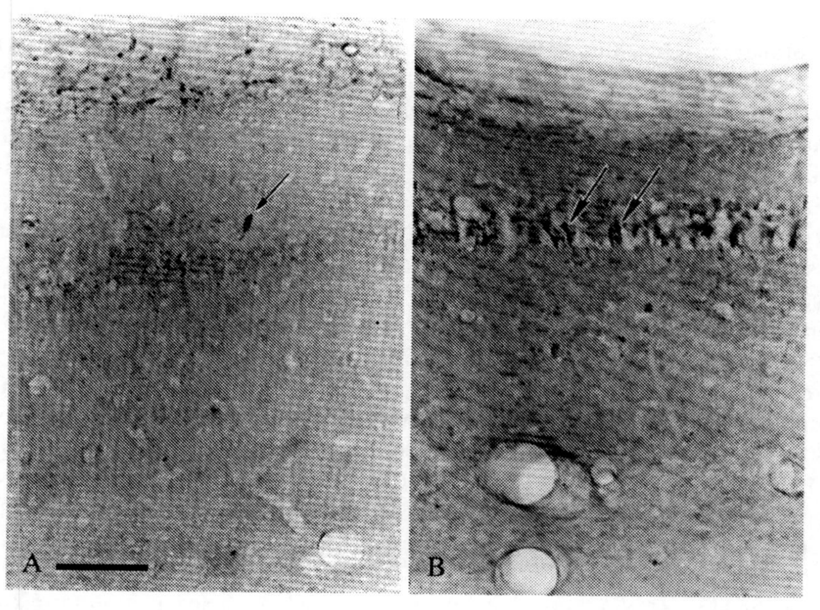

Fig. 6.3 Apolipoprotein D immunoreactivity in field CA1 of the hippocampus. A: section from a normal rat, showing occasional labeled neurons (arrow). B: section from a 1 day post-kainate inected rat, showing dense staining in pyramidal neurons (arrows). Scale = 160 μm. Reproduced with kind permission from Ong et al., 1997, Neuroscience 79:359–367. Pergamon

6.10 Growth Factor Expression in Neurotoxicity Mediated by Glutamate

Administration of glutamate and its analogs in rat brain produces a marked increase in the expression of mRNA for nerve growth factor (NGF), brain derived growth factor (BDNF), and insulin-like growth factor (IGF)-I, whilst neurotrophin-3 mRNA is decreased (Sperk, 1994; McCusker et al., 2006) (Table 6.3). The balance between the activation of the glutamatergic and GABAergic systems modulates the physiological levels of BDNF and NGF mRNAs in hippocampal neurons in vitro and *in vivo* (Zafra et al., 1991). Blockade of the glutamate receptors and/or stimulation of the GABAergic system retards synthesis of BDNF and NGF mRNAs in hippocampus and synthesis of NGF protein in hippocampus and septum. The inhibition of NGF synthesis in the septum reflects the diminished availability of NGF in the projection field of NGF-dependent septal cholinergic neurons. These neurons do not synthesize NGF themselves but accumulate it by retrograde axonal transport (Zafra et al., 1991). The modulation of BDNF and NGF synthesis by the glutamate and GABA transmitter systems suggests that BDNF and NGF play an important role in induction and maintenance of synaptic plasticity (Zafra et al., 1991).

KA administration increases the expression of glial derived neurotrophic factor, GDNF. However, the time course and pattern of GDNF mRNA expression is quite distinct from other neurotrophins. Thus, KA administration markedly and rapidly increases BDNF mRNA in hippocampus and in other cortical regions, but the increase in GDNF mRNA is slow and mainly restricted to the hippocampal formation. KA administration in rats also up-regulates the immunoreactivity of basic fibroblast growth factor-like in nuclei of astrocytes in several forebrain areas. This effect peaks 24 h after KA injection. Based on studies *in vivo*, basic fibroblast growth factor may play a neuroprotective role in excitotoxicity mediated by kainate, as seen from a massive and widespread astroglial increase in basic fibroblast growth factor immunoreactivity and messenger RNA (Humpel et al., 1993). The neuroprotective effect of BDNF is also observed on differentiating cells of retinal explants. A neutralizing antibody to BDNF, but not an antibody to neurotrophin-4, NT-4, can completely retard this effect (Martins et al., 2005). Consistently, chronic activation of NMDA receptor up-regulates the expression of BDNF and trkB mRNA, as well as BDNF protein content, but has no effect on the content of NT-4 mRNA in retinal tissue. Furthermore, inactivation of NMDA receptors *in vivo* by intraperitoneal injections of MK-801 increases the natural cell death of specific cell populations in the post-natal retina. Collectively, these studies indicate that chronic activation of NMDA receptors in vitro induces a BDNF-dependent neuroprotective state in differentiating retinal cells, and that NMDA receptor activation controls the programmed cell death of developing retinal neurons *in vivo*.

Endogenous BDNF-TrkB signaling in glutamate-mediated neurotransmission may be involved in consolidation of information at the synapse. Recent studies also indicate that sustained synthesis of the immediate early gene Arc, activity-regulated cytoskeleton-associated protein, during glutamate toxicity protracts the time-window required for hyperphosphorylation of actin-depolymerizing factor/cofilin and local expansion of the actin cytoskeleton *in vivo*. The synaptic consolidation mediated by Arc is upregulated in response to a brief infusion of BDNF (Soule et al., 2006). Genes regulated by BDNF that may co-operate with Arc during LTP maintenance were found with microarray expression profiling. In addition to regulating gene expression, BDNF signaling modulates the fine localization and biochemical activation of the translation machinery (Soule et al., 2006). Collective evidence suggests that by modulating the spatial and temporal translation of newly induced Arc, and constitutively expressed mRNA in neuronal dendrites, BDNF may effectively control the window of synaptic consolidation (Soule et al., 2006).

Thus alterations in growth factor expression due to exposure to glutamate and its analogs maintain neuronal circuitry, modulate synaptic efficacy, and consolidate information in developing and adult brain. Growth factors also protect hippocampal neurons from death induced by serum deprivation. Several molecular mechanisms are possible. These mechanisms include activation of phosphatidylinositol 3-kinase (PtdIns 3-K)/Akt (protein kinase B) kinase and the mitogen-activated protein kinase (MAPK) pathways by IGF-1 and BDNF (Zheng and Quirion, 2004) and enhancement of mitogen activated kinase (MEK)/extracellular signal regulated kinase (ERK) pathways. The phosphorylation of Akt and its downstream target, FKHRL1, a winged-helix family transcription factor, induced by IGF-1 is rapid and sustained whereas that of MAPK is transient. Furthermore, IGF-1 also mediates the tyrosine phosphorylation of insulin receptor substrate-1 (IRS-1) and its association with PtdIns 3-kinase.

In contrast, the tyrosine phosphorylation of Shc proteins is dramatically stimulated by BDNF, but not by IGF-1, suggesting different phosphorylation mechanisms. In hippocampal neurons only the inhibitor of the PtdIns 3-K/Akt pathway, LY294002, retards the survival effects of both IGF-1 and BDNF; an inhibitor of the MAPK pathway inhibitor, PD98059, has no effect (Zheng and Quirion, 2004). Taken together, these results indicate that the neuroprotective effects of both IGF-1 and BDNF against serum deprivation are mediated by the activation of the PtdIns 3-K/Akt, but not the MAPK pathway. Collective evidence suggests that growth factors protect neural cells against toxicity induced by glutamate. This effect is mediated primarily through the PtdIns 3-K/Akt and the MEK/ERK signaling pathways.

6.11 Cytokine Expression in Neurotoxicity Mediated by Glutamate

Injections of KA also induce the expression of mRNA for cytokines such as interleukin-1β (IL-1β), interleukin-6, tumor necrosis factor (TNF-α), and leukemia inhibitory factor in different regions of rat brain (Minami et al., 1991; Yabuuchi et al., 1993). Cytokines play important roles not only in neurodegeneration, but also in neuromodulation that occurs in brain tissue during recovery after neurotoxicity mediated by KA (Yabuuchi et al., 1993). KA injections also induce the synthesis of endogenous interleukin-1 receptor antagonist (IL-1ra) in rat brain (Chau and Tai, 1981). The induction of IL-1ra can be observed in the hippocampus, amygdale, thalamus, piriform cortex, entorhinal cortex, and some hypothalamic nuclei. The pattern and time-course of IL-1ra mRNA expression is closely associated with degenerative histopathological changes in brains of rats injected with KA (Eriksson et al., 1998). The molecular mechanism of induction of cytokine mRNA in neurotoxicity mediated by KA is not fully understood. However, activation of microglia mediated by KA may be associated with the expression of cytokines (Minami et al., 1991). These cytokines may mediate cellular intercommunication through autocrine, paracrine, or endocrine mechanisms (Wilson et al., 2002) following toxicity mediated by KA.

6.12 Regulation of NMDA and GABA Receptors in Neurotoxicity Mediated by Glutamate

During neurotoxicity induced by KA, glutamate release is stimulated (Stein-Behrens et al., 1992). The released glutamate may exert positive feedback to potentiate the action of KA by acting at the NMDA receptor, which produces a further increase in the entry of extracellular Ca^{2+} through receptor-operated ion channels, thus compounding the Ca^{2+}-mediated neuronal damage.

The expression of GABA transporters in the rat hippocampus was studied following KA-induced neuronal injury. A decrease in GAT-1 and GAT-3 immunostaining is observed in CA subfields but no change in expression of GAT-2 staining is observed in the degenerating CA subfields (Zhu and Ong, 2004) (Fig. 6.4). In contrast, increased BGT-1 immunoreactivity is observed in astrocytes following KA injection. BGT-1 is a weak transporter of GABA in comparison to other GABA transporters. The increased expression of BGT-1 in astrocytes may be a protective mechanism against increased osmotic stress known to occur after excitotoxic injury (Fig. 6.5). On the other hand, excessive or prolonged BGT-1 expression may cause astrocytic swelling following KA-mediated brain injury (Zhu and Ong, 2004).

KA receptor stimulation also induces a down-regulation of GABA-mediated transmission in the hippocampus (Rodríguez-Moreno and Lerma, 1998), thus inhibiting the release of GABA. This process involves the activation of phospholipase C (PLC) and protein kinase C through a pertussis toxin-sensitive G protein (Ziegra et al., 1992), indicating a metabotropic function of the KA receptor. Increased Ca^{2+} concentration can block KA inhibition of GABA release. This once again suggests the involvement of second messengers generated from glycerophospholipids in the regulation of Ca^{2+}, thus affecting processes mediated by KA receptors. More studies are required to understand the role of KA receptors in inhibiting GABA release at the presynaptic level.

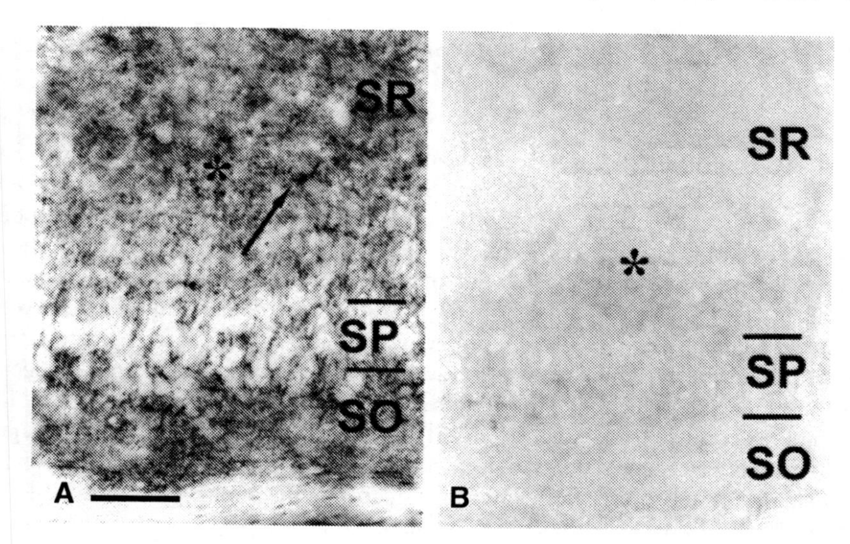

Fig. 6.4 GAT-3 immunoreactivity in field CA3 of the hippocampus. A: section from a normal rat, showing occasional labeled glial cell bodies (arrow) and large numbers of processes in the neuropil (asterisk). B: section from a 1 day post-kainate injected rat showing loss of labeling in an affected area (asterisk). Abbreviations: SO, stratum oriens; SP, stratum pyramidale; and SR stratum radiatum. Scale = 50 μm. Reproduced with kind permission from Zhu et al., 2004, Journal of Neuroscience Research 77:402–409. Wiley-Liss

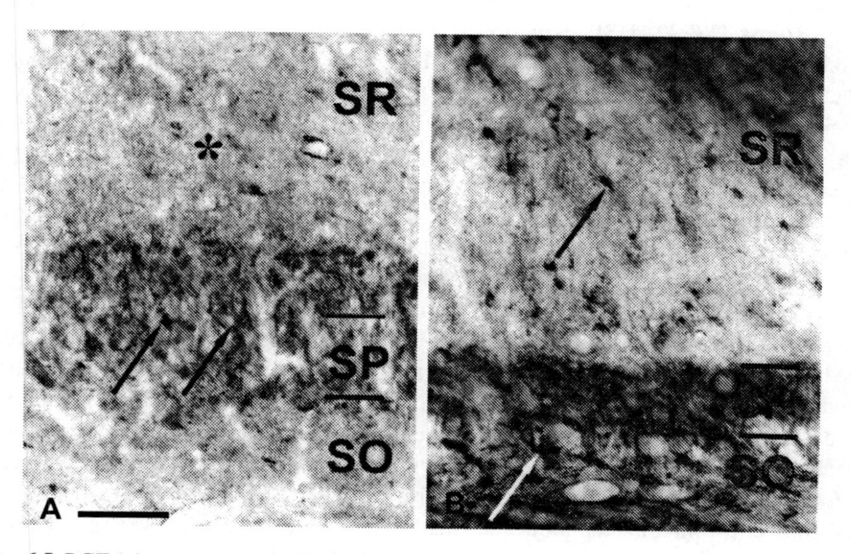

Fig. 6.5 BGT-1 immunoreactivity in field CA3 of the hippocampus. A: section from a normal rat, showing labeled pyramidal cell bodies (arrows), and the neuropil (asterisk). B: section from a 1 week post-kainate injected rat, showing loss of staining in neurons, but increased staining in astrocytes (arrows) in an affected area. Abbreviations: SO, stratum oriens; SP, stratum pyramidale; and SR stratum radiatum. Scale = 50 μm. Reproduced with kind permission from Zhu et al., 2004, Journal of Neuroscience Research 77:402–409. Wiley-Liss

6.13 Heat Shock Protein Expression in Neurotoxicity Mediated by Glutamate

Mammalian cells respond to stresses, heat shock, heavy metal toxicity, glutamate toxicity, and alcohol exposure, by inducing the expression of highly conserved proteins, the heat shock proteins and their mRNA (HSPs). These proteins include HSP-27, HSP-32, HSP-47, HSP-70, HSP-90, and heme oxygenase (HO). They serve as molecular chaperones. In brain, they are involved in the protection of neurons from various forms of stress. The heat shock genes contain highly conserved cis-acting elements (HSEs) in the 5' promoter regions upstream of the open reading frames of the genes that bind cellular transcription factors during heat shock (Sorger, 1991). Intracortical injection of the glutamate receptor agonist, NMDA, in 9-day-old rat brain produces considerable toxicity (Acarin et al., 2002). Excitotoxic damage results in primary cortical degeneration and secondary damage in the corresponding thalamus. In the injured cortex, reactive microglia/macrophages express HSP32 from 10 h until 14 days post-lesion, showing maximal levels at days 3–5.

In parallel, most cortical reactive astrocytes express HSP47 from 10 h until 14 days post-lesion. A population of them also expresses HSP27 labeling from 1 day post-lesion. In addition, some cortical reactive astrocytes display a temporary expression of HSP32 at day 1. In general, astroglial HSP expression in the cortex reaches maximal levels at days 3–5 post-lesion. In the damaged thalamus, HSP32 is not significantly induced, but reactive astrocytes express HSP47 and some of them also express HSP27. HSP induction in thalamic astroglia is transient peaking at 5 days post-lesion and returning to basal levels by day 14 (Acarin et al., 2002). The expression of HSP32, HSP27, and HSP47 mediated by glutamate analogs in glial cells may contribute to their protection and adaptation to damage, therefore playing an important role in the evolution of the glial response and the outcome of excitotoxic lesions. HSP32 may provide antioxidant protective mechanisms to microglia/macrophages, whereas HSP47 could contribute to extracellular matrix remodeling, and HSP27 may stabilize the astroglial cytoskeleton and participate in astroglial antioxidant mechanisms (Acarin et al., 2002).

In neurotoxicity mediated by KA, a moderate expression of HSP-70 mRNA is detected in pyramidal cell layers of CA1-3 and dentate gyrus of the hippocampus, the basolateral, lateral, central and medial amygdala, the piriform cortex, the central medial thalamic nucleus of rats, and in cultured hippocampal slices (Hashimoto et al., 1998; Akbar et al., 2001; Sato and Matsuki, 2002). Marked expression of HSP-70 mRNA is also detected in cingulate, parietal, somatosensory, insular, entorhinal, and piriform cortices of cerebral cortex, all regions of hippocampus, and the central medial thalamic nucleus of the rats that develop severe seizures. Collectively these studies suggest that the behavioral changes, seizure intensity, and body temperature, may be related to the early expression of various HSP mRNA by glial cells and that HSP have a protective effect against the vulnerability of CA1 and CA3 neurons and also with toxicity tolerance of neurons in the dentate gyrus (Hashimoto et al., 1998; Akbar et al., 2001; Sato and Matsuki, 2002).

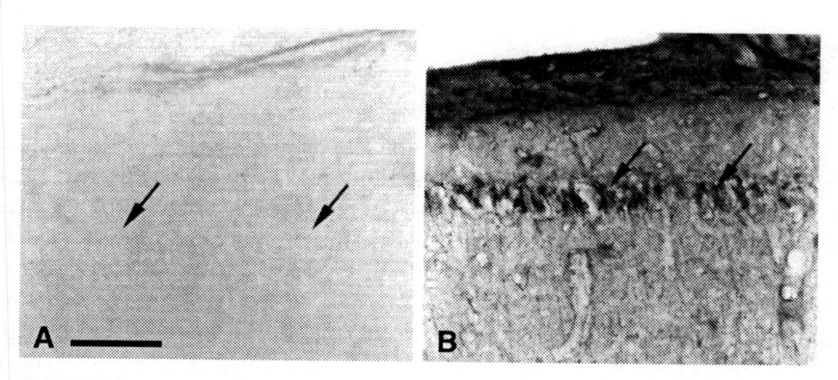

Fig. 6.6 HO-1 immunoreactivity in field CA1 of the hippocampus. A: section from a normal rat, showing very little HO-1 labeling. Arrows indicate pyramidal neurons. B: section from a 1 day post-kainate lesioned hippocampus, showing increased labeling in pyramidal neurons (arrows). Scale = 150 μm. Reproduced with kind permission from Huang et al., 2005, Journal of Neuroscience Research 80:268–278. Wiley-Liss

Of the various HSP, heme oxygenase-1, HO-1, by generating the vasoactive molecule carbon monoxide and the potent antioxidant bilirubin, represents a novel protective system potentially active against brain oxidative injury (Rossler et al., 2004). Kainate injection leads to an increased expression of the inducible form of HO-1 in the hippocampus (Huang et al., 2005) (Fig. 6.6), but it is not clear whether the HO-1 is enzymically active or not. Using monoclonal antibodies to bilirubin and HO-1 it was found that there is a close correlation between bilirubin and HO-1 expression. Both bilirubin and HO-1 are expressed in damaged neurons at early times, and astrocytes at later times, weeks, after kainate injection. These observations suggest that the increased HO-1 immunoreactivity in the hippocampus is enzymically active. To determine whether HO-1 activity after kainate may have a protective or, perhaps, destructive effect, kainate-injected rats were injected intraperitoneally with a blood-brain barrier-permeable inhibitor of HO, tin protoporphyrin (SnPP). SnPP treatment did not improve neuronal survival. Instead, increased mortality was observed in rats treated with SnPP. The surviving SnPP-treated rats show significantly less Nissl or MAP2 staining in the hippocampal field compared to the saline-injected rats. These results indicate that HO-1 induction has a net protective effect on neurons in the kainate model of excitotoxic injury (Huang et al., 2005).

The HO-1 gene is redox-regulated. Redox active compounds, including nutritional antioxidants, modulate its expression. Given the broad cytoprotective properties of the heat shock response, there is now strong interest in discovering and developing pharmacological agents capable of inducing the heat shock response. These findings have opened up new neuroprotective strategies, as molecules inducing this defense mechanism can be a therapeutic target to minimize the deleterious consequences associated with the accumulation of conformationally aberrant proteins during oxidative stress, such as in neurodegenerative disorders and brain aging, with resulting prolongation of a healthy life span (Rossler et al., 2004).

6.14 Cholesterol and Its Oxidation Products in Neurotoxicity Mediated by KA

Cholesterol is an integral component of neural membranes. It is crucial for the function of neuronal and glial cells in brain (Dietschy and Turley, 2001; Pfrieger, 2003). Cholesterol affects the physicochemical properties of neural membranes and regulates endocytosis, antigen expression, and activities of membrane-bound enzymes, receptors, and ion channels (Dietschy and Turley, 2001; Pfrieger, 2003). Neurons produce enough cholesterol to survive and grow, but the formation of numerous mature synapses during development demands additional amounts that must be provided by glia. Thus, the availability of cholesterol appears to limit synapse development. This may explain the delayed onset of CNS synaptogenesis after glial differentiation and neurobehavioral manifestations of defects in cholesterol or lipoprotein homeostasis. Collectively these studies indicate that neurons import glia-derived cholesterol via apolipoproteins to form numerous and efficient synaptic connections (Pfrieger, 2003).

Intraventricular injections of KA cause increased cholesterol staining in neurons of CA1 region of rat hippocampus (Fig. 6.7). Increased cholesterol staining is also observed in hippocampal slices and neuronal cultures after kainate treatment (Ong et al., 2003). No increase in cholesterol staining is observed in glial cells. In neuronal cultures, addition of lovastatin, an inhibitor of cholesterol synthesis, attenuates the increased filipin staining of cholesterol after kainate treatment, indicating that the increase in cholesterol may be due to an increase in cholesterol synthesis. The molecular mechanism involved in the upregulation of cholesterol synthesis in neurotoxicity induced by KA remains unknown. However, it may be the response to the oxidative stress of those pyramidal neurons that escape cell death. They then participate in compensatory reorganization of local networks and upregulate their cholesterol synthesizing machinery, not only for maintaining their own membrane

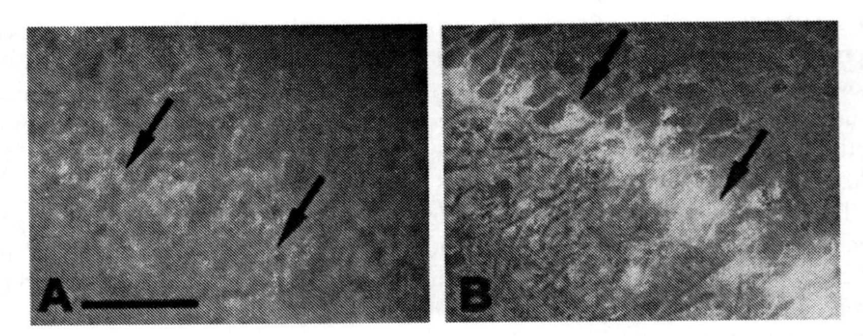

Fig. 6.7 Filipin staining for free cholesterol in field CA3 of the hippocampus. A: section from a normal rat, showing light staining for free cholesterol in pyramidal neurons (arrows). B: section from a rat that had been injected with kainate 1 day earlier, showing intense filipin staining in affected hippocampal neurons (arrows). Scale = 250 μm. Reproduced with kind permission from Ong et al., 2003, Brain Pathology 13:250–262. Blackwell Publishing

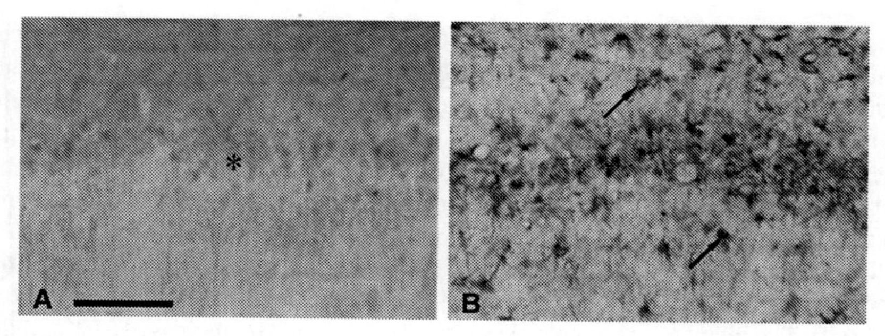

Fig. 6.8 Cholesterol 24-hydroxylase immunoreactivity in field CA1 of the hippocampus. A: section from a normal rat, showing light staining in neurons (asterisk). B: section from a rat that had been injected with kainate 2 weeks earlier, showing increased labeling in astrocytes (arrows). Scale = 50 μm. Reproduced with kind permission from He et al., 2006, Journal of Neuropathology and Experimental Neurology, 65: 652–663. Lippincott, Williams and Wilkins

integrity, but also for synthesizing new membranes (Vance et al., 1994; Perez et al., 1996; Dietschy and Turley, 2001).

KA treatment causes a marked increase in cholesterol 24-hydroxylase immunoreactivity in glial cells of the affected CA fields (Fig. 6.8). The number of glial cells positive for cholesterol 24-hydroxylase increases significantly from 1 week (1312 ± 195 cells / mm^2) to 2 weeks (1815 ± 225 cells / mm^2) after kainate injection, when compared to saline-injected rats (0 ± 0 cells / mm^2). The cells positive for cholesterol 24-hydroxylase were GFAP positive. No increase in cholesterol 24-hydroxylase immunoreactivity in glial cells was observed in the hippocampus or other parts of the brain that were not affected by the kainate injections.

Organotypic slices from the hippocampus of rats incubated with kainate plus lovastatin show significantly lower levels of cholesterol, 24-hydroxycholesterol, and 7-ketocholesterol, compared to those that have been treated with KA alone, suggesting that lovastatin modulated the loss of hippocampal neurons after KA treatment. The level of 24-hydroxycholesterol detected *in vivo* after KA treatment, >50 μM, is neurotoxic in hippocampal slice cultures. Brain-permeable statins such as lovastatin may have a neuroprotective effect by limiting the levels of oxysterols in brain areas undergoing neurodegeneration (Ong et al., 2003; He et al., 2006). Glutathione can attenuate neuronal injury produced by these cholesterol oxidation products. Collectively these studies suggest that cholesterol and cholesterol oxidation products in particular may be a factor in aggravating oxidative damage to neurons following neuronal injury induced by KA (Ong et al., 2003).

6.15 Ceramide in Neurotoxicity Mediated by KA

Ceramide or N-acylsphingosine is a component of microdomains or lipid rafts. These structures serve as platforms for neural cell signaling and events related to signal transduction. Ceramide is generated either by de novo synthesis or by hydrolysis of sphingomyelin (Marchesini and Hannun, 2004). Neurodegeneration mediated by kainate markedly increases ceramide levels in the rat hippocampus (Guan et al., 2006). This increase in ceramide levels may be due to increased sphingomyelin degradation and a significant increase in ceramide biosynthetic activity. The first reaction in the biosynthetic pathway involves the condensation of serine and palmitoyl CoA, a reaction catalyzed by serine palmitoyltransferase (SPT). This enzyme consists of heterodimers of 53-kDa SPT1 and 63-kDa SPT2 subunits that are bound to the endoplasmic reticulum (Hanada, 2003). KA-induced toxicity increases the expression of SPT in reactive astrocytes of the hippocampus after kainate injections (Fig. 6.9). The increase in enzyme expression is paralleled by increased SPT activity in the hippocampus at 2 weeks post-kainate injection (He et al., 2007). The treatment of hippocampal slice cultures with SPT inhibitor ISP-1, myriocin, or L-cycloserine modulated increases in 16:0, 18:0, and 20:0 ceramide species, and partially reduced cell death induced by kainate in vitro. Some decrease in the sphingomyelin content of the hippocampus is also associated with the increase in ceramide (Guan et al., 2006).

A possible role of ceramide biosynthesis in neuronal injury was also observed in slice cultures (He et al., 2007). Addition of kainate to slices results in a decrease in MAP2 staining and an increase in LDH release into the culture media indicating neural injury. Incubation of slices with serine palmitoyltransferase inhibitors (L-cycloserine or myriocin), after addition of kainate, decreases MAP2 staining and increases LDH release. In contrast to inhibition of ceramide biosynthesis by L-cycloserine or myriocin, the incubation of slices with an inhibitor of sphingomyelinase has no significant protective effect. Thus, inhibition of ceramide biosynthesis has a significant protective effect after kainate-mediated neural cell injury. It should be noted that light serine palmitoyltransferase immunoreactivity is observed in neurons up to 3 days after kainate lesions, whereas astrocytic immunoreactivity to SPT did not peak until 2 weeks after kainate injection (He et al., 2007).

Ceramide may act on neural cells in several ways. Ceramide mediates clustering of membrane receptors, thus enabling membrane domains to be involved in

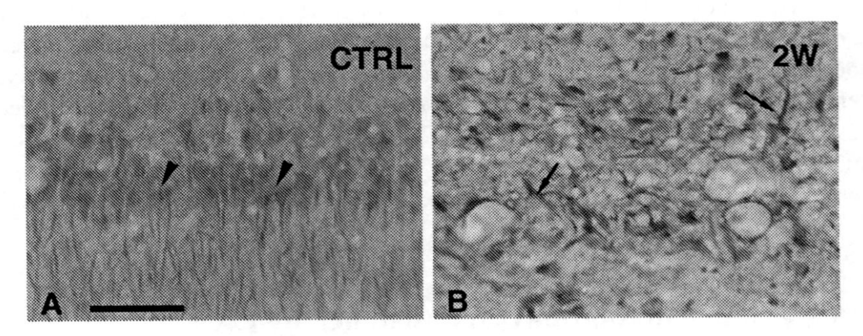

Fig. 6.9 Serine palmitoyltransferase (SPT) immunoreactivity in field CA1 of the hippocampus. A: section from a normal rat, showing light labeling in pyramidal neurons (arrowheads). B: section from a rat that had been injected with kainate 2 weeks earlier, showing loss of labeling in neurons, but increased labeling in astrocytes (arrows). Scale = 50 μm. Reproduced with kind permission from He et al., 2005, Journal of Neuroscience Research 85:423–432. Wiley-Liss

induction of apoptosis by death receptors (Dobrowsky and Carter, 1998; Gulbins and Li, 2006). Generation of ceramide mediates large protein permeable channels on the outer membrane of mitochondria (Siskind, 2005). This may facilitate the release of pro-apoptotic proteins including cytochrome c from mitochondria to the cytoplasm, and trigger induction of the mitochondrial-dependent intrinsic caspase pathway (Stoica et al., 2005). Ceramide-1-phosphate may facilitate translocation and activation of cytosolic phospholipase $A_2\alpha$ ($cPLA_2\alpha$) directly and by a PKC pathway (Nakamura et al., 2006). This may cause increased generation of arachidonic acid from membrane glycerophospholipids, and formation of toxic lipid peroxidation products such as 4-hydroxynonenal (Farooqui et al., 2004). Ceramide may also cause non-apoptotic, caspase-independent cell death by inducing generation of reactive oxygen species through the activation of $cPLA_2\alpha$ in A172 human glioma cells (Kim et al., 2005). Collectively these studies indicate that the generation of ceramide during neurotoxicity mediated by KA may facilitate neural cell death.

6.16 Uptake of Toxic Divalent Metal Ions in Neurotoxicity Induced by Kainate

Toxic divalent metal ions include lead, Pb^{2+}, cadmium, Cd^{2+}, mercuric, Hg^{2+}, and arsenic, As^{2+}. These ions enter cells via the divalent metal transporter 1 (DMT1). This transporter has an unusually broad substrate range that includes metal ions such as iron, zinc, manganese, cobalt, cadmium, copper, nickel, and lead (Gunshin et al., 1997). In normal brain, DMT1 is localized in both neurons and astrocytes (Burdo et al., 1999; Wang et al., 2001; Huang et al., 2004). The expression of DMT1 is increased in the hippocampus after neurotoxicity mediated by KA (Fig. 6.10) (Wang

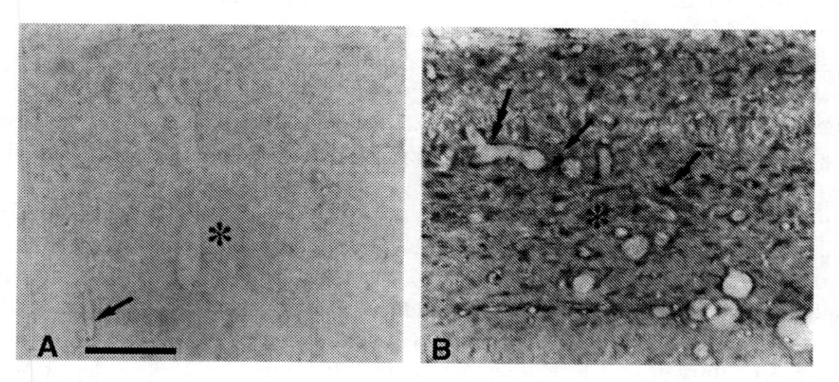

Fig. 6.10 Divalent metal transporter 1 (DMT1) immunoreactivity in field CA1 of the hippocampus. A: section from a normal rat, showing light staining (asterisk). B: section from a rat that had been injected with kainate 1 month earlier. Large numbers of DMT1 positive astrocytes (arrows) are present in the affected area (asterisk). Double arrow indicates glial end feet around blood vessels. Scale: 150 μm. Reproduced with kind permission from Wang et al., 2002, Experimental Neurology, 177:193–201. Elsevier

et al., 2002). This increased DMT1 expression correlates with increased Fe^{3+} (Wang et al., 2002), Pb^{2+}, and Cd^{2+} levels in rat brain after KA injection (Ong et al., 2006). Earlier studies suggested changes in Zn^{2+} distribution after injury mediated by KA (Kim et al., 2003), but more recent studies indicate that KA injections do not cause the accumulation of Zn^{2+} in the brain (Ong et al., 2006). These observations are consistent with the view that Zn^{2+} transport occurs through Zn^{2+} transporters (Kordas and Stoltzfus, 2004). Collectively these studies suggest that an increase in uptake of divalent metal ions into brain may be an important mechanism associated with neurodegeneration mediated by KA.

6.17 Other Neurochemical Changes in Neurotoxicity Mediated by Glutamate

A number of genes are upregulated following KA injections in adult rats, including immediate early genes like c-fos (Popovici et al., 1990; Shoham and Ebstein, 1997), Met-enkephalin and other neuropeptides (Gall, 1988), and neuropeptide Y (Vezzani et al., 1994). The significance and mechanism of upregulation of these genes is not known. However, alterations in intracellular Ca^{2+} play an important role in activating these genes (Liu et al., 1999). These genes are closely associated with synaptic reorganization in brain tissue after seizures and neurotoxicity induced by KA. Systemic KA injections also decrease β-amyloid precursor protein, β-APP, immunoreactivity in neurons, whereas astrocytic and microglial β-APP immunoreactivity is markedly increased (Shoham and Ebstein, 1997). The increase in β-APP immunoreactivity was more extensive in adult brain than young rat brain. The molecular mechanism of β-APP up-regulation remains unknown. However, activation of microglia and reactive astrocytes and their migration to sites of brain injury suggests that the increased β-APP synthesis in toxicity induced by KA may be associated with cell adhesion and gliosis. KA injections up-regulate several proteins of astrocytic origin, including GFAP. Once again, this indicates the formation of glial scars in brain tissue (Sohl et al., 2000). p53, a tumor-suppressor gene that encodes a nuclear phosphoprotein that plays an important role in regulation of cell proliferation and DNA damage repair, is up-regulated after KA-induced toxicity (Sakhi et al., 1994). Mice deficient in p53 are protected from KA-induced cell death (Morrison et al., 1996). Involvement of p53 in KA toxicity suggests that cell cycle components may also contribute to KA-induced cell death.

Administration of KA also produces profound alterations in energy metabolites in the striatum within 2 h of injection (Retz and Coyle, 1982). The depletion in ATP levels may result from either an inhibition of ATP synthesis or an increase in ATP consumption in toxicity induced by KA. Collective evidence suggests that the reduction in glucose levels and concomitant increase in lactate after intrastriatal injections of KA result from an increase in energy consumption during toxicity. It has been proposed that in toxicity induced by KA, ATP depletion to a critical level invariably precedes the onset of cell death (Coyle, 1983).

References

Acarin L., Paris J., González B., and Castellano B. (2002). Glial expression of small heat shock proteins following an excitotoxic lesion in the immature rat brain. Glia 38:1–14.

Akbar M. T., Wells D. J., Latchman D. S., and De Belleroche J. (2001). Heat shock protein 27 shows a distinctive widespread spatial and temporal pattern of induction in CNS glial and neuronal cells compared to heat shock protein 70 and caspase 3 following kainate administration. Brain Res. Mol. Brain Res. 93: 148–163.

Al Noori S. and Swann J. W. (2000). A role for sodium and chloride in kainic acid-induced beading of inhibitory interneuron dendrites. Neuroscience 101:337–348.

Almeida A., Heales S. J., Bolanos J. P., and Medina J. M. (1998). Glutamate neurotoxicity is associated with nitric oxide-mediated mitochondrial dysfunction and glutathione depletion. Brain Res. 790:209–216.

Atlante A., Gagliardi S., Minervini G. M., Ciotti M. T., Marra E., and Calissano P. (1997). Glutamate neurotoxicity in rat cerebellar granule cells: A major role for xanthine oxidase in oxygen radical formation. J. Neurochem. 68:2038–2045.

Atlante A., Calissano P., Bobba A., Azzariti A., Marra E., and Passarella S. (2000). Cytochrome c is released from mitochondria in a reactive oxygen species (ROS)-dependent fashion and can operate as a ROS scavenger and as a respiratory substrate in cerebellar neurons undergoing excitotoxic death. J. Biol. Chem. 275:37159–37166.

Beckman J. S., Ischiropoulos H., Zhu L., van der Woerd M., Smith C., Chen J., Harrison J., Martin J. C., and Tsai M. (1992). Kinetics of superoxide dismutase- and iron-catalyzed nitration of phenolics by peroxynitrite. Arch. Biochem. Biophys. 298:438–445.

Blanc E. M., Kelly J. F., Mark R. J., Wäg G., and Mattson M. P. (1997). 4-hydroxynonenal, an aldehydic product of lipid peroxidation, impairs signal transduction associated with muscarinic acetylcholine and metabotropic glutamate receptors: Possible action on $G\alpha q/11$. J. Neurochem. 69:570–580.

Bolanos J. P., Almeida A., Stewart V., Peuchen S., Land J. M., Clark J. B., and Heales S. J. (1997). Nitric oxide-mediated mitochondrial damage in the brain: mechanisms and implications for neurodegenerative diseases. J. Neurochem. 68:2227–2240.

Boonplueang R., Akopian G., Stevenson F. F., Kuhlenkamp J. F., Lu S. C., Walsh J. P., and Andersen J. K. (2005). Increased susceptibility of glutathione peroxidase-1 transgenic mice to kainic acid-related seizure activity and hippocampal neuronal cell death. Exp. Neurol. 192:203–214.

Boschert U., Merlo-Pich E., Higgins G., Roses A. D., and Catsicas S. (1999). Apolipoprotein E expression by neurons surviving excitotoxic stress. Neurobiol. Dis. 6:508–514.

Brorson J. R., Manzolillo P. A., and Miller R. J. (1994). Ca^{2+} entry via AMPA/KA receptor and excitotoxicity in cultured cerebellar Purkinje cells. J. Neurosci. 14:187–197.

Burdo J. R., Martin J., Menzies S. L., Dolan K. G., Romano M. A., Fletcher R. J., Garrick M. D., Garrick L. M., and Connor J. R. (1999). Cellular distribution of iron in the brain of the Belgrade rat. Neuroscience 93:1189–1196.

Camandola S., Poli G., and Mattson M. P. (2000). The lipid peroxidation product 4-hydroxy-2,3-nonenal increases AP-1-binding activity through caspase activation in neurons. J. Neurochem. 74:159–168.

Castagne V., Gautschi M., Lefevre K., Posada A., and Clarke P. G. H. (1999). Relationships between neuronal death and the cellular redox status. Focus on the developing nervous system. Prog. Neurobiol. 59:397–423.

Chau L. Y. and Tai H. H. (1981). Release of arachidonate from diglyceride in human platelet requires the sequential action of a diacylglycerol lipase and a monoglyceride lipase. Biochem. Biophys. Res. Commun. 100:1688–1695.

Ciani E. and Contestabile A. (1993). Ornithine decarboxylase is differentially induced by kainic acid during brain development in the rat. Brain Res. Dev. Brain Res. 71:258–260.

Codazzi F., Di Cesare A., Chiulli N., Albanese A., Meyer T., Zacchetti D., and Grohovaz F. (2006). Synergistic control of protein kinase $C\gamma$ activity by ionotropic and metabotropic glutamate receptor inputs in hippocampal neurons. J. Neurosci. 26:3404–3411.

Cohen M. R., Ramchand C. N., Sailer V., Fernandez M., McAmis W., Sridhara N., and Alston C. (1987). Detoxification enzymes following intrastriatal kainic acid. Neurochem. Res. 12: 425–429.

Coyle J. T. (1983). Neurotoxic action of kainic acid. J. Neurochem. 41:1–11.

Cruise L., Ho L. K., Veitch K., Fuller G., and Morris B. J. (2000). Kainate receptors activate NF-κB via MAP kinase in striatal neurones. NeuroReport 11:395–398.

Davis K. E., Straff D. J., Weinstein E. A., Bannerman P. G., Correale D. M., Rothstein J. D., and Robinson M. B. (1998). Multiple signaling pathways regulate cell surface expression and activity of the excitatory amino acid carrier 1 subtype of Glu transporter in C6 glioma. J. Neurosci. 18:2475–2485.

de Vera N., Artigas F., Serratosa J., and Martínez E. (1991). Changes in polyamine levels in rat brain after systemic kainic acid administration: relationship to convulsant activity and brain damage. J. Neurochem. 57:1–8.

Dell'Acqua M. L., Smith K. E., Gorski J. A., Horne E. A., Gibson E. S., and Gomez L. L. (2006). Regulation of neuronal PKA signaling through AKAP targeting dynamics. Eur. J. Cell Biol. 85:627–633.

Dietschy J. M. and Turley S. D. (2001). Cholesterol metabolism in the brain. Curr. Opin. Lipidol. 12:105–112.

Djebaili M., Rondouin G., Baille V., and Bockaert J. (2000). p53 and Bax implication in NMDA induced-apoptosis in mouse hippocampus. NeuroReport 11:2973–2976.

Djebaili M., De Bock F., Baille V., Bockaert J., and Rondouin G. (2002). Implication of p53 and caspase-3 in kainic acid but not in *N*-methyl-d-aspartic acid-induced apoptosis in organotypic hippocampal mouse cultures. Neurosci. Lett. 327:1–4.

Dobrowsky R. T. and Carter B. D. (1998). Coupling of the p75 neurotrophin receptor to sphingolipid signaling. Ann. N.Y Acad. Sci. 845:32–45.

Enslen H., Tokumitsu H., Stork P. J., Davis R. J., and Soderling T. R. (1996). Regulation of mitogen-activated protein kinases by a calcium/calmodulin-dependent protein kinase cascade. Proc. Natl. Acad. Sci. USA 93:10803–10808.

Eriksson C., Winblad B., and Schultzberg M. (1998). Kainic acid-induced expression of interleukin-1 receptor antagonist mRNA in the rat brain. Brain Res. Mol. Brain Res. 58: 195–208.

Fage D., Voltz C., Scatton B., and Carter C. (1992). Selective release of spermine and spermidine from the rat striatum by N-methyl-D-aspartate receptor activation *in vivo*. J. Neurochem. 58:2170–2175.

Faherty C. J., Xanthoudakis S., and Smeyne R. J. (1999). Caspase-3-dependent neuronal death in the hippocampus following kainic acid treatment. Brain Res. Mol. Brain Res. 70:159–163.

Farooqui A. A. and Horrocks L. A. (1991). Excitatory amino acid receptors, neural membrane phospholipid metabolism and neurological disorders. Brain Res. Rev. 16:171–191.

Farooqui A. A. and Horrocks L. A. (1994). Excitotoxicity and neurological disorders: involvement of membrane phospholipids. Int. Rev. Neurobiol. 36:267–323.

Farooqui A. A. and Horrocks L. A. (2006). Phospholipase A_2-generated lipid mediators in the brain: the good, the bad, and the ugly. Neuroscientist 12:245–260.

Farooqui A. A., Anderson D. K., and Horrocks L. A. (1993). Effect of glutamate and its analogs on diacylglycerol and monoacylglycerol lipase activities of neuron-enriched cultures. Brain Res. 604:180–184.

Farooqui A. A., Horrocks L. A., and Farooqui T. (2000). Deacylation and reacylation of neural membrane glycerophospholipids. J. Mol. Neurosci. 14:123–135.

Farooqui A. A., Ong W. Y., and Horrocks L. A. (2004). Biochemical aspects of neurodegeneration in human brain: involvement of neural membrane phospholipids and phospholipases A_2. Neurochem. Res. 29:1961–1977.

Furukawa K. and Mattson M. P. (1998). The transcription factor NF-κB mediates increases in calcium currents and decreases in NMDA- and AMPA/kainate-induced currents induced by tumor necrosis factor-α in hippocampal neurons. J. Neurochem. 70:1876–1886.

Gall C. (1988). Seizures induce dramatic and distinctly different changes in enkephalin, dynorphin, and CCK immunoreactivities in mouse hippocampal mossy fibers. J. Neurosci. 8:1852–1862.

González M. I., Susarla B. T. S., and Robinson M. B. (2005). Evidence that protein kinase Cα interacts with and regulates the glial glutamate transporter GLT-1. J. Neurochem. 94: 1180–1188.

Grilli M., Pizzi M., Memo M., and Spano P. (1996). Neuroprotection by aspirin and sodium salicylate through blockade of NF-κB activation. Science 274:1383–1385.

Guan X. L., He X., Ong W. Y., Yeo W. K., Shui G. H., and Wenk M. R. (2006). Non-targeted profiling of lipids during kainate-induced neuronal injury. FASEB J. 20:1152–1161.

Guglielmetti F., Rattray M., Baldessari S., Butelli E., Samanin R., and Bendotti C. (1997). Selective up-regulation of protein kinase C epsilon in granule cells after kainic acid-induced seizures in rat. Brain Res. Mol. Brain Res. 49:188–196.

Gulbins E. and Li P. L. (2006). Physiological and pathophysiological aspects of ceramide. Am. J. Physiol. Regul. Integr. Comp. Physiol. 290:R11–R26.

Gunshin H., Mackenzie B., Berger U. V., Gunshin Y., Romero M. F., Boron W. F., Nussberger S., Gollan J. L., and Hediger M. A. (1997). Cloning and characterization of a mammalian proton-coupled metal-ion transporter. Nature 388:482–488.

Hanada K. (2003). Serine palmitoyltransferase, a key enzyme of sphingolipid metabolism. Biochim. Biophys. Acta 1632:16–30.

Hashimoto K., Watanabe K., Nishimura T., Iyo M., Shirayama Y., and Minabe Y. (1998). Behavioral changes and expression of heat shock protein hsp-70 mRNA, brain-derived neurotrophic factor mRNA, and cyclooxygenase-2 mRNA in rat brain following seizures induced by systemic administration of kainic acid. Brain Res. 804:212–223.

He X., Jenner A. M., Ong W. Y., Farooqui A. A., and Patel S. C. (2006). Lovastatin modulates increased cholesterol and oxysterol levels and has a neuroprotective effect on rat hippocampal neurons after kainate injury. J. Neuropathol. Exp. Neurol. 65:652–663.

He X., Guan X. L., Ong W. Y., Farooqui A. A., and Wenk M. R. (2007). Expression, activity, and role of serine palmitoyltransferase in the rat hippocampus after kainate injury. J. Neurosci. Res. 85:423–432.

Hjelle O. P., Chaudhry F. A., and Ottersen O. P. (1994). Antisera to glutathione: characterization and immunocytochemical application to the rat cerebellum. Eur. J. Neurosci. 6:793–804.

Huang J. and Philbert M. A. (1995). Distribution of glutathione and glutathione-related enzyme systems in mitochondria and cytosol of cultured cerebellar astrocytes and granule cells. Brain Res. 680:16–22.

Huang E., Ong W. Y., and Connor J. R. (2004). Distribution of divalent metal transporter-1 in the monkey basal ganglia. Neuroscience 128:487–496.

Huang E., Ong W. Y., Go M. L., and Garey L. J. (2005). Heme oxygenase-1 activity after excitotoxic injury: immunohistochemical localization of bilirubin in neurons and astrocytes and deleterious effects of heme oxygenase inhibition on neuronal survival after kainate treatment. J. Neurosci. Res. 80:268–278.

Humpel C., Lippoldt A., Chadi G., Ganten D., Olson L., and Fuxe K. (1993). Fast and widespread increase of basic fibroblast growth factor messenger RNA and protein in the forebrain after kainate-induced seizures. Neuroscience 57:913–922.

Janaky R., Ogita K., Pasqualotto B. A., Bains J. S., Oja S. S., Yoneda Y., and Shaw C. A. (1999). Glutathione and signal transduction in the mammalian CNS. J. Neurochem. 73:889–902.

Jenkinson A. M., Collins A. R., Duthie S. J., Wahle K. W. J., and Duthie G. G. (1999). The effect of increased intakes of polyunsaturated fatty acids and vitamin E on DNA damage in human lymphocytes. FASEB J. 13:2138–2142.

Kim H., Bing G., Jhoo W., Ko K. H., Kim W. K., Suh J. H., Kim S. J., Kato K., and Hong J. S. (2000a). Changes of hippocampal Cu/Zn-superoxide dismutase after kainate treatment in the rat. Brain Res. 853:215–226.

Kim H. C., Jhoo W. K., Kim W. K., Suh J. H., Shin E. J., Kato K., and Ho K. K. (2000b). An immunocytochemical study of mitochondrial manganese-superoxide dismutase in the rat hippocampus after kainate administration. Neurosci. Lett. 281:65–68.

Kim H. C., Bing G., Kim S. J., Jhoo W. K., Shin E. J., Bok W. M., Ko K. H., Kim W. K., Flanders K. C., Choi S. G., and Hong J. S. (2002). Kainate treatment alters TGF-β3 gene expression in the rat hippocampus. Brain Res. Mol. Brain Res. 108:60–70.

Kim D., Kim E. H., Kim C., Sun W., Kim H. J., Uhm C. S., Park S. H., and Kim H. (2003). Differential regulation of metallothionein-I, II, and III mRNA expression in the rat brain following kainic acid treatment. NeuroReport 14:679–682.
Kim S. Y., Min D. S., Choi J. S., Choi Y. S., Park H. J., Sung K. W., Kim J., and Lee M. Y. (2004). Differential expression of phospholipase D isozymes in the hippocampus following kainic acid-induced seizures. J. Neuropathol. Exp. Neurol. 63:812–820.
Kim W. H., Choi C. H., Kang S. K., Kwon C. H., and Kim Y. K. (2005). Ceramide induces non-apoptotic cell death in human glioma cells. Neurochem. Res. 30:969–979.
Ko H. W., Park K. Y., Kim H., Han P. L., Kim Y. U., Gwag B. J., and Choi E. J. (1998). Ca^{2+} -mediated activation of c-Jun N-terminal kinase and nuclear factor κB by NMDA in cortical cell cultures. J. Neurochem. 71:1390–1395.
Kondo T., Kakegawa W., and Yuzaki M. (2005). Induction of long-term depression and phosphorylation of the δ2 glutamate receptor by protein kinase C in cerebellar slices. Eur. J. Neurosci. 22:1817–1820.
Kordas K. and Stoltzfus R. J. (2004). New evidence of iron and zinc interplay at the enterocyte and neural tissues. J. Nutr. 134:1295–1298.
Kotecha S. A. and MacDonald J. F. (2003). Signaling molecules and receptor transduction cascades that regulate NMDA receptor-mediated synaptic transmission. In: Bradley R. J., Harris R. A., and Jenner P. (eds.), *International Review of Neurobiology, Vol 54.* International Review of Neurobiology Academic Press Inc, San Diego, pp. 53–108.
Lauderback C. M., Hackett J. M., Huang F. F., Keller J. N., Szweda L. I., Markesbery W. R., and Butterfield D. A. (2001). The glial glutamate transporter, GLT-1, is oxidatively modified by 4-hydroxy-2-nonenal in the Alzheimer's disease brain: the role of Aβ1-42. J. Neurochem. 78:413–416.
Lee M. C., Ban S. S., Woo Y. J., and Kim S. U. (2001a). Calcium/calmodulin kinase II activity of hippocampus in kainate-induced epilepsy. J. Korean Med. Sci. 16:643–648.
Lee M. C., Rho J. L., Kim M. K., Woo Y. J., Kim J. H., Nam S. C., Suh J. J., Chung W. K., Moon J. D., and Kim H. I. (2001b). c-JUN expression and apoptotic cell death in kainate-induced temporal lobe epilepsy. J. Korean Med. Sci. 16:649–656.
Lerea L. S., Carlson N. G., and McNamara J. O. (1995). N-methyl-D-aspartate receptors activate transcription of c-fos and NGFI-A by distinct phospholipase A_2-requiring intracellular signaling pathways. Molec. Pharmacol. 47:1119–1125.
Leslie S. W., Brown L. M., Trent R. D., Lee Y. H., Morris J. L., Jones T. W., Randall P. K., Lau S. S., and Monks T. J. (1992). Stimulation of N-methyl-D-aspartate receptor-mediated calcium entry into dissociated neurons by reduced and oxidized glutathione. Mol. Pharmacol. 41:308–314.
Liang L. P., Ho Y. S., and Patel M. (2000). Mitochondrial superoxide production in kainate-induced hippocampal damage. Neuroscience 101:563–570.
Liu H. N., Larocca J. N., and Almazan G. (1999). Molecular pathways mediating activation by kainate of mitogen-activated protein kinase in oligodendrocyte progenitors. Brain Res. Mol. Brain Res. 66:50–61.
Liu W., Liu R., Schreiber S. S., and Baudry M. (2001). Role of polyamine metabolism in kainic acid excitotoxicity in organotypic hippocampal slice cultures. J. Neurochem. 79:976–984.
Lombardi G., Szekely A. M., Bristol L. A., Guidotti A., and Manev H. (1993). Induction of ornithine decarboxylase by N-methyl-D-aspartate receptor activation is unrelated to potentiation of glutamate excitotoxicity by polyamines in cerebellar granule neurons. J. Neurochem. 60:1317–1324.
Lu C., Chan S. L., Haughey N., Lee W. T., and Mattson M. P. (2001a). Selective and biphasic effect of the membrane lipid peroxidation product 4-hydroxy-2,3-nonenal on *N*-methyl-D-aspartate channels. J. Neurochem. 78:577–589.
Lu X. R., Ong W. Y., Halliwell B., Horrocks L. A., and Farooqui A. A. (2001b). Differential effects of calcium-dependent and calcium-independent phospholipase A_2 inhibitors on kainate-induced neuronal injury in rat hippocampal slices. Free Radical Biol. Med. 30:1263–1273.
Mahley R. W. (1988). Apoprotein E: cholesterol transport protein with expanding role in cell biology. Science 240:622–630.

Manev H., Uz T., and Qu T. Y. (2000). 5-Lipoxygenase and cyclooxygenase mRNA expression in rat hippocampus: early response to glutamate receptor activation by kainate. Exp. Gerontol. 35:1201–1209.

Marchesini N. and Hannun Y. A. (2004). Acid and neutral sphingomyelinases: roles and mechanisms of regulation. Biochem. Cell Biol. 82:27–44.

Mark R. J., Lovell M. A., Markesbery W. R., Uchida K., and Mattson M. P. (1997). A role for 4-hydroxynonenal, an aldehydic product of lipid peroxidation, in disruption of ion homeostasis and neuronal death induced by amyloid β-peptide. J. Neurochem. 68:255–264.

Martins R. A., Silveira M. S., Curado M. R., Police A. I., and Linden R. (2005). NMDA receptor activation modulates programmed cell death during early post-natal retinal development: a BDNF-dependent mechanism. J. Neurochem. 95:244–253.

Matute C., Domercq M., and Sánchez-Gómez M. V. (2006). Glutamate-mediated glial injury: Mechanisms and clinical importance. Glia 53:212–224.

Maybodi L., Pow D. V., Kharazia V. N., and Weinberg R. J. (1999). Immunocytochemical demonstration of reduced glutathione in neurons of rat forebrain. Brain Res. 817:199–205.

McCusker R. H., McCrea K., Zunich S., Dantzer R., Broussard S. R., Johnson R. W., and Kelley K. W. (2006). Insulin-like growth factor-I enhances the biological activity of brain-derived neurotrophic factor on cerebrocortical neurons. J. Neuroimmunol. 179:186–190.

McInnis J., Wang C., Anastasio N., Hultman M., Ye Y., Salvemini D., and Johnson K. M. (2002). The role of superoxide and nuclear factor-κB signaling in *N*-methyl-D-aspartate-induced necrosis and apoptosis. J. Pharmacol. Exp. Ther. 301:478–487.

McNamara R. K. and Lenox R. H. (2000). Differential regulation of primary protein kinase C substrate (MARCKS, MLP, GAP-43, RC3) mRNAs in the hippocampus during kainic acid-induced seizures and synaptic reorganization. J. Neurosci. Res. 62:416–426.

McNamara R. K., Wees E. A., and Lenox R. H. (1999). Differential subcellular redistribution of protein kinase C isozymes in the rat hippocampus induced by kainic acid. J. Neurochem. 72:1735–1743.

Milatovic D., Gupta R. C., and Dettbarn W. D. (2002). Involvement of nitric oxide in kainic acid-induced excitotoxicity in rat brain. Brain Res. 957:330–337.

Minami M., Kuraishi Y., and Satoh M. (1991). Effects of kainic acid on messenger RNA levels of IL-1β, IL-6, TNF-α and LIF in the rat brain. Biochem. Biophys. Res. Commun. 176: 593–598.

Morais Cabral J. H., Atkins G. L., Sánchez L. M., López-Baodo Y. S., López-Oton C., and Sawyer L. (1995). Arachidonic acid binds to apolipoprotein D: Implications for the protein's function. FEBS Lett. 366:53–56.

Moreno J. J. and Pryor W. A. (1992). Inactivation of alpha 1-proteinase inhibitor by peroxynitrite. Chem. Res. Toxicol. 5:425–431.

Moriguchi S., Han F., Nakagawasai O., Tadano T., and Fukunaga K. (2006). Decreased calcium/calmodulin-dependent protein kinase II and protein kinase C activities mediate impairment of hippocampal long-term potentiation in the olfactory bulbectomized mice. J. Neurochem. 97:22–29.

Morrison R. S., Wenzel H. J., Kinoshita Y., Robbins C. A., Donehower L. A., and Schwartzkroin P. A. (1996). Loss of the p53 tumor suppressor gene protects neurons from kainate-induced cell death. J. Neurosci. 16:1337–1345.

Nadkarni D. V. and Sayre L. M. (1995). Structural definition of early lysine and histidine adduction chemistry of 4-hydroxynonenal. Chem. Res. Toxicol. 8:284–291.

Najm I., el-Skaf G., Massicotte G., Vanderklish P., Lynch G., and Baudry M. (1992). Changes in polyamine levels and spectrin degradation following kainate-induced seizure activity: effect of difluoromethylornithine. Exp. Neurol. 116:345–354.

Nakamura H., Hirabayashi T., Shimizu M., and Murayama T. (2006). Ceramide-1-phosphate activates cytosolic phospholipase A2α directly and by PKC pathway. Biochem. Pharmacol. 71:850–857.

Ong W. Y., Hu C. Y., Hjelle O. P., Ottersen O. P., and Halliwell B. (2000a). Changes in glutathione in the hippocampus of rats injected with kainate: depletion in neurons and upregulation in glia. Exp. Brain Res. 132:510–516.

Ong W. Y., Lu X. R., Hu C. Y., and Halliwell B. (2000b). Distribution of hydroxynonenal-modified proteins in the kainate-lesioned rat hippocampus: evidence that hydroxynonenal formation precedes neuronal cell death. Free Radic. Biol. Med. 28:1214–1221.

Ong W. Y., Goh E. W. S., Lu X. R., Farooqui A. A., Patel S. C., and Halliwell B. (2003). Increase in cholesterol and cholesterol oxidation products, and role of cholesterol oxidation products in kainate-induced neuronal injury. Brain Path. 13:250–262.

Ong W. Y., He X., Chua L. H., and Ong C. N. (2006). Increased uptake of divalent metals lead and cadmium into the brain after kainite-induced neuronal injury. Exp. Brain Res. 173:468–474.

Oyama Y., Sadakata C., Chikahisa L., Nagano T., and Okazaki E. (1997). Flow-cytometric analysis on kainate-induced decrease in the cellular content of non-protein thiols in dissociated rat brain neurons. Brain Res. 760:277–280.

Paradis É., Clavel S., Julien P., Murthy M. R. V., de Bilbao F., Arsenijevic D., Giannakopoulos P., Vallet P., and Richard D. (2004). Lipoprotein lipase and endothelial lipase expression in mouse brain: regional distribution and selective induction following kainic acid-induced lesion and focal cerebral ischemia. Neurobiol. Dis. 15:312–325.

Parihar M. S. and Hemnani T. (2003). Phenolic antioxidants attenuate hippocampal neuronal cell damage against kainic acid induced excitotoxicity. J. Biosci. 28:121–128.

Parihar M. S. and Hemnani T. (2004). Experimental excitotoxicity provokes oxidative damage in mice brain and attenuation by extract of *Asparagus racemosus*. J. Neural Transm. 111:1–12.

Park D. S., Obeidat A., Giovanni A., and Greene L. A. (2000). Cell cycle regulators in neuronal death evoked by excitotoxic stress: implications for neurodegeneration and its treatment. Neurobiol. Aging 21:771–781.

Patel S. C., Asotra K., Patel Y. C., McConathy W. J., Patel R. C., and Suresh S. (1995). Astrocytes synthesize and secrete the lipophilic ligand carrier apolipoprotein D. NeuroReport 6:653–657.

Pepicelli O., Fedele E., Bonanno G., Raiteri M., Ajmone-Cat M. A., Greco A., Levi G., and Minghetti L. (2002). *In vivo* activation of *N*-methyl-D-aspartate receptors in the rat hippocampus increases prostaglandin E_2 extracellular levels and triggers lipid peroxidation through cyclooxygenase-mediated mechanisms. J. Neurochem. 81:1028–1034.

Perez Y., Morin F., Beaulieu C., and Lacaille J. C. (1996). Axonal sprouting of CA1 pyramidal cells in hyperexcitable hippocampal slices of kainate-treated rats. Eur. J. Neurosci. 8:736–748.

Pfrieger F. W. (2003). Outsourcing in the brain: do neurons depend on cholesterol delivery by astrocytes? BioEssays 25:72–78.

Phillis J. W., Horrocks L. A., and Farooqui A. A. (2006). Cyclooxygenases, lipoxygenases, and epoxygenases in CNS: Their role and involvement in neurological disorders. Brain Res. Rev. 52:201–243.

Popovici T., Represa A., Crepel V., Barbin G., Beaudoin M., and Ben Ari Y. (1990). Effects of kainic acid-induced seizures and ischemia on c-fos-like proteins in rat brain. Brain Res. 536:183–194.

Pryor W. A. and Squadrito G. L. (1995). The chemistry of peroxynitrite: a product from the reaction of nitric oxide with superoxide. Am. J. Physiol 268:L699–L722.

Qi W., Reiter R. J., Tan D. X., Manchester L. C., Siu A. W., and Garcia J. J. (2000). Increased levels of oxidatively damaged DNA induced by chromium(III) and H_2O_2: protection by melatonin and related molecules. J. Pineal Res. 29:54–61.

Qin Z. H., Chen R. W., Wang Y., Nakai M., Chuang D. M., and Chase T. N. (1999). Nuclear factor kB nuclear translocation upregulates c-Myc and p53 expression during NMDA receptor-mediated apoptosis in rat striatum. J. Neurosci. 19:4023–4033.

Radi R., Beckman J. S., Bush K. M., and Freeman B. A. (1991a). Peroxynitrite oxidation of sulfhydryls. The cytotoxic potential of superoxide and nitric oxide. J. Biol. Chem. 266: 4244–4250.

Radi R., Beckman J. S., Bush K. M., and Freeman B. A. (1991b). Peroxynitrite-induced membrane lipid peroxidation: the cytotoxic potential of superoxide and nitric oxide. Arch. Biochem. Biophys. 288:481–487.

Reed L. J. and De Belleroche J. (1990). Induction of ornithine decarboxylase in cerebral cortex by excitotoxin lesion of nucleus basalis: association with postsynaptic responsiveness and N-methyl-D-aspartate receptor activation. J. Neurochem. 55:780–787.

Retz K. C. and Coyle J. T. (1982). Effects of kainic acid on high-energy metabolites in the mouse striatum. J. Neurochem. 38:196–203.

Rodríguez-Moreno A. and Lerma J. (1998). Kainate receptor modulation of GABA release involves a metabotropic function. Neuron 20:1211–1218.

Rossler O. G., Bauer I., Chung H. Y., and Thiel G. (2004). Glutamate-induced cell death of immortalized murine hippocampal neurons: neuroprotective activity of heme oxygenase-1, heat shock protein 70, and sodium selenite. Neurosci. Lett. 362:253–257.

Rothstein J. D. and Kuncl R. W. (1995). Neuroprotective strategies in a model of chronic glutamate-mediated motor neuron toxicity. J. Neurochem. 65:643–651.

Rusanescu G., Qi H., Thomas S. M., Brugge J. S., and Halegoua S. (1995). Calcium influx induces neurite growth through a Src-Ras signaling cassette. Neuron 15:1415–1425.

Sakhi S., Bruce A., Sun N., Tocco G., Baudry M., and Schreiber S. S. (1994). p53 induction is associated with neuronal damage in the central nervous system. Proc. Natl. Acad. Sci. USA 91:7525–7529.

Sandhya T. L., Ong W. Y., Horrocks L. A., and Farooqui A. A. (1998). A light and electron microscopic study of cytoplasmic phospholipase A_2 and cyclooxygenase-2 in the hippocampus after kainate lesions. Brain Res. 788:223–231.

Sato K. and Matsuki N. (2002). A 72 kDa heat shock protein is protective against the selective vulnerability of CA1 neurons and is essential for the tolerance exhibited by CA3 neurons in the hippocampus. Neuroscience 109:745–756.

Shoham S. and Ebstein R. P. (1997). The distribution of β-amyloid precursor protein in rat cortex after systemic kainate-induced seizures. Exp. Neurol. 147:361–376.

Siskind L. J. (2005). Mitochondrial ceramide and the induction of apoptosis. J. Bioenerg. Biomembr. 37:143–153.

Sohl G., Guldenagel M., Beck H., Teubner B., Traub O., Gutierrez R., Heinemann U., and Willecke K. (2000). Expression of connexin genes in hippocampus of kainate-treated and kindled rats under conditions of experimental epilepsy. Brain Res. Mol. Brain Res. 83:44–51.

Sola C., Tusell J. M., and Serratosa J. (1997). Calmodulin is expressed by reactive microglia in the hippocampus of kainic acid-treated mice. Neuroscience 81:699–705.

Sorger P. K. (1991). Heat shock factor and the heat shock response. Cell 65:363–366.

Soule J., Messaoudi E., and Bramham C. R. (2006). Brain-derived neurotrophic factor and control of synaptic consolidation in the adult brain. Biochem. Soc. Trans. 34:600–604.

Sperk G. (1994). Kainic acid seizures in the rat. Prog. Neurobiol. 42:1–32.

Stein-Behrens B. A., Elliott E. M., Miller C. A., Schilling J. W., Newcombe R., and Sapolsky R. M. (1992). Glucocorticoids exacerbate kainic acid-induced extracellular accumulation of excitatory amino acids in the rat hippocampus. J. Neurochem. 58: 1730–1735.

Stoica B. A., Movsesyan V. A., Knoblach S. M., and Faden A. I. (2005). Ceramide induces neuronal apoptosis through mitogen-activated protein kinases and causes release of multiple mitochondrial proteins. Mol. Cell Neurosci. 29:355–371.

Sun A. Y., Cheng Y., Bu Q., and Oldfield F. (1992). The biochemical mechanisms of the excitotoxicity of kainic acid. Free radical formation. Mol. Chem. Neuropathol. 17:51–63.

Tamagno E., Robino G., Obbili A., Bardini P., Aragno M., Parola M., and Danni O. (2003). H_2O_2 and 4-hydroxynonenal mediate amyloid beta-induced neuronal apoptosis by activating JNKs and $p38^{MAPK}$. Exp. Neurol. 180:144–155.

Tatsukawa T., Chimura T., Miyakawa H., and Yamaguchi K. (2006). Involvement of basal protein kinase C and extracellular signal-regulated kinase 1/2 activities in constitutive internalization of AMPA receptors in cerebellar Purkinje cells. J. Neurosci. 26:4820–4825.

Thorburne S. K. and Juurlink B. H. (1996). Low glutathione and high iron govern the susceptibility of oligodendroglial precursors to oxidative stress. J. Neurochem. 67:1014–1022.

van der Brug M. P., Goodenough S., and Wilce P. (2002). Kainic acid induces 14-3-3 ζ expression in distinct regions of rat brain. Brain Res. 956:110–115.

Vance J. E., Pan D., Campenot R. B., Bussière M., and Vance D. E. (1994). Evidence that the major membrane lipids, except cholesterol, are made in axons of cultured rat sympathetic neurons. J. Neurochem. 62:329–337.

Varga V., Jenei Z., Janaky R., Saransaari P., and Oja S. S. (1997). Glutathione is an endogenous ligand of rat brain N-methyl-D-aspartate (NMDA) and 2-amino-3-hydroxy-5-methyl-4-isoxazolepropionate (AMPA) receptors. Neurochem. Res. 22:1165–1171.

Verdaguer E., García-Jordà E., Canudas A. M., Domínguez E., Jiménez A., Pubill D., Escubedo E., Camarasa Pallàs Mercè J., and Camins A. (2002). Kainic acid-induced apoptosis in cerebellar granule neurons: an attempt at cell cycle re-entry. NeuroReport 13:413–416.

Vezzani A., Civenni G., Rizzi M., Monno A., Messali S., and Samanin R. (1994). Enhanced neuropeptide Y release in the hippocampus is associated with chronic seizure susceptibility in kainic acid treated rats. Brain Res. 660:138–143.

Wang X. S., Ong W. Y., and Connor J. R. (2001). A light and electron microscopic study of the iron transporter protein DMT-1 in the monkey cerebral neocortex and hippocampus. J. Neurocytol. 30:353–360.

Wang X. S., Ong W. Y., and Connor J. R. (2002). A light and electron microscopic study of divalent metal transporter-1 distribution in the rat hippocampus, after kainate-induced neuronal injury. Exp. Neurol. 177:193–201.

Wang J. Q., Arora A., Yang L., Parelkar N. K., Zhang G., Liu X., Choe E. S., and Mao L. (2005a). Phosphorylation of AMPA receptors: mechanisms and synaptic plasticity. Mol. Neurobiol. 32:237–249.

Wang Q., Yu S., Simonyi A., Sun G. Y., and Sun A. Y. (2005b). Kainic acid-mediated excitotoxicity as a model for neurodegeneration. Mol. Neurobiol. 31:3–16.

Wang J. Q., Liu X., Zhang G., Parelkar N. K., Arora A., Haines M., Fibuch E. E., and Mao L. (2006). Phosphorylation of glutamate receptors: a potential mechanism for the regulation of receptor function and psychostimulant action. J. Neurosci. Res. 84:1621–1629.

Wang J. Q., Fibuch E. E., and Mao L. (2007). Regulation of mitogen-activated protein kinases by glutamate receptors. J. Neurochem. 100:1–11.

Weber G. F. (1999). Final common pathways in neurodegenerative diseases: regulatory role of the glutathione cycle. Neurosci. Biobehav. Rev. 23:1079–1086.

Wilson C. J., Finch C. E., and Cohen H. J. (2002). Cytokines and cognition—the case for a head-to-toe inflammatory paradigm. J. Am. Geriatr. Soc. 50:2041–2056.

Xia J., Chung H. J., Wihler C., Huganir R. L., and Linden D. J. (2000). Cerebellar long-term depression requires PKC-regulated interactions between GluR2/3 and PDZ domain-containing proteins. Neuron 28:499–510.

Yabuuchi K., Minami M., Katsumata S., and Satoh M. (1993). In situ hybridization study of interleukin-1β mRNA induced by kainic acid in the rat brain. Brain Res. Mol. Brain Res. 20:153–161.

Yoshikawa K., Kita Y., Kishimoto K., and Shimizu T. (2006). Profiling of eicosanoid production in the rat hippocampus during kainic acid-induced seizure - Dual phase regulation and differential involvement of COX-1 and COX-2. J. Biol. Chem. 281:14663–14669.

Zafra F., Castrén E., Thoenen H., and Lindholm D. (1991). Interplay between glutamate and γ-aminobutyric acid transmitter systems in the physiological regulation of brain-derived neurotrophic factor and nerve growth factor synthesis in hippocampal neurons. Proc. Natl. Acad. Sci. USA 88:10037–10041.

Zaleska M. M. and Wilson D. F. (1989). Lipid hydroperoxides inhibit reacylation of phospholipids in neuronal membranes. J. Neurochem. 52:255–260.

Zheng W. H. and Quirion R. (2004). Comparative signaling pathways of insulin-like growth factor-1 and brain-derived neurotrophic factor in hippocampal neurons and the role of the PI3 kinase pathway in cell survival. J. Neurochem. 89:844–852.

Zhu X. M. and Ong W. Y. (2004). Changes in GABA transporters in the rat hippocampus after kainate-induced neuronal injury: decrease in GAT-1 and GAT-3 but upregulation of betaine/GABA transporter BGT-1. J. Neurosci. Res. 77:402–409.

Ziegra C. J., Willard J. M., and Oswald R. E. (1992). Coupling of a purified goldfish brain kainate receptor with a pertussis toxin-sensitive G protein. Proc. Natl. Acad. Sci. USA 89:4134–4138.

Chapter 7
Possible Mechanisms of Neural Injury Caused by Glutamate and Its Receptors

7.1 Excitotoxicity

Glutamate serves as the primary excitatory neurotransmitter in the mammalian CNS. It contributes to neuronal communication and can be involved in neuropathological damage through the activation of excitatory amino acid (EAA) receptors (Farooqui and Horrocks, 1994a, 2007; Salinska et al., 2005; Wang et al., 2005). Excitotoxicity, a term coined by Dr. John Olney, is the process by which high levels of glutamate and its analogs excite neurons and bring about their demise (Olney et al., 1979; Choi, 1988). It is also widely used for glial cells that do not exhibit excitability due to their absent action potential. However, glial cells express functional AMPA and KA receptors (Matute et al., 2006).

Ca^{2+} influx mediates excitotoxicity with a cascade of events involving free radical generation, mitochondrial dysfunction, and the activation of many enzymes (Table 7.1), including those involved in the generation and metabolism of arachidonic acid (Farooqui and Horrocks, 1991, 1994b; Farooqui et al., 2001; Wang et al., 2005). These enzymes include isoforms of PLA_2, cyclooxygenase-2 (COX-2), and lipoxygenases (LOX) (Phillis et al., 2006; Wolf and Eastman, 1999). Increased Ca^{2+} influx into the cytoplasm produces accumulation of Ca^{2+} inside mitochondria. Ca^{2+} is sequestered into the mitochondrial matrix through a proton electrochemical gradient generated by the electron transport chain. This influx of Ca^{2+} decreases the electrochemical gradient and subsequently decreases ATP synthesis. Furthermore, the synthesis of reactive oxygen species (ROS) in mitochondria inhibits Ca^{2+}-ATPase and hence reduces the ability of neural membranes to expel Ca^{2+} ions. This process results in a further increase in mitochondrial Ca^{2+}. When a threshold concentration is reached, mitochondrial permeability transition pores open irreversibly (Shi et al., 2006). Then a further release of mitochondrial Ca^{2+}, cytochrome c (Table 7.2), and PAF from mitochondria leads to uncoupling of mitochondrial electron transfer from ATP synthesis. The release of Ca^{2+} from mitochondria also causes activation of calpains, endonucleases, and Ca^{2+}-calmodulin-dependent nitric oxide synthase (NOS). Nitric oxide generated by NOS reacts with the superoxide radical to form peroxynitrite, which may cause DNA fragmentation. Finally, a high concentration of Ca^{2+} stimulates protein kinases including protein kinase C and Jun amino-terminal kinase (Farooqui and Horrocks, 1991, 1994a). Thus the accumulation of oxygenated

Table 7.1 Excitotoxicity-mediated alterations in enzymic activities in brain tissues

Enzyme	Effect	Reference
Phospholipase A_2	Increased	(Bartolomeo et al., 1997; Kim et al., 1995)
Phospholipase D	Increased	(Kim et al., 2004)
Lipoprotein lipase	Increased	(Paradis et al., 2004)
Cyclooxygenase	Increased	(Adams et al., 1996)
Lipoxygenase	Increased	(Manev et al., 1998)
Nitric oxide synthase	Increased	(Milatovic et al., 2002)
Calpain	Increased	(Ong et al., 1997)
Diacylglycerol lipase	Increased	(Farooqui et al., 1993)
Monoacylglycerol lipase	Increased	(Farooqui et al., 1993)
Protein kinases	Increased	(Vaccarino et al., 1987; Manev et al., 1989)
Endonucleases	Increased	(Siesjö, 1990)

arachidonic acid metabolites along with abnormal ion homeostasis, activation of NOS, and protein kinases, alterations in cellular redox, and lack of energy generation, is associated with neural cell injury.

Glutamate-mediated damage to glial cells does not involve the activation of glutamate receptors, but rather glutamate uptake (Oka et al., 1993; Matute et al., 2006). Glutamate uptake from the extracellular space by specific glutamate transporters is essential for maintaining excitatory postsynaptic currents (Auger and Attwell, 2000) and for blocking excitotoxic death due to overstimulation of glutamate receptors (Rothstein et al., 1996) (Chapter 4). Exposure of astroglial, oligodendroglial, and microglial cell cultures to glutamate produces glial cell demise by a transporter-related mechanism involving the inhibition of cystine uptake, which causes a decrease in glutathione and makes glial cells vulnerable to toxic free radicals (Oka et al., 1993; Matute et al., 2006). The accumulation of extracellular glutamate at the synapse is toxic to neurons. Astrocytes protect neurons by removing glutamate from the extracellular space through glutamate transporters. If removal of glutamate from the synapse does not keep pace with accumulation, neuronal damage may occur. Thus, by regulating the levels of extracellular glutamate that have access to these receptors, glutamate uptake systems have the potential to affect both normal synaptic signaling and the abnormal over-activation of the receptors that can trigger excitotoxic cell death.

Neural injury mediated by glutamate involves two important components: neuroinflammation, a neuroprotective mechanism whose prolonged presence is injurious to neurons, and oxidative stress, cytotoxic consequences produced by oxy-

Table 7.2 Excitotoxicity-mediated changes in neurochemical parameters involved in oxidative stress

Neurochemical parameter	Effect	Reference
Free radicals and lipid peroxides	Increased	(Wang et al., 2005)
4-Hydroxynonenal	Increased	(Ong et al., 2001)
Glutathione	Increased	(Ong et al., 2001)
Cytochrome c	Increased	(Garrido et al., 2001)
F2-isoprostane	Increased	(Pepicelli et al., 2005)

gen free radicals. Although neuroinflammation and oxidative stress may occur independently, growing evidence indicates that ROS formation may be a specific consequence of glutamate receptor activation, and may partly mediate excitotoxic neuronal injury (Olney et al., 1979; Choi, 1988). Neuroinflammation and oxidative stress are interrelated processes that may bring about neural cell demise independently or synergistically.

It remains controversial whether inflammation and oxidative stress are the cause or consequence of neural injury (Andersen, 2004; Juranek and Bezek, 2005). Similarly, very little information is available on neural injury and clinical expression of inflammation and oxidative stress with trauma and neurodegenerative diseases involving neural cell death mediated by glutamate through apoptosis as well as necrosis (Farooqui et al., 2004). Thus discovering the molecular mechanisms of brain damage in acute neural trauma, ischemia, and the neurodegenerative diseases AD and PD remains a most challenging area of neuroscience research (Graeber and Moran, 2002).

7.2 Glutamate-Mediated Inflammation and Neural Cell Injury

Inflammation is a protective mechanism that isolates the damaged brain tissue from uninjured areas, destroys affected cells, and repairs the extracellular matrix (Correale and Villa, 2004). Inflammation is a "double-edged sword". On one hand the continuous antigenic challenge, duration, and intensity of inflammation damages neural cells through the synthesis of pro-inflammatory lipid mediators. On the other hand immune-mediated processes induce neuroprotection and repair through the generation of anti-inflammatory lipid mediators and repair proteins. Without a strong inflammatory response, brain tissue would be highly susceptible to acute neural trauma, neurodegenerative diseases, and microbial, viral, and prion infections.

All neural cells, including microglia, astrocytes, neurons, and oligodendrocytes, participate in inflammatory responses. Morphologically in brain tissue, major hallmarks of inflammatory reaction are phenotypic changes of glial cells, mainly activation and transformation of microglial cells into phagocytic cells, and to a lesser extent, reactive astrocytosis. Although the chemical nature of signals that initiate the activation of microglial cells responding to inflammation remains unknown, neuronal depolarization following injury combined with extracellular ion changes, metabolic perturbations, and alterations in acid-base balance may be the major stimuli (Block and Hong, 2005).

The identity of factors released from damaged neurons to signal microglial cell activation may depend upon which type of neural cell is damaged, neuron versus glial, and on the the toxin or stimulus, glutamate versus β-amyloid versus α-synuclein, and the nature of cellular death, apoptosis versus necrosis. Similarly, the molecular mechanisms and internal and external factors that modulate the dynamic aspects of acute and chronic inflammation in cell injury mediated by glutamate remain unclear. It also remains unclear to what extent inflammation is beneficial for the injured neurons and glial cells in brain tissue and how the inflammation contributes to secondary brain injury and progressive neuronal loss.

A variety of immune system modulators, including complement proteins, adhesion molecules, inflammatory cytokines such as interleukin-1 alpha (IL-1α), interleukin-1 beta (IL-1β), interleukin-3 (IL-3), interleukin-6 (IL-6), interleukin-17 (IL-17), interleukin-18 (IL-18), tumor necrosis factor-alpha (TNF-α), colony-stimulating factor-1, and tumor and growth factors (TGF-α and β), are made and secreted by both microglia and astrocytes (Sun et al., 2004; Kim et al., 2001; Hays, 1998; Noda et al., 2006; Minghetti et al., 2005; Drew et al., 2005; Wu et al., 1998; Block and Hong, 2005). These factors propagate and maintain inflammation by a number of mechanisms, including the activation of multiple forms of PLA_2, cyclooxygenases (COX), and lipoxygenases (LOX), causing the release of non-esterified AA from neural membrane phospholipids and generating lyso-glycerophospholipids, platelet-activating factor (PAF), pro-inflammatory prostaglandins, and reactive oxygen species (ROS) (Lin et al., 2004; Moses et al., 2006; Phillis et al., 2006).

TNF-α and IL-1β facilitate enhanced arachidonic acid metabolism. In turn, arachidonic acid metabolites increase TNF-α and IL-1β production by blood macrophages and microglial cells, implying that this positive feed-forward loop may amplify events mediated by arachidonic acid metabolites in the brain and spinal cord (Genis et al., 1992). Furthermore, the release of arachidonic acid also exacerbates neural injury by increasing extracellular levels of glutamate through the inhibition of Na^+-dependent uptake and stimulation of exocytosis of glutamate. Furthermore, microglia, astrocytes, neurons, endothelial cells, and oligodendrocytes also produce complement proteins that may have pro-inflammatory as well as anti-inflammatory effects on brain tissue (Hosokawa et al., 2003).

At the molecular level, neurotoxic concentrations of glutamate and its analogs interact with their receptors and induce the expression of cytokines and chemokines in brain tissue (Minami et al., 1991; Strijbos and Rothwell, 1995; Allen and Attwell, 2001). These cytokines include TNF-α, IL-1β, IL-6, and IL-17. It is proposed that cytokines intensify glutamate-mediated neural injury by acting synergistically. Thus, TNF-α potentiates glutamate-induced neurotoxicity in an in vitro model of rat spinal cord injury. In this model, cytokines may act on AMPA receptors in microglia. This process may induce glial cells to release more glutamate and cytokines, thus exacerbating the outcome of neural injury (Hermann et al., 2001; Gomes-Leal et al., 2004).

Glutamate-mediated ROS generation and NF-κB activation are mediated by a direct interaction with NADPH oxidase 4 (Nox4), a protein related to gp91^{phox} (Nox2) of phagocytic cells. This enzyme catalyzes the production of superoxide radical by one-electron reduction of oxygen, using NADPH as the electron donor. NADPH oxidase plays a pivotal role in the inflammatory response. Studies on LPS-induced toxicity indicate that inflammation-mediated neurodegeneration is less pronounced in NADPH oxidase-deficient (PHOX$^{-/-}$) mice compared to control (PHOX$^{+/+}$) mice (Qin et al., 2004). Dopaminergic neurons in primary mesencephalic neuron-glia cultures from PHOX$^{+/+}$ mice are more sensitive to LPS-induced neurodegeneration in vitro when compared with PHOX$^{-/-}$ mice.

Furthermore, microglia depleted $PHOX^{+/+}$ neuron-glia cultures fail to show dopaminergic neurotoxicity with the addition of LPS. Similarly, neuron-enriched cultures from both $PHOX^{+/+}$ mice and $PHOX^{-/-}$ mice do not show any direct dopaminergic neurotoxicity induced by LPS. However, the addition of $PHOX^{+/+}$ microglia to neuron-enriched cultures from either strain results in reinstatement of LPS-induced dopaminergic neurotoxicity. These observations indicate that microglial cells are the primary source of NADPH oxidase-induced neurotoxicity. Collectively these studies suggest the dual neurotoxic functions of microglial NADPH oxidase not only in production of extracellular ROS but also in amplification of pro-inflammatory gene expression and associated neurotoxicity (Qin et al., 2004).

Glutamate-mediated Ca^{2+} entry through NMDA at the plasma membrane level and mobilization of $Ca^{2+}$from intracellular stores through PLC-mediated generation of PtdIns-3*P* is indispensable for the basal NF-κB activity. Three cytosolic Ca^{2+} sensors, calmodulin, protein kinases C (PKC), and the p21(ras)/phosphatidylinositol 3-kinase (PtdIns-3K)/Akt pathways, are simultaneously involved in the steps linking the Ca^{2+} to NF-κB activity (Lilienbaum and Israel, 2003; Marchetti et al., 2004; Lubin et al., 2005). Calmodulin modulates calcineurin, a Ca^{2+}-dependent protein phosphatase, which plays a role in the basal NF-κB activity, whilst stimulation of both the calmodulin kinase II and Akt kinase pathways results in the up-regulation of the transcriptional potential of the p65 subunit of NF-κB (Lilienbaum and Israel, 2003). In primary cultures of neonatal cerebellar granule neurons, all Ca^{2+} sensors, calmodulin, protein kinases C (PKC), and the p21(ras)/phosphatidylinositol 3-kinase (PtdIns-3K)/Akt pathway, converge towards NF-κB at the levels of nuclear translocation as well as transcription. The duration of NF-κB activation is a critical determinant for sensitivity toward excitotoxic stress and is dependent on the different upstream and downstream signaling associated with various kinases. This is in contrast to studies in non-neuronal cells, which either do not respond to Ca^{2+} or do not simultaneously activate all three cascades (Lilienbaum and Israel, 2003). Collective evidence suggests that brain inflammatory processes differ from systemic inflammation not only in the involvement of various types of neural cells but also in differences in response to second messengers.

In neural cells NF-κB is present in the cytoplasm in an inhibitory form attached to its inhibitory protein, I-κB. NF-κB activity is tightly controlled by the IκB kinase complex, consisting of IκB kinases I-κKα, I-κKβ, and I-κKγ. I-κKβ is essential for the inflammatory cytokine-mediated activation of NF-κB (Yamamoto and Gaynor, 2004). NF-κB is stimulated by multiple stimuli within stressed cells of different origin to form part of a protective response (Baichwal and Baeuerle, 1997) (Table 7.3). Upon stimulation IκB is rapidly phosphorylated, ubiquinated, and then degraded by proteasomes resulting in the release and subsequent nuclear translocation of active NF-κB.

Treatment of hippocampal slices with kainate produces a time-dependent increase in NF-κB levels in areas CA3 and CA1, but not dentate gyrus, compared with controls (Lubin et al., 2005). The kainate-mediated NF-κB complex that binds to DNA is composed of p65, p50, and c-Rel subunits. Pharmacological studies indicate that extracellular signal-regulated protein kinase (ERK) and PtdIns-3K

Table 7.3 Factors stimulating NF-κB in brain tissue

Factor	Effect	Reference
ROS	Stimulation	(Mazière et al., 1999)
H_2O_2	Stimulation	(Schmidt et al., 1995)
Ca^{2+}	Stimulation	(Lilienbaum and Israel, 2003)
Cytokines	Stimulation	(Minghetti et al., 2005; Drew et al., 2005)
Nerve growth factor	Stimulation	(Maggirwar et al., 1998)
Free fatty acids	Stimulation	(Lin et al., 2004; Moses et al., 2006)
Glutamate	Stimulation	(O'Riordan et al., 2006)
Mitogens	Stimulation	(Gabriel et al., 1999)
UV radiations	Stimulation	(Gabriel et al., 1999)
Viruses and bacteria	Stimulation	(Schmidt et al., 1995)
β-Amyloid (Aβ)	Stimulation	(Schmidt et al., 1995)

are coupled to basal and kainate-mediated NF-κB DNA binding activity in the CA3 area of hippocampus. Kainate induces a decrease in total and increase in phospho-inhibitor-κBα (I-κBα), suggesting that kainate-mediated activation of NF-κB is via the classical I-κB kinase pathway. Interestingly, inhibition of ERK but not PtdIns-3K inhibits the kainate-mediated increase in phospho-I-κBα. These findings support a role for the ERK and PtdIns-3K pathways in kainate-mediated NF-κB activation in the CA3 area of hippocampus. It is also likely that these kinases may target the NF-κB pathway at different loci (Lubin et al., 2005).

In the nucleus, NF-κB mediates the transcription of more than 150 genes that not only influence the survival of neural cells but also maintain their normal functional integrity. NF-κB also induces many genes implicated in inflammation, oxidative stress, and immune responses (Fig. 7.1). These genes code for enzymes, intracellular adhesion molecule-1 (ICAM-1), vascular adhesion molecule-1 (VCAM-1), E-selectin, cytokines, and matrix metalloproteinases (MMP) (Table 7.4). Activation of NF-κB also leads to the local generation of more cytokines, which in turn promulgate inflammatory signals (Block and Hong, 2005).

The lack of NF-κB in neurons may be linked to an increased susceptibility to glutamate-mediated cell death (Yu et al., 1999), β-amyloid neurotoxicity (Kaltschmidt et al., 1999), and neural cell death associated with traumatic brain injury (Sullivan et al., 1999). Thus mice deficient in the p50 subunit of NF-κB show increased damage to hippocampal pyramidal neurons after administration of kainate. Gel-shift analyses indicate that p50 is required for the majority of κB DNA-binding activity in hippocampus. Intraventricular administration of κB decoy DNA before kainate administration in wild-type mice produces an enhancement of damage to hippocampal pyramidal neurons, indicating that reduced NF-κB activity is sufficient to account for the enhanced excitotoxic neuronal injury in p50$(^{-/-})$ mice (Yu et al., 1999). Cultured hippocampal neurons from p50$(^{-/-})$ mice show not only a marked increase in intracellular Ca^{2+} levels but also a significant increase in oxidative stress after exposure to glutamate. This suggests that neurons from p50$(^{-/-})$ mice are more vulnerable to excitotoxicity than those from p50$(^{+/+})$ and p50$(^{+/-})$ mice (Yu et al., 1999). Collectively, these studies indicate the importance of the p50 subunit of NF-κB in protecting neurons against excitotoxic cell death.

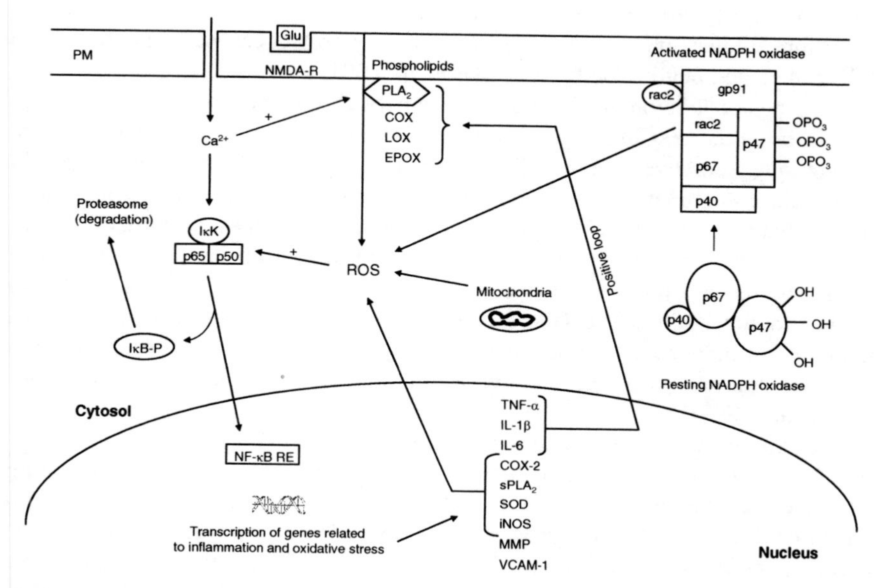

Fig. 7.1 Glutamate-mediated generation of ROS and its effect on NF-κB activation in brain tissue. Glu, glutamate; NMDA-R, NMDA receptor; PLA_2, phospholipase A_2; COX, cyclooxygenase; LOX, lipoxygenase; EPOX, epoxygenase; ROS, reactive oxygen species; NF-κB, nuclear factor κB; NF-κB-RE, nuclear factor κB-response element; IκB, inhibitory subunit of NF-κB; IκK, IκB kinase; TNF-α, tumor necrosis factor-α; IL-1β, interleukin-1β; IL-6, interleukin-6; COX-2, cyclooxygenase-2; iNOS, inducible nitric oxide synthase; MMPs, matrix metalloproteinases; VCAM-1, vascular adhesion molecule-1; $cPLA_2$, cytosolic phospholipase A_2; $sPLA_2$, secretory phospholipase A_2; Positive sign (+) indicates stimulation (pro-inflammatory) and negative sign (–) indicates inhibition (anti-inflammatory)

Table 7.4 NF-κB mediates the transcription of genes implicated in inflammation, oxidative stress, and immune responses

Genes	Effect	Reference
$sPLA_2$	Upregulated	(Yedgar et al., 2006)
COX-2	Upregulated	(Minghetti and Levi, 1998)
INOS	Upregulated	(Minghetti and Levi, 1998; Bal-Price et al., 2002)
SOD	Upregulated	(Noh and Koh, 2000; Block and Hong, 2005)
NADPH oxidase	Upregulated	(Qin et al., 2004; Hilburger et al., 2005)
ICAM-1	Upregulated	(Basta et al., 2005; Block and Hong, 2005)
VCAM-1	Upregulated	(Basta et al., 2005; Block and Hong, 2005)
E-selectin	Upregulated	(Block and Hong, 2005)
TNF-α	Upregulated	(Block and Hong, 2005)
IL-1β	Upregulated	(Block and Hong, 2005)
IL-6	upregulated	(Block and Hong, 2005)
IL-17	Upregulated	(Witowski et al., 2004)
MMP	Upregulated	(Farooqui et al., 2007; Block and Hong, 2005)

$sPLA_2$, secretory phospholipase A_2; COX-2, inducible nitric oxide synthase; iNOS, inducible nitric oxide synthase; SOD, superoxide dismutase; ICAM-1, intracellular adhesion molecule-1; VCAM-1, vascular adhesion molecule-1; TNF-α, tumor necrosis factor-α; IL, interleukin; and MMP, matrix metalloproteinases.

The associated molecular mechanism is not fully understood. However, based on cell culture studies, the p52 subunit of NF-κB may protect cerebellar neurons from TNF-α-mediated neurotoxicity (Nicholas et al., 2001).

Two types of inflammatory processes, namely chronic and acute, are known to occur in brain tissue. Chronic neuroinflammation is associated with slow progressive neurodegenerative diseases (Block and Hong, 2005; Farooqui et al., 2007). Acute neuroinflammation is involved in ischemia, head injury, and spinal cord trauma (Farooqui et al., 2007). Acute neuroinflammation is a short-lived process. In contrast, chronic inflammation is a long-lasting phenomenon associated with mononuclear infiltration, tissue hyperplasia, progressive cavitation, and glial scarring in the brain tissue (Fitch et al., 1999). Time-lapse video analyses of inflammation-induced cavitation show astrocyte abandonment of neuronal processes and neurite stretching. These processes are associated with secondary injury (Fitch et al., 1999).

There are two phases in inflammatory responses: one at the onset for the generation of pro-inflammatory lipid mediators such as eicosanoids and platelet activating factor and the other at the onset of resolution. Resolution, a turning off mechanism by neural cells to limit tissue injury, involves the synthesis of pro-resolving and anti-inflammatory eicosanoids. The molecular mechanism of resolution remains elusive (Serhan and Savill, 2005). However, lipoxins (Chiang et al., 2006), PGD_2 (Gilroy et al., 2004), and docosanoids, the resolvins and neuroprotectins (Serhan, 2005b; Bazan, 2005c) (Fig. 7.2), may play an important role during resolution. Lipoxins are potent anti-inflammatory and pro-resolving molecules that act through specific G protein-coupled receptors, the ALX and LXA receptors (Norel and Brink, 2004; Chiang et al., 2005). The activation of these receptors triggers the expression of a suppressor of cytokine signaling (SOCS-2). SOCS-2-deficient mice show uncontrolled synthesis of pro-inflammatory cytokines, aberrant leukocyte infiltration, and increased mortality (Machado et al., 2006). In the absence of biosynthetic pathways for LXA_4, the resulting uncontrolled inflammation can become lethal, despite pathogen clearance (Machado et al., 2006). Collectively these studies suggest that lipoxins regulate cellular activities associated with inflammation and resolution (Serhan, 2005a; Chiang et al., 2005, 2006).

Resolvins and neuroprotectins (Fig. 7.2), metabolic products of docosahexaenoic acid, slow down the inflammatory cycle (Serhan, 2005b; Bazan, 2005a). Thus the infusion of neuroprotectin D1 (NPD1), following ischemic reperfusion injury or during oxidative stress in cell culture, down-regulates oxidative stress and apoptotic DNA damage. NPD1 also up-regulates the anti-apoptotic Bcl-2 proteins, Bcl-2 and bclxL, and decreases the expression of the pro-apoptotic proteins, Bax and Bad (Mukherjee et al., 2004; Bazan, 2005c). Collectively these studies suggest that the generation of resolvins and docosatrienes is an internal neuroprotective mechanism for preventing brain damage (Lukiw et al., 2005; Bazan, 2005b; Serhan, 2005b). The molecular mechanism by which NPD1 exerts its neuroprotective effect remains unknown. However, these lipid mediators may restore signal transduction processes that are disturbed from the harmful effects of inflammation.

The generation of oxidized glycerophospholipids during acute inflammation also plays an important role in the resolution of inflammation and adaptive immune

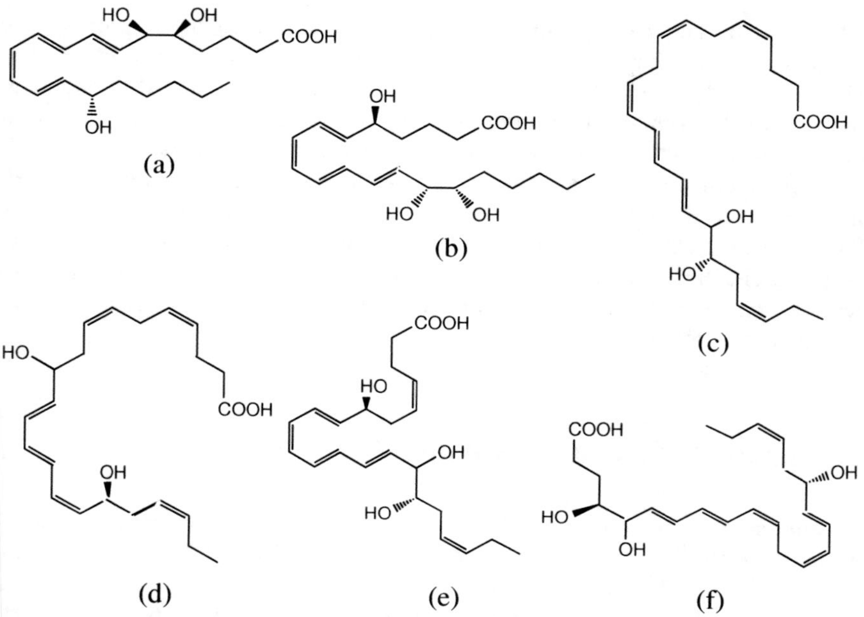

Fig. 7.2 Chemical structures of lipoxins, resolvins and neuroprotectins. Lipoxin A4 (a), Lipoxin B4 (b); 16,17S-docosatriene (c); 10,17S-docosatriene (d); 7,16,17S-resolvin (e); and 4S5,17S-resolvin (f). These metabolites retard the actions of arachidonic acid and its metabolites (eicosanoids)

responses (Bochkov and Leitinger, 2003). This process may include defense strategies such as (a) induction of signaling pathways leading to the upregulation of anti-inflammatory genes, (b) inhibition of signaling pathways coupled to the expression of pro-inflammatory genes, and (c) prevention of the interaction of pro-inflammatory bacterial products with host cells (Bochkov and Leitinger, 2003).

Beside the above mediators, in Parkinson disease (PD) the damaged dopaminergic neurons release a protein called neuromelanin (Zecca et al., 2003). This protein activates microglia and is responsible for the self-propelling cycle of microgliosis. The molecular mechanism of neuromelanin action is not known. However, neuromelanin may scavenge reactive metal ions and neurotoxins to form stable adducts. Degenerating neurons that release neuromelanin in PD may induce a vicious cycle of chronic neuroinflammation and neuronal loss (Zecca et al., 2003). The addition of neuromelanin to microglial cultures also results in activation of NF-κB via phosphorylation and degradation of the inhibitor protein κB (I-κB) and induction of TNF-α, IL-6, and nitric oxide synthesis (Wilms et al., 2003). Neuromelanin also activates p38 mitogen-activated protein kinase (MAPK). The inhibition of this pathway by SB203580 reduces the phosphorylation of the transactivation domain of the p65 subunit of NF-κB suggesting that neuromelanin modulates NF-κB activity. Collective evidence suggests that neuromelanin plays a protective role during chronic neuroinflammation and neuronal loss in PD.

In the optic nerve during inflammatory reactions, macrophages promote the secretion of a Ca^{2+}-binding protein known as oncomodulin (mol mass of 30 kDa) (Yin et al., 2004; Filbin, 2006). This protein in combination with cAMP and mannose effectively promotes regeneration in crushed optic nerve beyond the site of optic nerve injury. The effect of oncomodulin may involve downstream signaling via Ca^{2+}/calmodulin kinase and gene transcription. Studies *in vivo* show that oncomodulin released from microspheres promotes regeneration in mature rat optic nerve (Yin et al., 2004). Oncomodulin also stimulates outgrowth from peripheral sensory neurons. Collectively these studies indicate that oncomodulin is a new growth factor secreted by neurons of the mature central and peripheral nervous systems to protect the optic nerve from inflammatory processes.

7.3 Glutamate-Mediated Oxidative Stress in Brain

Oxidative stress is a cytotoxic process that occurs in the cell when antioxidant mechanisms are overwhelmed by reactive oxygen species (ROS) (Porter et al., 1995; Halliwell, 2006). Thus, oxidative stress is a threshold phenomenon characterized by a major increase in the amount of oxidized cellular components. ROS include superoxide anions, hydroxyl, alkoxyl, and peroxyl radicals, and hydrogen peroxide (Table 7.5). The major source of ROS is the mitochondrial respiratory chain. A number of cellular enzymes, including enzymes in the mitochondrial respiratory chain, xanthine/xanthine oxidase, myeloperoxidase, cytochrome P450 in cell cytoplasm, COX, LOX, nitric oxide synthase, and NADPH oxidase in plasma membranes, utilize molecular oxygen and produce reactive oxygen species (ROS), e.g. superoxide anion (O_2^{-}) and H_2O_2. Superoxide is rapidly converted to H_2O_2 by superoxide dismutase (SOD), and in turn, H_2O_2 is converted to H_2O by catalase (Lambeth, 2004). In the presence of metal ions, such as Fe^{2+} and Cu^{2+}, H_2O_2 can be further converted to hydroxyl radical ($\cdot OH$) through the Fenton reaction. Hydroxyl radicals can attack polyunsaturated fatty acids in membrane phospholipids forming the peroxyl radical ($\cdot ROO$) and then propagate the chain reaction of lipid peroxidation.

Table 7.5 Radical reactive oxygen species and non-radical oxygen species found in brain tissue

Name	Chemical formula
Molecular oxygen	$:O_2$
Superoxide anion	$\cdot O^{-2}$
Hydroxyl radical	$\cdot OH$
Alkoxyl radical	$RO\cdot^{-}$
Peroxyl radical	$RO\cdot$
Nitric oxide radical	$NO\cdot$
Peroxynitrite anion	$ONOO^{-}$
Singlet oxygen	$^{1}O_2$
Hypochlorous acid	HOCl
Hydrogen peroxide	H_2O_2

The treatment of mixed hippocampal cell cultures with glutamate and its analogs also results in generation of ROS via PLA_2, COX, and LOX-catalyzed reactions (Bondy and Lee, 1993; Lafon-Cazal et al., 1993; Gunasekar et al., 1995; Kahlert et al., 2005). Glutamate treatment produces a marked rise in generation of ROS in neurons, but not in astrocytes. However, in both cell types, the mitochondrial potential is increased in response to glutamate challenge. It is proposed that in neurons, Ca^{2+} influx accounts for the increased ROS generation in response to glutamate. This observation may explain the high vulnerability of neurons to glutamate challenge, compared to the vulnerability of astrocytes. The high resistance of astrocytes is accompanied by an efficient buffering of cytosolic Ca^{2+}, which is not found in neurons. Mitochondria play a vital role in buffering the cytosolic Ca^{2+} overload in stimulated neurons. Collectively these studies indicate that mitochondria play a vital role in buffering the cytosolic Ca^{2+} overload in neurons stimulated by glutamate.

N-methyl-D-aspartate (NMDA) receptor activation results in production of reactive oxygen species (ROS) and activation of extracellular signal-regulated kinase (ERK) in hippocampal area CA1 (Kishida et al., 2005). The NMDA receptor-dependent activation of ERK is not only attenuated by a number of antioxidants but also by the NADPH oxidase inhibitor diphenylene iodonium (DPI). No attenuation of ERK is observed in mice that lack the $p47^{phox}$ subunit of NADPH oxidase. This suggests that NADPH oxidase is closely associated with NMDA-mediated neurotoxicity. Collective evidence suggests that ROS production, especially superoxide production via NADPH oxidase, is required for NMDA receptor-dependent activation of ERK in hippocampal area CA1 (Kishida et al., 2005).

NADPH oxidase also plays an important role in glutamate-induced cell death in SH-SY5Y human neuroblastoma cells (Nikolova et al., 2005). One of the downstream targets of NADPH oxidase-derived superoxide radicals is the transcription factor NF-κB, which controls the expression of a large array of genes involved in immune function, inflammation, and cell survival. NF-κB itself is a key factor in controlling NADPH oxidase expression and function (Anrather et al., 2006). In monocytic and microglial cell lines LPS-mediated induction of NADPH oxidase subunit $gp91^{phox}$ is inhibited by the expression of I-κBκ. Similarly the generation of ROS is retarded in monocytic and microbial cells over expressing I-κBκ. Expression of $gp91^{phox}$ is very low in RelA($^{-/-}$) fibroblasts but can be induced by reconstituting these cells with p65/RelA (Anrather et al., 2006). Thus, $gp91^{phox}$ expression is dependent on the presence of p65/RelA. It is also shown that $gp91^{phox}$ transcription is dependent on NF-κB. Two potential cis-acting elements in the murine $gp91^{phox}$ promoter that control NF-κB-dependent regulation were identified. The findings raise the possibility of a positive feedback loop in which NF-κB activation by oxidative stress leads to further radical production via NADPH oxidase (Anrather et al., 2006).

The chemical reactivity of ROS varies from the very toxic hydroxyl [$^{\cdot}OH^-$] to the less reactive superoxide radical [$O_2{}^-$]. H_2O_2, although less reactive than $O^-{}_2$, is more highly diffusible and can cross the plasma membrane. ROS inactivates membrane proteins and DNA (Berlett and Stadtman, 1997). Various processes generate ROS both extracellularly and intracellularly. ROS can directly oxidize and damage

Table 7.6 Effect of ROS on cellular components, lipids, proteins, and nucleic acids

Component	Effect	Reference
Lipids	Peroxidation of fatty acids, decrease in neural membrane fluidity, permeability, microviscosity, and activities of membrane-bound enzymes	(Farooqui and Horrocks, 2007)
Proteins	Oxidation of SH groups, chemical cross-linking of membrane proteins and lipids, and inhibition of enzymic activities	(Farooqui and Horrocks, 2007)
Nucleic acids	DNA strand scission, consumption of NAD and impairments in ATP synthesis, and generation of 8-hydroxy-2-deoxyguanosine	(Farooqui and Horrocks, 2007)

macromolecules such as DNA, proteins, and lipids. The polyunsaturated fatty acid at the *sn*-2 position of the glycerophospholipid molecule is most susceptible to free radical attack at the α-methylene carbon in the alkyl chain of the fatty acid that is adjacent to the carbon-carbon double bond (Table 7.6). Under aerobic conditions a polyunsaturated fatty acid with an unpaired electron undergoes a molecular rearrangement by reaction with O_2 to generate a peroxyl radical. The peroxyl radical captures hydrogen atoms from the adjacent fatty acids to form a lipid hydroperoxide (Fig. 7.3). The lipid hydroperoxides thus formed are not completely stable *in vivo* and, in the presence of iron, can further decompose to radicals that can propagate the chain reactions started by an initial free radical attack.

The damage to neural membranes induced by ROS has many potential consequences. These include: (a) changes in physicochemical properties of neural membranes (microviscosity) resulting in alterations in the orientation of optimal domains for the interaction of functional membrane proteins such as receptors, enzymes, and ion-channels, (b) changes in the number of receptors and their affinity

$$
\begin{aligned}
&(1)\quad \text{Lipid-H} + {}^{\cdot}\text{OH} \longrightarrow \text{H}_2\text{O} + {}^{\cdot}\text{Lipid}\\
&(2)\quad {}^{\cdot\cdot}\text{Lipid} + \text{O}_2 \longrightarrow \text{lipid-OO}^{\cdot}\\
&(3)\quad \text{Lipid-OO}^{\cdot} + \text{Lipid-H} \longrightarrow \text{Lipid-OOH} + {}^{\cdot}\text{lipid}\\
&(4)\quad {}^{\cdot}\text{NO} + {}^{\cdot}\text{O}_2^- \longrightarrow \text{ONOO}^-\\
&(5)\quad \text{ONOO}^- + \text{H}^+ \longrightarrow \text{ONOOH}\\
&(6)\quad \text{ONOOH} \longrightarrow {}^{\cdot}\text{OH} + \text{NO}_2^{\cdot}\\
&(7)\quad \text{H}_2\text{O}_2 + \text{H}^+ + \text{Cl}^- \longrightarrow \text{H}_2\text{O} + \text{HOCl}\\
&(8)\quad \text{O}_2^- + \text{O}_2^- + 2\text{H}^+ \longrightarrow \text{H}_2\text{O}_2\ \text{O}_2\\
&(9)\quad 2\text{H}_2\text{O}_2 \longrightarrow 2\text{H}_2 + \text{O}_2\\
&(10)\quad \text{H}_2\text{O}_2 + 2\text{GSH} \longrightarrow 2\text{H}_2\text{O} + \text{GSSG}
\end{aligned}
$$

Fig. 7.3 Reactions showing the generation of ROS during lipid peroxidation and oxidative stress. Hydroxyl radical ($^{\cdot}OH$); lipid radical ($^{\cdot}$lipid), peroxyl radical (lipid-OO$^{\cdot}$); lipid peroxide (lipid-OOH); nitric oxide ($^{\cdot}NO$); nitrogen dioxide (NO_2); peroxynitrite anion ($ONOO^-$); hypochlorous acid (HOCl), and hydrogen peroxide (H_2O_2)

for neurotransmitters and drugs, and (c) inhibition of ion pump operation resulting in changes in ion homeostasis. The presence of peroxidized glycerophospholipids in neural membranes may also produce a membrane-packing defect, making the *sn*-2 ester bond more accessible to the action of PLA_2. In fact, glycerophospholipid hydroperoxides are better substrates for PLA_2 than are the native glycerophospholipids. Oxidative modification of neural membrane glycerophospholipids leads to the formation and accumulation of biologically active lipid hydroperoxides and other oxidation products that induce specific cellular reactions (Bochkov and Leitinger, 2003) such as inhibition of reacylation of phospholipids in neuronal membranes (Zaleska and Wilson, 1989). This inhibition may constitute another mechanism whereby oxidative processes contribute to neural cell death. The hydrolysis of peroxidized glycerophospholipids results in removal of peroxidized fatty acyl chains, which are reduced and re-esterified under normal conditions (McLean et al., 1993). Collective evidence suggests that lipid hydroperoxides are another type of ROS whose biological function has not yet been fully understood. However, 1-palmitoyl-2-(5)oxovaleroyl-*sn*-glycerophosphocholine acts through G protein coupled receptors to produces an up-regulation of the cAMP/PKC-mediated pathway and down-regulation of NF-κB-dependent transcription (Leitinger, 2003; Furnkranz and Leitinger, 2004).

The cellular response to oxidized glycerophospholipids depends not only on their concentration, but also the extent of their oxidation. Lower concentrations of moderately oxidized glycerophospholipids do not induce cell death, but instead induce an adaptive response to the stress of a subsequent exposure to ROS. In non-neural systems, the oxidized glycerophospholipids also induce monocyte-endothelial cell interactions that are independent of receptors for platelet-activating factor (Leitinger et al., 1997; Leitinger, 2003). These glycerophospholipids may also be involved in the induction of vascular endothelial growth factor expression (McIntyre et al., 1999; Leitinger, 2005). In non-neural cells, high concentrations of fully oxidized glycerophospholipids and their metabolites produce neural degeneration (Farooqui et al., 1997).

The detoxification of glycerophospholipid hydroperoxides can be accomplished through the combined enzymic activity of PLA_2 and reduction of the released fatty acid hydroperoxides with phospholipid hydroperoxide glutathione peroxidase (van Kuijk et al., 1987; Imai and Nakagawa, 2003). The latter enzyme acts on membranes and reduces glycerophospholipid hydroperoxides to the non-toxic hydroxyl derivatives (Fisher et al., 1999; Nakagawa, 2004). Phospholipid hydroperoxide glutathione peroxidase is different from the classic glutathione peroxidase, which mainly reduces hydrogen peroxide. The restoration of membrane integrity by the reaction catalyzed by phospholipid hydroperoxide glutathione peroxidase is achieved by the reinsertion of non-oxidized fatty acyl groups by the deacylation/reacylation cycle (Farooqui et al., 2000).

The reaction between ROS and proteins (Table 7.6) leads to a chemical cross-linking of membrane proteins and lipids and a reduction in membrane unsaturation (Pamplona et al., 2005). This depletion of unsaturation in membrane lipids is associated with decreased membrane fluidity and decreases in the activity of membrane-bound enzymes, ion-channels, and receptors (Ray et al., 1994). Neurons are most susceptible to ROS-mediated oxidative injury. ROS also contribute to brain damage by activating a number of cellular pathways resulting in the expression of stress-sensitive genes and proteins associated with oxidative injury (Wang et al., 2006). Furthermore, oxidative stress also activates mechanisms that result in a glia-mediated inflammation that also causes secondary neuronal damage (Farooqui et al., 2007).

ROS-mediated hydroxylation of nucleic acid bases generates 8-hydroxy-2-deoxyguanosine (8-OHdG) from deoxyguanosine and alters DNA (Table 7.6). DNA alterations are associated with aging and Alzheimer disease (AD). In fact 8-OHdG is commonly used as a marker for DNA damage in AD. The concentration of 8-OHdG increases several-fold in excitotoxicity. It is present in significant amounts in the mitochondrial and nuclear DNA of AD brains as compared to control cases (Farooqui et al., 2001). The altered DNA can be repaired by DNA glycosylase. However, if the intensity of glutamate-mediated oxidative stress is high, then DNA repair does not occur and the neural cell dies by either apoptosis or necrosis (Fraga et al., 1990).

7.4 Glutamate-Mediated Energy Status of Degenerating Neurons

Brain has a very high metabolic rate that it must maintain for its normal function. This requires an uninterrupted supply of both glucose and oxygen, which are used by brain tissue for producing ATP that maintains the high metabolic rate. Due to the high rate of oxygen consumption, the enrichment of polyunsaturated fatty acids in membrane phospholipids, and the relatively low abundance of anti-oxidative defense enzymes, neural cells are highly susceptible to oxidative stress (Halliwell, 2006). Underlying this unusual high-energy demand is the need for not only maintaining the appropriate ionic gradients across the neural membranes, but also creating the proper cellular redox potentials.

Full and transient deficits in glucose and oxygen can rapidly compromise ATP generation and threaten cellular integrity by either not maintaining or abnormally modulating ion homeostasis and cellular redox. The initial response to a transient insufficiency of energy is depolarization resulting in Na^+ influx. This relieves the voltage-dependent Mg^{2+} block of NMDA receptor channels, produces the opening of voltage-dependent Ca^{2+} channels, and reverses the Na^+/Ca^{2+} antiporter system in such a way that Ca^{2+} enters the neural cell as Na^+ is deported. The decreased ATP level also prevents ATP-dependent extrusion of Ca^{2+}. Thus, prolonged ATP insufficiency results in a massive influx and accumulation of Ca^{2+} that facilitates neural cell injury.

Treatment of neuronal cultures with glutamate is accompanied by the loss of neuronal viability, reduced neuronal energy state (ATP level and mitochondrial membrane potential), and increased cytoplasmic mitochondrial Ca^{2+}. This decrease in ATP results in failure of ionic pumps and increase in extracellular K^+. These processes cause neuronal depolarization and release of glutamate, which over-stimulates NMDA receptors. The NMDA antagonist MK-801 retards these effects, suggesting involvement of glutamate in neuronal cell death (Rundén-Pran et al., 2005). Addition of U0126, a selective inhibitor of MAPK kinase, also blocks glutamate-mediated neuronal cell death (Franceschini et al., 2006). This suggests the involvement of protein kinases in glutamate-mediated neural injury.

The mode of cell death in glutamate toxicity depends on the intracellular energy charge (Sastry and Rao, 2000; Farooqui et al., 2004). In neurons, the intracellular energy levels and mitochondrial function are rapidly compromised in necrosis, but not in apoptosis, suggesting that the ATP level is a prominent factor in determining the neurochemical events associated with apoptotic and necrotic neural cell death (Nicotera and Leist, 1997). Apoptosis and necrosis represent the extreme ends of the spectrum of mechanisms for neural cell death (Williamson and Schlegel, 2002; Leist and Nicotera, 1998). Nuclear degradation and the expression of annexin V, two important markers for apoptosis, require sufficient ATP generation. In the face of a marked reduction in ATP, the caspase cascade is blocked with subsequent neurodegeneration occurring through necrosis (Denecker et al., 2001).

7.5 Glutamate-Mediated Alterations in Cellular Redox Status

In neural cells, the redox status is controlled by the thioredoxin (Trx) and glutathione (GSH) systems that scavenge harmful intracellular ROS. Thioredoxins are antioxidants that serve as a general protein disulphide oxidoreductase (Saitoh et al., 1998). They interact with a broad range of proteins by a redox mechanism based on the reversible oxidation of 2 cysteine thiol groups to a disulphide, accompanied by the transfer of 2 electrons and 2 protons. These proteins maintain their reduced state through the thioredoxin system, which consists of NADPH, thioredoxin reductase (TR), and thioredoxin (Trx) (Williams, Jr. et al., 2000; Saitoh et al., 1998). The thioredoxin system is a system inducible by oxidative stress that reduces the disulfide bond in proteins (Fig. 7.4). It is a major cellular redox system that maintains cysteine residues in the reduced state in numerous proteins.

Glutathione (GSH) is a tripeptide that acts as a non-enzymic antioxidant in brain tissue (Bains and Shaw, 1997). It exists in at least two pools, 85–90% in

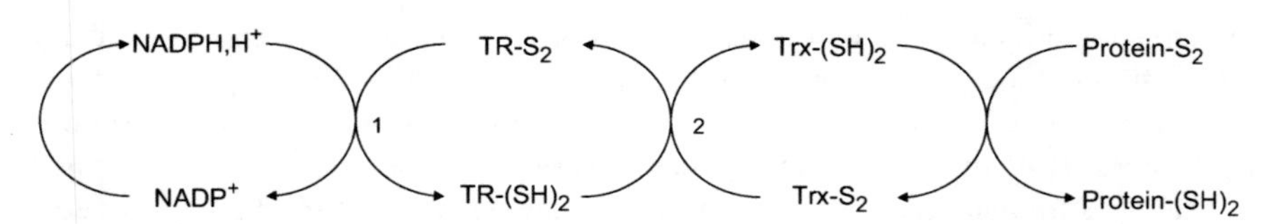

Fig. 7.4 Reactions associated with the thioredoxin system. Thioredoxin is a redox-regulating protein with a redox-active disulfide/dithiol within the conserved active site sequence –Cys-Gly-pro-Cys-. Thioredoxin reductase, a 55 kDa flavoprotein that catalyzes the NADPH-dependent reduction of thioredoxin (1) and thioredoxin oxidase (2), a flavin-dependent sulfhydryl oxidase that catalyzes the oxidative protein folding with the generation of disulfides

the cytoplasm and 10–15% in the mitochondria. Mitochondrial GSH, which detoxifies H_2O_2, is vital for cellular defense against endogenous or exogenous oxidative stress. Complete depletion of mitochondrial GSH causes a gradual loss of neural cell viability. Thus limitations of GSH synthesis and transmembrane transport in mitochondria suggest that optimal functioning of the mitochondrial GSH cycle and maintenance of adequate thiol-disulfide redox tone is essential to protect against glutamate-mediated ROS generation. Oxygen radicals generated during glutamate-mediated oxidative stress are detoxified by glutathione (GSH). In brain tissue the concentration of GSH ranges from 1 to 3 mM. Astrocytes and nerve terminals also contain high levels of GSH whereas a neuronal cell body has limited amounts of GSH (Chang et al., 1997). GSH depletion not only enhances oxidative stress but also increases the levels of excitotoxic molecules. These processes initiate neural injury in distinct neuronal populations of brain tissue (Bains and Shaw, 1997).

NMDA receptor channels contain a "redox modulatory site" formed from two cysteine residues in the NMDA receptor sequence. These residues are oxidized to disulfide bonds when the NMDA receptor is oxidized. The reduction of disulfide bonds to thiol groups in the reduced state is a molecular switch that turns on or turns off conformational changes in the NMDA receptor. This process may modulate Ca^{2+} influx through the NMDA channel. Thus the treatment of neural cells with DTT potentiates NMDA-mediated Ca^{2+} influx whilst exposure to DTNB or ROS inhibits Ca^{2+} influx. Moreover, the NMDA redox-sensitive site is also a target for nitric oxide. The interaction of nitric oxide with the redox site results in decreased Ca^{2+} influx through the NMDA receptor. Glutathione may act as an endogenous ligand at NMDA receptors and thus can modulate NMDA receptor channel activity and cellular redox under normal and pathological conditions. Glutamate-mediated neurotoxicity also results in the inhibition of high-affinity cystine uptake by neurons through the cystine/glutamate antiporter. This inhibition decreases the availability of cystine for glutathione synthesis, thus weakening neuronal antioxidant defenses (Dubovsky et al., 1989; Ratan et al., 1994). Collective evidence suggests that glutamate markedly affects the redox status of neural cells in brain tissue.

7.6 Glutamate-Mediated Alterations in Gene Expression

Glutamate and its analogs modulate de novo protein synthesis of immediate early genes through the expression of transcription factor AP1 (Pennypacker et al., 1994; Ogita and Yoneda, 1994; Yoneda et al., 1999). Under normal conditions the expression of immediate early genes is very low. Injections of glutamate and its analogs stimulate the expression of immediate early genes (Ogita and Yoneda, 1994; Yoneda et al., 1999). During this process transcription factor CREB migrates from the cytoplasm to the nucleus where it is phosphorylated at serine 133. The phosphorylation of CREB activates transcription of the c-fos gene to result in expression of c-Fos protein. This expressed c-Fos protein then heterodimerizes with c-Jun protein to form AP1 which binds to the consensus element TGACTCA upstream of the inducible target genes (Yoneda et al., 1999). Thus increased binding activity to the cAMP response element (CRE), a response element present in the promoter region of c-fos, has been detected in the extract derived from nuclear fractions of glutamate-treated rat brain (Ogita and Yoneda, 1994; Yoneda et al., 1999).

Glutamate-mediated signaling *in vivo* involves the participation of the antioxidant-response element (ARE). In nuclear extracts from the hippocampus of mice injected with kainate (Ogita et al., 2004), ARE binding is upregulated from 2 h to 3 days after kainate treatment. Supershift analysis suggests the involvement of Nrf2 (NF-E2-related factor 2), Fos-B, and c-Fos in ARE binding in hippocampal nuclear extracts from kainate-treated animals. Kainate treatment also produces a marked increase in levels of c-Fos and Fos-B, without markedly affecting that of Nrf2 in nuclear extracts from the hippocampus. RT-PCR analysis indicates that kainate treatment increases the glutathione-S-transferase mRNA level in the hippocampus (Ogita et al., 2004). Collectively these studies suggest that kainate-mediated oxidative stress enhances nuclear ARE binding through an interaction between constitutive Nrf2 with inducible Fos-B expressed in murine hippocampus.

In cultured cerebellar granular neurons, glutamate signals lead not only to neuronal cell death, but also to expression of both the c-fos gene and c-Fos protein with marked potentiation of DNA binding activity of the transcription factor activator protein-1 (AP-1) that consists of Fos and Jun family member proteins (Szekely et al., 1989). Increases in AP1 binding activity after glutamate treatment suggest that the AP-1 transcription factor-mediated modulation of downstream gene expression may be an important component of glutamate-mediated neurotoxicity. Glutamate-mediated regulation of AP-1, NF-κB, and other redox-sensitive transcription factors may interfere with the regulation of expression of a number of genes that have consensus sequences for these peptides in their promoter regions. In brain tissue immediate early genes mediate gene regulation in response to trophic signals, neurotransmitters, and oxidative stress. Some of these genes are associated with apoptotic cell death in brain tissue (Koistinaho and Hökfelt, 1997).

NMDA receptor antagonists reduce the expression of immediate early genes, strongly supporting the view that glutamate may be involved early in the signaling pathway (Hisanaga et al., 1992). In contrast, the AMPA/KA receptor antagonist, 6-cyano-7-nitroquinoxaline-2,3-dione (CNQX), shows no protection against excitotoxicity and has no significant effect on the Glu-induced delay in c-fos mRNA expression. The Q240/30 c-fos mRNA ratio may be a predictive index for excitotoxic neuronal death and provide information on the identity of the receptor subtype mediating excitotoxicity in different brain cell types. The Q240/30 c-fos mRNA ratio may also aid in establishing the role of excitotoxicity during the development of neurons in vitro (Griffiths et al., 1997). RT-PCR studies also support the proposal that c-fos mRNA can be used as a specific biomarker of excitotoxicity. Furthermore, these studies have potential for screening newly-designed excitatory amino acid receptor antagonists in the search for clinically relevant drugs to treat acute neural trauma and neurodegenerative diseases.

References

Adams J., Collaco-Moraes Y., and De Belleroche J. (1996). Cyclooxygenase-2 induction in cerebral cortex: An intracellular response to synaptic excitation. J. Neurochem. 66:6–13.

Allen N. J. and Attwell D. (2001). A chemokine-glutamate connection. Nat. Neurosci. 4:676–678.

Andersen J. K. (2004). Oxidative stress in neurodegeneration: cause or consequence? Nature Rev. Neurosci.S18–S25.

Anrather J., Racchumi G., and Iadecola C. (2006). NF-κB regulates phagocytic NADPH oxidase by inducing the expression of gp91^{phox}. J. Biol. Chem. 281:5657–5667.

Auger C. and Attwell D. (2000). Fast removal of synaptic glutamate by postsynaptic transporters. Neuron 28:547–558.

Baichwal V. R. and Baeuerle P. A. (1997). Activate NF-κB or die? Curr. Biol. 7:R94–R96.

Bains J. S. and Shaw C. A. (1997). Neurodegenerative disorders in humans: the role of glutathione in oxidative stress-mediated neuronal death. Brain Res. Rev. 25:335–358.

Bal-Price A., Matthias A., and Brown G. C. (2002). Stimulation of the NADPH oxidase in activated rat microglia removes nitric oxide but induces peroxynitrite production. J. Neurochem. 80:73–80.

Bartolomeo A. C., Morris H., and Boast C. A. (1997). Arecoline via miniosmotic pump improves AF64A-impaired radial maze performance in rats: A possible model of Alzheimer's disease. Neurobiol. Learn. Mem. 68:333–342.

Basta G., Lazzerini G., Del Turco S., Ratto G. M., Schmidt A. M., and De Caterina R. (2005). At least 2 distinct pathways generating reactive oxygen species mediate vascular cell adhesion molecule-1 induction by advanced glycation end products. Arterioscler. Thromb. Vasc. Biol. 25:1401–1407.

Bazan N. G. (2005a). Lipid signaling in neural plasticity, brain repair, and neuroprotection. Mol. Neurobiol. 32:89–103.

Bazan N. G. (2005b). Neuroprotectin D1 (NPD1): A DHA-derived mediator that protects brain and retina against cell injury-induced oxidative stress. Brain Path. 15:159–166.

Bazan N. G. (2005c). Synaptic signaling by lipids in the life and death of neurons. Mol. Neurobiol. 31:219–230.

Berlett B. S. and Stadtman E. R. (1997). Protein oxidation in aging, disease, and oxidative stress. J. Biol. Chem. 272:20313–20316.

Block M. L. and Hong J. S. (2005). Microglia and inflammation-mediated neurodegeneration: multiple triggers with a common mechanism. Prog. Neurobiol. 76:77–98.

Bochkov V. N. and Leitinger N. (2003). Anti-inflammatory properties of lipid oxidation products. J. Mol. Med. 81:613–626.

Bondy S. C. and Lee D. K. (1993). Oxidative stress induced by glutamate receptor agonists. Brain Res. 610:229–233.

Chang M. L., Klaidman L. K., and Adams J. D., Jr. (1997). The effects of oxidative stress on *in vivo* brain GSH turnover in young and mature mice. Mol. Chem. Neuropathol. 30:187–197.

Chiang N., Arita M., and Serhan C. N. (2005). Anti-inflammatory circuitry: Lipoxin, aspirin-triggered lipoxins and their receptor ALX. Prostaglandins Leukot. Essent. Fatty Acids 73: 163–177.

Chiang N., Serhan C. N., Dahlén S. E., Drazen J. M., Hay D. W., Rovati G. E., Shimizu T., Yokomizo T., and Brink C. (2006). The lipoxin receptor ALX: potent ligand-specific and stereoselective actions *in vivo*. Pharmacol. Rev. 58:463–487.

Choi D. W. (1988). Glutamate neurotoxicity and diseases of the nervous system. Neuron 1: 628–634.

Correale J. and Villa A. (2004). The neuroprotective role of inflammation in nervous system injuries. J. Neurol. 251:1304–1316.

Denecker G., Vercammen D., Declercq W., and Vandenabeele P. (2001). Apoptotic and necrotic cell death induced by death domain receptors. Cell Mol. Life Sci. 58:356–370.

Drew P. D., Storer P. D., Xu J. H., and Chavis J. A. (2005). Hormone regulation of microglial cell activation: relevance to multiple sclerosis. Brain Res. Rev. 48:322–327.

Dubovsky S. L., Christiano J., Daniell L. C., Franks R. D., Murphy J., Adler L., Baker N., and Harris R. A. (1989). Increased platelet intracellular calcium concentration in patients with bipolar affective disorders. Arch. Gen. Psychiatr. 46:632–638.

Farooqui A. A. and Horrocks L. A. (1991). Excitatory amino acid receptors, neural membrane phospholipid metabolism and neurological disorders. Brain Res. Rev. 16:171–191.

Farooqui A. A. and Horrocks L. A. (1994a). Excitotoxicity and neurological disorders: involvement of membrane phospholipids. Int. Rev. Neurobiol. 36:267–323.

Farooqui A. A. and Horrocks L. A. (1994b). Involvement of glutamate receptors, lipases, and phospholipases in long-term potentiation and neurodegeneration. J. Neurosci. Res. 38:6–11.

Farooqui A. A. and Horrocks L. A. (2007). *Glycerophospholipids in the Brain: Phospholipases A_2 in Neurological Disorders*, pp. 1–394. Springer, New York, in press.

Farooqui A. A., Anderson D. K., and Horrocks L. A. (1993). Effect of glutamate and its analogs on diacylglycerol and monoacylglycerol lipase activities of neuron-enriched cultures. Brain Res. 604:180–184.

Farooqui A. A., Yang H.-C., and Horrocks L. A. (1997). Involvement of phospholipase A_2 in neurodegeneration. Neurochem. Int. 30:517–522.

Farooqui A. A., Horrocks L. A., and Farooqui T. (2000). Deacylation and reacylation of neural membrane glycerophospholipids. J. Mol. Neurosci. 14:123–135.

Farooqui A. A., Ong W. Y., Lu X. R., Halliwell B., and Horrocks L. A. (2001). Neurochemical consequences of kainate-induced toxicity in brain: involvement of arachidonic acid release and prevention of toxicity by phospholipase A_2 inhibitors. Brain Res. Rev. 38: 61–78.

Farooqui A. A., Ong W. Y., and Horrocks L. A. (2004). Biochemical aspects of neurodegeneration in human brain: involvement of neural membrane phospholipids and phospholipases A_2. Neurochem. Res. 29:1961–1977.

Farooqui A. A., Horrocks L. A., and Farooqui T. (2007). Modulation of inflammation in brain: a matter of fat. J. Neurochem. 101:577–599.

Filbin M. T. (2006). How inflammation promotes regeneration. Nat. Neurosci. 9:715–717.

Fisher A. B., Dodia C., Manevich Y., Chen J. W., and Feinstein S. I. (1999). Phospholipid hydroperoxides are substrates for non-selenium glutathione peroxidase. J. Biol. Chem. 274:21326–21334.

Fitch M. T., Doller C., Combs C. K., Landreth G. E., and Silver J. (1999). Cellular and molecular mechanisms of glial scarring and progressive cavitation: *in vivo* and in vitro analysis of inflammation-induced secondary injury after CNS trauma. J. Neurosci. 19:8182–8198.

Fraga C. G., Shigenaga M. K., Park J. W., Degan P., and Ames B. N. (1990). Oxidative damage to DNA during aging: 8-hydroxy-2′-deoxyguanosine in rat organ DNA and urine. Proc. Natl. Acad. Sci. U. S. A 87:4533–4537.

Franceschini D., Giusti P., and Skaper S. D. (2006). MEK inhibition exacerbates ischemic calcium imbalance and neuronal cell death in rat cortical cultures. Eur. J. Pharmacol. 553: 18–27.

Furnkranz A. and Leitinger N. (2004). Regulation of inflammatory responses by oxidized phospholipids structure-function relationships. Curr. Pharmaceut. Design 10:915–921.

Gabriel C., Justicia C., Camins A., and Planas A. M. (1999). Activation of nuclear factor-κB in the rat brain after transient focal ischemia. Brain Res. Mol. Brain Res. 65:61–69.

Garrido R., Mattson M. P., Hennig B., and Toborek M. (2001). Nicotine protects against arachidonic-acid-induced caspase activation, cytochrome c release and apoptosis of cultured spinal cord neurons. J. Neurochem. 76:1395–1403.

Genis P., Jett M., Bernton E. W., Boyle T., Gelbard H. A., Dzenko K., Keane R. W., Resnick L., Mizrachi Y., Volsky D. J., Epstein L. G., and Gendelman H. E. (1992). Cytokines and arachidonic metabolites produced during human immunodeficiency virus (HIV)-infected macrophage-astroglia interactions: implications for the neuropathogenesis of HIV disease. J. Exp. Med. 176:1703–1718.

Gilroy D. W., Newson J., Sawmynaden P. A., Willoughby D. A., and Croxtall J. D. (2004). A novel role for phospholipase A_2 isoforms in the checkpoint control of acute inflammation. FASEB J. 18:489–498.

Gomes-Leal W., Corkill D. J., Freire M. A., Picanço-Diniz C. W., and Perry V. H. (2004). Astrocytosis, microglia activation, oligodendrocyte degeneration, and pyknosis following acute spinal cord injury. Exp. Neurol. 190:456–467.

Graeber M. B. and Moran L. B. (2002). Mechanisms of cell death in neurodegenerative diseases: fashion, fiction, and facts. Brain Pathol. 12:385–390.

Griffiths R., Malcolm C., Ritchie L., Frandsen A., Schousboe A., Scott M., Rumsby P., and Meredith C. (1997). Association of c-fos mRNA expression and excitotoxicity in primary cultures of mouse neocortical and cerebellar neurons. J Neurosci. Res. 48:533–542.

Gunasekar P. G., Kanthasamy A. G., Borowitz J. L., and Isom G. E. (1995). NMDA receptor activation produces concurrent generation of nitric oxide and reactive oxygen species: implication for cell death. J. Neurochem. 65:2016–2021.

Halliwell B. (2006). Oxidative stress and neurodegeneration: where are we now? J. Neurochem. 97:1634–1658.

Hays S. J. (1998). Therapeutic approaches to the treatment of neuroinflammatory diseases. Curr. Pharm. Des. 4:335–348.

Hermann G. E., Rogers R. C., Bresnahan J. C., and Beattie M. S. (2001). Tumor necrosis factor-α induces cFOS and strongly potentiates glutamate-mediated cell death in the rat spinal cord. Neurobiol. Dis. 8:590–599.

Hilburger E. W., Conte E. J., McGee D. W., and Tammariello S. P. (2005). Localization of NADPH oxidase subunits in neonatal sympathetic neurons. Neurosci. Lett. 377:16–19.

Hisanaga K., Sagar S. M., and Sharp F. R. (1992). N-methyl-D-aspartate antagonists block fos-like protein expression induced via multiple signaling pathways in cultured cortical neurons. J. Neurochem. 58:1836–1844.

Hosokawa M., Klegeris A., Maguire J., and McGeer P. L. (2003). Expression of complement messenger RNAs and proteins by human oligodendroglial cells. Glia 42:417–423.

Imai H. and Nakagawa Y. (2003). Biological significance of phospholipid hydroperoxide glutathione peroxidase (PHGPx, GPx4) in mammalian cells. Free Radical Biol. Med. 34: 145–169.

Juranek I. and Bezek S. (2005). Controversy of free radical hypothesis: Reactive oxygen species-Cause or consequence of tissue injury? Gen. Physiol. Biophys. 24:263–278.

Kahlert S., Zundorf G., and Reiser G. (2005). Glutamate-mediated influx of extracellular Ca^{2+} is coupled with reactive oxygen species generation in cultured hippocampal neurons but not in astrocytes. J. Neurosci. Res. 79:262–271.

Kaltschmidt B., Uherek M., Wellmann H., Volk B., and Kaltschmidt C. (1999). Inhibition of NF-κB potentiates amyloid β-mediated neuronal apoptosis. Proc. Natl. Acad. Sci. USA 96: 9409–9414.

Kim D. K., Rordorf G., Nemenoff R. A., Koroshetz W. J., and Bonventre J. V. (1995). Glutamate stably enhances the activity of two cytosolic forms of phospholipase A_2 in brain cortical cultures. Biochem. J. 310:83–90.

Kim G. M., Xu J., Xu J. M., Song S. K., Yan P., Ku G., Xu X. M., and Hsu C. Y. (2001). Tumor necrosis factor receptor deletion reduces nuclear factor-kappa B activation, cellular inhibitor of apoptosis protein 2 expression, and functional recovery after traumatic spinal cord injury. J. Neurosci. 21:6617–6625.

Kim S. Y., Min D. S., Choi J. S., Choi Y. S., Park H. J., Sung K. W., Kim J., and Lee M. Y. (2004). Differential expression of phospholipase D isozymes in the hippocampus following kainic acid-induced seizures. J. Neuropathol. Exp. Neurol. 63:812–820.

Kishida K. T., Pao M., Holland S. M., and Klann E. (2005). NADPH oxidase is required for NMDA receptor-dependent activation of ERK in hippocampal area CA1. J. Neurochem. 94: 299–306.

Koistinaho J. and Hökfelt T. (1997). Altered gene expression in brain ischemia. NeuroReport 8: i–viii.

Lafon-Cazal M., Pietri S., Culcasi M., and Bockaert J. (1993). NMDA-dependent superoxide production and neurotoxicity. Nature 364:535–537.

Lambeth J. D. (2004). NOX enzymes and the biology of reactive oxygen. Nat. Rev. Immunol. 4:181–189.

Leist M. and Nicotera P. (1998). Apoptosis, excitotoxicity, and neuropathology. Exp. Cell Res. 239:183–201.

Leitinger N. (2003). Oxidized phospholipids as modulators of inflammation in atherosclerosis. Curr. Opin. Lipidol. 14:421–430.

Leitinger N. (2005). Oxidized phospholipids as triggers of inflammation in atherosclerosis. Mol. Nutr. Food Res. 49:1063–1071.

Leitinger N., Watson A. D., Faull K. F., Fogelman A. M., and Berliner J. A. (1997). Monocyte binding to endothelial cells induced by oxidized phospholipids present in minimally oxidized low density lipoprotein is inhibited by a platelet activating factor receptor antagonist. Adv. Exp. Med. Biol. 433:379–382.

Lilienbaum A. and Israel A. (2003). From calcium to NF-kappa B signaling pathways in neurons. Mol. Cell. Biol. 23:2680–2698.

Lin T. N., Wang Q., Simonyi A., Chen J. J., Cheung W. M., He Y. Y., Xu J., Sun A. Y., Hsu C. Y., and Sun G. Y. (2004). Induction of secretory phospholipase A_2 in reactive astrocytes in response to transient focal cerebral ischemia in the rat brain. J. Neurochem. 90:637–645.

Lubin F. D., Johnston L. D., Sweatt J. D., and Anderson A. E. (2005). Kainate mediates nuclear factor-kappa B activation in hippocampus via phosphatidylinositol-3 kinase and extracellular signal-regulated protein kinase. Neuroscience 133:969–981.

Lukiw W. J., Cui J. G., Marcheselli V. L., Bodker M., Botkjaer A., Gotlinger K., Serhan C. N., and Bazan N. G. (2005). A role for docosahexaenoic acid-derived neuroprotectin D1 in neural cell survival and Alzheimer disease. J. Clin. Invest 115:2774–2783.

Machado F. S., Johndrow J. E., Esper L., Dias A., Bafica A., Serhan C. N., and Aliberti J. (2006). Anti-inflammatory actions of lipoxin A_4 and aspirin-triggered lipoxin are SOCS-2 dependent. Nature Med. 12:330–334.

Maggirwar S. B., Sarmiere P. D., Dewhurst S., and Freeman R. S. (1998). Nerve growth factor-dependent activation of NF-κB contributes to survival of sympathetic neurons. J. Neurosci. 18:10356–10365.

Manev H., Favaron M., Guidotti A., and Costa E. (1989). Delayed increase in Ca^2+ influx elicited by glutamate: Role in neuronal death. Mol. Pharmacol. 36:106–112.

Manev H., Uz T., and Qu T. (1998). Early upregulation of hippocampal 5-lipoxygenase following systemic administration of kainate to rats. Restor. Neurol. Neurosci. 12:81–85.

Marchetti L., Klein M., Schlett K., Pfizenmaier K., and Eisel U. L. M. (2004). Tumor necrosis factor (TNF)-mediated neuroprotection against glutamate-induced excitotoxicity is enhanced by N-methyl-D-aspartate receptor activation-Essential role of a TNF receptor 2-mediated phosphatidylinositol 3-kinase-dependent NF-κB pathway. J. Biol. Chem. 279: 32869–32881.

Matute C., Domercq M., and Sánchez-Gómez M. V. (2006). Glutamate-mediated glial injury: Mechanisms and clinical importance. Glia 53:212–224.

Mazière C., Conte M. A., Degonville J., Ali D., and Mazière J. C. (1999). Cellular enrichment with polyunsaturated fatty acids induces an oxidative stress and activates the transcription factors AP1 and NFκB. Biochem. Biophys. Res. Commun. 265:116–122.

McIntyre T. M., Zimmerman G. A., and Prescott S. M. (1999). Biologically active oxidized phospholipids. J. Biol. Chem. 274:25189–25192.

McLean L. R., Hagaman K. A., and Davidson W. S. (1993). Role of lipid structure in the activation of phospholipase A_2 by peroxidized phospholipids. Lipids 28:505–509.

Milatovic D., Gupta R. C., and Dettbarn W. D. (2002). Involvement of nitric oxide in kainic acid-induced excitotoxicity in rat brain. Brain Res. 957:330–337.

Minami M., Kuraishi Y., and Satoh M. (1991). Effects of kainic acid on messenger RNA levels of IL-1β, IL-6, TNF-α and LIF in the rat brain. Biochem. Biophys. Res. Commun. 176: 593–598.

Minghetti L. and Levi G. (1998). Microglia as effector cells in brain damage and repair: focus on prostanoids and nitric oxide. Prog. Neurobiol. 54:99–125.

Minghetti L., Ajmone-Cat M. A., De Berardinis M. A., and De Simone R. (2005). Microglial activation in chronic neurodegenerative diseases: roles of apoptotic neurons and chronic stimulation. Brain Res. Rev. 48:251–256.

Moses G. S. D., Jensen M. D., Lue L. F., Walker D. G., Sun A. Y., Simonyi A., and Sun G. Y. (2006). Secretory PLA2-IIA: a new inflammatory factor for Alzheimer's disease. J. Neuroinflammation 3:28.

Mukherjee P. K., Marcheselli V. L., Serhan C. N., and Bazan N. G. (2004). Neuroprotectin D1: A docosahexaenoic acid-derived docosatriene protects human retinal pigment epithelial cells from oxidative stress. Proc. Natl. Acad. Sci. USA 101:8491–8496.

Nakagawa Y. (2004). Role of mitochondrial phospholipid hydroperoxide glutathione peroxidase (PHGPx) as an antiapoptotic factor. Biol. Pharm. Bull. 27:956–960.

Nicholas R. S. J., Compston A., and Brown D. R. (2001). Inhibition of tumour necrosis factor-α (TNFα)-induced NF-κB p52 converts the metabolic effects of microglial-derived TNFα on mouse cerebellar neurones to neurotoxicity. J. Neurochem. 76:1431–1438.

Nicotera P. and Leist M. (1997). Energy supply and the shape of death in neurons and lymphoid cells. Cell Death Differ. 4:435–442.

Nikolova S., Lee Y. S., Lee Y. S., and Kim J. A. (2005). Rac1-NADPH oxidase-regulated generation of reactive oxygen species mediates glutamate-induced apoptosis in SH-SY5Y human neuroblastoma cells. Free Radic. Res. 39:1295–1304.

Noda M., Kettenmann H., and Wada K. (2006). Anti-inflammatory effects of kinins via microglia in the central nervous system. Biol. Chem. 387:167–171.

Noh K. M. and Koh J. Y. (2000). Induction and activation by zinc of NADPH oxidase in cultured cortical neurons and astrocytes. J. Neurosci. 20:RC111.

Norel X. and Brink C. (2004). The quest for new cysteinyl-leukotriene and lipoxin receptors: recent clues. Pharmacol. Ther. 103:81–94.

Ogita K. and Yoneda Y. (1994). Selective potentiation of DNA binding activities of both activator protein 1 and cyclic AMP response element binding protein through *in vivo* activation of N-methyl-D-aspartate receptor complex in mouse brain. J. Neurochem. 63:525–534.

Ogita K., Kubo M., Nishiyama N., Watanabe M., Nagashima R., and Yoneda Y. (2004). Enhanced binding activity of nuclear antioxidant-response element through possible formation of Nrf2/Fos-B complex after *in vivo* treatment with kainate in murine hippocampus. Neuropharmacology 46:580–589.

Oka A., Belliveau M. J., Rosenberg P. A., and Volpe J. J. (1993). Vulnerability of oligodendroglia to glutamate: pharmacology, mechanisms, and prevention. J. Neurosci. 13:1441–1453.

Olney J. W., Fuller T., and de Gubareff T. (1979). Acute dendrotoxic changes in the hippocampus of kainate treated rats. Brain Res. 176:91–100.

Ong W. Y., He Y., Suresh S., and Patel S. C. (1997). Differential expression of apolipoprotein D and apolipoprotein E in the kainic acid-lesioned rat hippocampus. Neuroscience 79:359–367.

Ong W. Y., Kumar U., Switzer R. C., Sidhu A., Suresh G., Hu C. Y., and Patel S. C. (2001). Neurodegeneration in Niemann-Pick type C disease mice. Exp. Brain Res. 141: 218–231.

O'Riordan K. J., Huang I. C., Pizzi M., Spano P., Boroni F., Egli R., Desai P., Fitch O., Malone L., Ahn H. J., Liou H. C., Sweatt J. D., and Levenson J. M. (2006). Regulation of nuclear factor κB in the hippocampus by group I metabotropic glutamate receptors. J. Neurosci. 26:4870–4879.

Pamplona R., Dalfó E., Ayala V., Bellmunt M. J., Prat J., Ferrer I., and Portero-Otín M. (2005). Proteins in human brain cortex are modified by oxidation, glycoxidation, and lipoxidation. Effects of Alzheimer disease and identification of lipoxidation targets. J. Biol. Chem. 280: 21522–21530.

Paradis É., Clavel S., Julien P., Murthy M. R. V., de Bilbao F., Arsenijevic D., Giannakopoulos P., Vallet P., and Richard D. (2004). Lipoprotein lipase and endothelial lipase expression in mouse brain: regional distribution and selective induction following kainic acid-induced lesion and focal cerebral ischemia. Neurobiol. Dis. 15:312–325.

Pennypacker K. R., Thai L., Hong J. S., and McMillian M. K. (1994). Prolonged expression of AP-1 transcription factors in the rat hippocampus after systemic kainate treatment. J. Neurosci. 14:3998–4006.

Pepicelli O., Fedele E., Berardi M., Raiteri M., Levi G., Greco A., Ajmone-Cat M. A., and Minghetti L. (2005). Cyclo-oxygenase-1 and-2 differently contribute to prostaglandin E-2 synthesis and lipid peroxidation after *in vivo* activation of N-methyl-D-aspartate receptors in rat hippocampus. J. Neurochem. 93:1561–1567.

Phillis J. W., Horrocks L. A., and Farooqui A. A. (2006). Cyclooxygenases, lipoxygenases, and epoxygenases in CNS: their role and involvement in neurological disorders. Brain Res. Rev. 52:201–243.

Porter N. A., Caldwell S. E., and Mills K. A. (1995). Mechanisms of free radical oxidation of unsaturated lipids. Lipids 30:277–290.

Qin L., Liu Y., Wang T., Wei S. J., Block M. L., Wilson B., Liu B., and Hong J. S. (2004). NADPH oxidase mediates lipopolysaccharide-induced neurotoxicity and proinflammatory gene expression in activated microglia. J. Biol. Chem. 279:1415–1421.

Ratan R. R., Murphy T. H., and Baraban J. M. (1994). Oxidative stress induces apoptosis in embryonic cortical neurons. J. Neurochem. 62:376–379.

Ray P., Ray R., Broomfield C. A., and Berman J. D. (1994). Inhibition of bioenergetics alters intracellular calcium, membrane composition, and fluidity in a neuronal cell line. Neurochem. Res. 19:57–63.

Rothstein J. D., Dykes-Hoberg M., Pardo C. A., Bristol L. A., Jin L., Kuncl R. W., Kanai Y., Hediger M. A., Wang Y., Schielke J. P., and Welty D. F. (1996). Knockout of glutamate transporters reveals a major role for astroglial transport in excitotoxicity and clearance of glutamate. Neuron 16:675–686.

Rundén-Pran E., TansøR., Haug F. M., Ottersen O. P., and Ring A. (2005). Neuroprotective effects of inhibiting *N*-methyl-D-aspartate receptors, P2X receptors and the mitogen-activated protein kinase cascade: a quantitative analysis in organotypical hippocampal slice cultures subjected to oxygen and glucose deprivation. Neuroscience 136:795–810.

Saitoh M., Nishitoh H., Fujii M., Takeda K., Tobiume K., Sawada Y., Kawabata M., Miyazono K., and Ichijo H. (1998). Mammalian thioredoxin is a direct inhibitor of apoptosis signal-regulating kinase (ASK) 1. EMBO J. 17:2596–2606.

Salinska E., Danysz W., and Lazarewicz J. W. (2005). The role of excitotoxicity in neurodegeneration. Folia Neuropathol. 43:322–339.

Sastry P. S. and Rao K. S. (2000). Apoptosis and the nervous system. J. Neurochem. 74:1–20.

Schmidt K. N., Amstad P., Cerutti P., and Baeuerle P. A. (1995). The roles of hydrogen peroxide and superoxide as messengers in the activation of transcription factor NF-κB. Chemistry & Biology 2:13–22.

Serhan C. N. (2005a). Lipoxins and aspirin-triggered 15-epi-lipoxins are the first lipid mediators of endogenous anti-inflammation and resolution. Prostaglandins Leukot. Essent. Fatty Acids 73:141–162.

Serhan C. N. (2005b). Novel ω-3-derived local mediators in anti-inflammation and resolution. Pharmacol. Ther. 105:7–21.

Serhan C. N. and Savill J. (2005). Resolution of inflammation: The beginning programs the end. Nature Immunol. 6:1191–1197.

Shi H. L., Liu S. M., Miyake M., and Liu K. J. (2006). Ebselen induced C6 glioma cell death in oxygen and glucose deprivation. Chem. Res. Toxicol. 19:655–660.

Siesjö B. K. (1990). Calcium in the brain under physiological and pathological conditions. Eur. Neurol. 30:3–9.

Strijbos P. J. and Rothwell N. J. (1995). Interleukin-1 beta attenuates excitatory amino acid-induced neurodegeneration in vitro: involvement of nerve growth factor. J. Neurosci. 15: 3468–3474.

Sullivan P. G., Bruce-Keller A. J., Rabchevsky A. G., Christakos S., Clair D. K., Mattson M. P., and Scheff S. W. (1999). Exacerbation of damage and altered NF-κB activation in mice lacking tumor necrosis factor receptors after traumatic brain injury. J. Neurosci. 19:6248–6256.

Sun D., Newman T. A., Perry V. H., and Weller R. O. (2004). Cytokine-induced enhancement of autoimmune inflammation in the brain and spinal cord: implications for multiple sclerosis. Neuropathol. Appl. Neurobiol. 30:374–384.

Szekely A. M., Barbaccia M. L., Alho H., and Costa E. (1989). In primary cultures of cerebellar granule cells the activation of N-methyl-D-aspartate-sensitive glutamate receptors induces c-fos mRNA expression. Molec. Pharmacol. 35:401–408.

Vaccarino F., Guidotti A., and Costa E. (1987). Ganglioside inhibition of glutamate-mediated protein kinase C translocation in primary cultures of cerebellar neurons. Proc. Natl. Acad. Sci. USA 84:8707–8711.

van Kuijk F. J. G. M., Sevanian A., Handelman G. J., and Dratz E. A. (1987). A new role for phospholipase A_2: protection of membranes from lipid peroxidation damage. Trends Biochem. Sci. 12:31–34.

Wang Q., Yu S., Simonyi A., Sun G. Y., and Sun A. Y. (2005). Kainic acid-mediated excitotoxicity as a model for neurodegeneration. Mol. Neurobiol. 31:3–16.

Wang J. Y., Wen L. L., Huang Y. N., Chen Y. T., and Ku M. C. (2006). Dual effects of antioxidants in neurodegeneration: Direct neuroprotection against oxidative stress and indirect protection via suppression of glia-mediated inflammation. Curr. Pharmaceut. Design 12:3521–3533.

Williams C. H., Jr., Arscott L. D., Müller S., Lennon B. W., Ludwig M. L., Wang P. F., Veine D. M., Becker K., and Schirmer R. H. (2000). Thioredoxin reductase two modes of catalysis have evolved. Eur. J. Biochem. 267:6110–6117.

Williamson P. and Schlegel R. A. (2002). Transbilayer phospholipid movement and the clearance of apoptotic cells. Biochim. Biophys. Acta Mol. Cell Biol. Lipids 1585:53–63.

Wilms H., Rosenstiel P., Sievers J., Deuschl G., Zecca L., and Lucius R. (2003). Activation of microglia by human neuromelanin is NF-κB dependent and involves p38 mitogen-activated protein kinase: implications for Parkinson's disease. FASEB J. 17:500–502.

Witowski J., Ksiazek K., and Jorres A. (2004). Interleukin-17: a mediator of inflammatory responses. Cell Mol. Life Sci. 61:567–579.

Wolf C. M. and Eastman A. (1999). The temporal relationship between protein phosphatase, mitochondrial cytochrome c release, and caspase activation in apoptosis. Exp. Cell Res. 247: 505–513.

Wu S. M., Patel D. D., and Pizzo S. V. (1998). Oxidized α_2-macroglobulin (α_2M) differentially regulates receptor binding by cytokines/growth factors: implications for tissue injury and repair mechanisms in inflammation. J. Immunol. 161:4356–4365.

Yamamoto Y. and Gaynor R. B. (2004). IκB kinases: key regulators of the NF-κB pathway. Trends Biochem. Sci. 29:72–79.

Yedgar S., Cohen Y., and Shoseyov D. (2006). Control of phospholipase A_2 activities for the treatment of inflammatory conditions. Biochim. Biophys. Acta 1761:1373–1382.

Yin H. Y., Morrow J. D., and Porter N. A. (2004). Identification of a novel class of endoperoxides from arachidonate autoxidation. J. Biol. Chem. 279:3766–3776.

Yoneda Y., Ogita K., Azuma Y., Kuramoto N., Manabe T., and Kitayama T. (1999). Predominant expression of nuclear activator protein-1 complex with DNA binding activity following systemic administration of N-methyl-D-aspartate in dentate granule cells of murine hippocampus. Neuroscience 93:19–31.

Yu Z., Zhou D., Bruce-Keller A. J., Kindy M. S., and Mattson M. P. (1999). Lack of the p50 subunit of nuclear factor-κB increases the vulnerability of hippocampal neurons to excitotoxic injury. J. Neurosci. 19:8856–8865.

Zaleska M. M. and Wilson D. F. (1989). Lipid hydroperoxides inhibit reacylation of phospholipids in neuronal membranes. J. Neurochem. 52:255–260.

Zecca L., Zucca F. A., Wilms H., and Sulzer D. (2003). Neuromelanin of the substantia nigra: a neuronal black hole with protective and toxic characteristics. Trends Neurosci. 26:578–580.

Chapter 8
Glutamate Receptors and Neurological Disorders

Glutamate, the major excitatory neurotransmitter in brain, plays a critical role in acute neural trauma, ischemia, spinal cord trauma, and head injury, and neurodegenerative diseases, Alzheimer disease, amyotrophic lateral sclerosis, and Huntington disease). Following neuronal insult, the enhanced glutamate efflux and impaired cellular reuptake result in a rapid buildup of glutamate in the extracellular space (Matute et al., 2006). This prolonged and excessive accumulation of glutamate causes over-stimulation of glutamate receptors and a sustained increase in intracellular calcium concentration not only through NMDA receptor channels, but also through glutamate transporters operating in the reverse mode. These processes are sufficient to trigger downstream pathophysiological mechanisms associated with neural cell death. Nevertheless, a direct link between the enhanced glutamate release and the neural cell injury has not yet been fully established (Doble, 1999). In neurological disorders neuronal demise displays "selective vulnerability". Neurons in certain regions of the brain, such as the hippocampus in ischemia, the nucleus basalis in Alzheimer disease, and in the substantia nigra pars compacta in Parkinson disease, display a lower threshold for death compared to other regions of brain. At present mechanisms associated with "selective vulnerability" of neurons are not understood. Morphologically glutamate-mediated cell death is characterized by somatodendritic swelling, chromatin condensation into irregular clumps, and organelle damage.

There are many neurochemical similarities between acute neural trauma and neurodegenerative diseases. For example, in acute neural trauma and neurodegenerative diseases, neuronal and glial cells generate ROS and activities of specific mitochondrial enzymic complexes, such as cytochrome oxidase, are reduced (Fiskum et al., 1999; Schinder et al., 1996). The overexpression of endogenous mitochondrial uncoupling proteins (UCP) follows neuronal injury (Sullivan et al., 2004). These proteins decrease the mitochondrial membrane potential and increase neuronal cell death following oxidative stress. The increase in UCP activity reduces glutamate-mediated ROS generation and cell death whereas a decrease in UCP levels increases susceptibility to neuronal injury (Sullivan et al., 2004). Furthermore, activated microglia and astrocytes impose inflammation on neurons by virtue of their production of cytokines such as tumor necrosis factor-α (TNF-α) that may have paracrine effects on neurons. Elevated cytokine production is a characteristic feature of acute neural trauma as well as neurodegenerative diseases (McGeer and McGeer, 1998; Bramlett and Dietrich, 2004; Farooqui and Horrocks, 2007a). The consequences of lack of oxygen and mitochondrial dysfunction in acute neural trauma (ischemia, spinal cord trauma, and head injury) also include failure of ATP production, rapid generation of ROS, exacerbation of excitotoxicity, and induction of apoptosis through the release of cytochrome c (Sullivan et al., 2004; Farooqui et al., 2004). These processes result in a rapid loss of ion homeostasis and neuronal demise.

In neurodegenerative diseases, Alzheimer, Parkinson, and Huntington diseases, on the other hand, glucose metabolism slows down, but mitochondria, in spite of their dysfunction, are still capable of generating some ATP for maintaining ion homeostasis to a limited extent. This process results in a cumulative slow brain damage that takes a much longer time to develop. Thus, many neurodegenerative diseases occur later in life and their onset is consistent with prolonged exposure to excitotoxicity, oxidative stress, and neuroinflammation. Importantly, neurogenesis, a process associated with birth and maturation of functional new hippocampal neurons, is impaired by interplay among excitotoxicity, oxidative stress, and neuroinflammation accounting for brain atrophy in patients with neurodegenerative diseases.

Furthermore, in neurodegenerative diseases, neurons increase their defenses by developing compensatory responses, oxidative strength (Moreira et al., 2005a; Numazawa et al., 2003), aimed to avoid or at least reduce cellular damage. These studies are supported by the view that Aβ deposition may not be the initiator of AD pathogenesis, but rather a downstream protective adaptation mechanism developed by cells in response to coordinated and upregulated interplay among excitotoxicity, oxidative stress, and neuroinflammation (Numazawa et al., 2003; Lee et al., 2004; Moreira et al., 2005a,b). This protective role of Aβ explains why many aged individuals, despite having a high number of senile plaques in their brain, show little or no cognitive loss.

8.1 Glutamate in Ischemic Injury

Ischemic insult is caused by a shortage of oxygen and substrate and failure to remove metabolic waste products (Auer and Siesjö, 1988). Following ischemic injury, a marked accumulation of glutamate occurs (Benveniste et al., 1989; Phillis and O'Regan, 1996). The source of this extracellular glutamate is the Ca^{2+}-stimulated release of neurotransmitter-containing presynaptic vesicles and depolarization-induced reversal of the Na^{+}-dependent high-affinity acidic amino acid plasma membrane transporter (Szatkowski and Attwell, 1994). In vitro a brief exposure of neurons to glutamate and its analogs causes widespread neuronal death that can be attenuated by the removal of extracellular Ca^{2+} (Choi, 1988). This finding suggests that Ca^{2+} entry via the NMDA type of glutamate receptor channel and the intracellular accumulation of Ca^{2+} is responsible for neuronal cell death.

A speculative description indicates that glutamate toxicity may have three stages: induction, amplification, and expression (Choi, 1990). Neurochemical events in induction involve the over-stimulation of glutamate receptors leading to a set of immediate cellular derangements. These derangements include an accumulation of intracellular Ca^{2+}, Na^{+}, Cl^{-}, water and generation of diacylglycerol and InsP_3 resulting in neuronal swelling and subsequent delayed neuronal disintegration. Amplification involves the postsynaptic cascade of processes that intensify initial derangements including more accumulation of intracellular Ca^{2+} and significant elevations in Na^{+}, InsP_3, diacylglycerol, and activation of protein kinase C. These processes promote the spread of excitotoxicity to other neurons. Thus the main neurochemical processes in amplification include accumulation of Ca^{2+} and the stimulation of Ca^{2+}-dependent enzymes (Table 8.1). Expression includes the destruction cascade directly responsible for neuronal cell degeneration (Choi, 1990).

The release of glutamate following ischemic insult decreases ATP through several neurophysiological and neurochemical mechanisms. Thus Na^{+}-dependent uptake of glutamate not only increases the rate of membrane depolarization and hence increased ATP consumption through Na^{+}, K^{+}-ATPase but also potentiates the lifting of the voltage-dependent Mg^{2+} blockade of NMDA receptors. This process results in a massive Ca^{2+} influx, thereby enhancing ATP depletion (Love, 1999).

Glutamate-mediated ischemic injury results in either immediate neuronal death within the core of infarct or delayed cell death in the surrounding penumbra. Neurons continue to die in stroke patients up to 10 days after infarction (Saunders et al., 1995). Neurons within the core undergo necrosis, whereas neurons in the penumbra may undergo either necrotic or apoptotic cell death (Hill et al., 1995). The extent of ischemic injury varies according to the age of animals. Thus 10 and 21 day-old rats develop greater damage from ischemic insult than 6 week, 9 week, and 6 month-old rats (Yager et al., 1996; Yager and Thornhill, 1997). Younger rats may be more susceptible to ischemic insult because of an unbalanced maturation of excitatory versus inhibitory neurotransmitter systems (Hattori and Wasterlain, 1990).

Glial cell death mediated by glutamate does not involve glutamate receptor activation, but rather glutamate uptake (Oka et al., 1993; Matute et al., 2006). Glutamate

Table 8.1 Stimulatory effect of Ca^{2+} on enzymic activities associated with excitotoxicity

Enzyme	Reference
Phospholipase A_2	(Edgar et al., 1982; Saluja et al., 1999)
Phospholipase C	(Ming et al., 2006)
Diacylglycerol lipase	(Farooqui et al., 1993)
Cyclooxygenase	(Phillis et al., 2006)
Lipoxygenase	(Phillis et al., 2006)
Epoxygenase	(Phillis et al., 2006)
Calpain	(Siman and Noszek, 1988)
Protein kinase C	(Nakagawa et al., 1985; Saluja et al., 1999)
Calmodulin-dependent protein kinase	(Chin et al., 1985)
Guanylate cyclase	(Novelli et al., 1987)
Nitric oxide synthase	(Gally et al., 1990)
Calcineurin	(Halpain et al., 1990)
Endonuclease	(Siesjö, 1990)

uptake from the extracellular space by specific glutamate transporters is essential for maintaining excitatory postsynaptic currents (Auger and Attwell, 2000) and for blocking excitotoxic death due to over-stimulation of glutamate receptors (Rothstein et al., 1996). As stated in Chapter 4, out of 5 glutamate transporters cloned from brain tissue, at least two transporters, namely excitatory amino acid transporter E1 (EAAT1) and excitatory amino acid transporter E2 (EAAT2), are expressed in astrocytes, oligodendrocytes, and microglial cells (Matute et al., 2006). Exposure of astroglial, oligodendroglial, and microglial cell cultures to glutamate produces glial cell demise by a transporter-related mechanism involving the inhibition of cystine uptake, which causes a decrease in glutathione and makes glial cells vulnerable to toxic free radicals (Oka et al., 1993; Matute et al., 2006).

Neuropathological consequences of glutamate-mediated alterations in glycerophospholipid metabolism include alterations in membrane fluidity and permeability, alterations in ion homeostasis, and changes in activities of membrane bound-enzymes, receptors, and ion channels, and increased oxidative stress. Many of these lipid mediators are pro-inflammatory. Their effects are accompanied by the activation of astrocytes and microglia and the release of inflammatory cytokines. These cytokines in turn propagate and intensify neuroinflammation by a number of mechanisms including further upregulation of PLA_2 isoforms, generation of platelet-activating factor, stimulation of nitric oxide synthase (NOS), and calpain activation (Farooqui and Horrocks, 1991, 2006; Farooqui et al., 2000, 2002; Ray et al., 2003).

Collectively, these studies suggest that increased glycerophospholipid degradation through the activation of PLA_2 isoforms leads to changes in membrane permeability and stimulation of many enzymes associated with lipolysis, proteolysis, and disaggregation of microtubules with a disruption of cytoskeleton and membrane structure and sets the stage for increased production of reactive oxygen species (ROS) through the arachidonic acid cascade (Farooqui et al., 1997). At low levels, under normal conditions ROS function as signaling intermediates in the regulation of fundamental cell activities such as growth and adaptation responses. At higher concentrations, ROS contribute to neural membrane damage when the balance between reducing and oxidizing (redox) forces shifts toward oxidative stress (Juranek and Bezek, 2005; Farooqui and Horrocks, 1994, 2006). ROS also up regulate the gene expression of cytokines through transcription factors in the nucleus (Gabriel et al., 1999; Schneider et al., 1999). This process leads to the intensification of glutamate-mediated neural injury in ischemia.

Stimulation of NMDA receptors results in the synthesis of nitric oxide (NO) in a Ca^{2+}/calmodulin-dependent manner through the activation of neuronal NOS. NO is a second messenger that activates soluble guanylyl cyclase and participates in a signal transduction pathway involving cyclic GMP. All forms of NOS, NOS 1 or nNOS, NOS 2 or iNOS, and NOS 3 or eNOS, are stimulated following ischemic insult (Crocker et al., 1991; Samdani et al., 1997). NOS 1 and 3 are stimulated within minutes whereas NOS 2 is stimulated after several hours. NO also binds to cytochrome oxidase, inhibits cellular respiration, and releases superoxide anion from the mitochondrial respiratory chain. It reacts with superoxide to produce peroxynitrite. This metabolite damages lipids, proteins, and nucleic acids and also produces apoptotic cell death. In cell cultures, excitotoxic injury mediated by glutamate can be reduced by nitric oxide synthase inhibitors (Samdani et al., 1997), suggesting that nitric oxide production is linked to the NMDA type of glutamate receptors.

8.2 Glutamate in Spinal Cord Injury

Traumatic injury to spinal cord consists of two broadly defined components: a primary component, attributable to the mechanical insult itself, and a secondary component, attributable to the series of systemic and local neurochemical and pathophysiological changes that occur in spinal cord after the initial traumatic insult (Klussmann and Martin-Villalba, 2005). The primary injury causes a rapid deformation of spinal cord tissues, leading to rupture of neural cell membranes and release of intracellular contents. In contrast, secondary injury to the spinal cord includes glial cell reactions involving both activated microglia and astroglia with synthesis of cytokines and chemokines and demyelination involving oligodendroglia (Beattie et al., 2000).

Neurochemically, secondary injury is characterized by the release of glutamate from intracellular stores (Demediuk et al., 1988; Panter et al., 1990; Sundström and Mo, 2002) and the over-expression of cytokines (Hayes et al., 2002; Ahn et al., 2004) (Table 8.2). In spinal cord tissue, glutamate produces neural cell death through several mechanisms. Glutamate overstimulates both NMDA and AMPA types of glutamate receptors resulting in the influx of Na^{+}, efflux of K^{+}, and a large Ca^{2+} influx into neurons (Katayama et al., 1990). This process leads to an uncontrolled and sustained increase in cytosolic Ca^{2+} levels, which is different from the transient increase in Ca^{2+} that occurs during receptor stimulation. This increase in intracellular Ca^{2+} may be responsible not only for the uncoupling of mitochondrial electron transport, but also for the stimulation of many Ca^{2+}-dependent enzymes including lipases, phospholipases, calpains, nitric oxide synthase, protein phosphatases, matrix metalloproteinases, and various protein kinases (Bazan et al., 1995; Pavel et al., 2001; Ray et al., 2003; Ellis et al., 2004; Arundine and Tymianski, 2004; Xu et al., 2006; Wells et al., 2003). Stimulation of these enzymes, along with a rapid decrease in ATP level, changes in ion homeostasis, alterations in cellular redox, and induction of inflammation, results in neural cell death in spinal cord trauma. These processes are accompanied by the activation of proinflammatory transcription factors (NF-κB and AP-1) (Table 8.3). Thus inflammatory reactions and oxidative stress are major components of secondary injury (Farooqui et al., 2004, 2007). Both these processes play a central role in regulating the pathogenesis of acute and chronic spinal cord trauma.

Glutamate also produces its toxicity through delayed post-traumatic white matter degeneration (Park et al., 2004). This process involves glutamate release by the reversal of Na^{+}-dependent glutamate transport with subsequent activation of AMPA receptors and oligodendrocyte death (Li and Stys, 2000). Thus, the general response of spinal cord tissue to traumatic injury is blood-brain barrier disruption,

Table 8.2 Involvement of glutamate and its analogs in neurological and neuropsychiatric disorders

Disorder	Excitotoxin	Glutamate receptor involved	Reference
Ischemia	Glutamate	NMDA, AMPA	(Haba et al., 1991; Phillis and O'Regan, 1996; Olney et al., 1997)
Spinal cord trauma	Glutamate	NMDA, AMPA	(Faden and Simon, 1988)
Head injury	Glutamate	NMDA, AMPA	(Hayes et al., 1992)
Epilepsy	Glutamate	NMDA, AMPA	(McDonald et al., 1991)
Alzheimer disease	Glutamate	NMDA, AMPA	(Dewar et al., 1991)
Parkinson disease	Glutamate	NMDA	(Katsuki et al., 2001)
Huntington disease	Glutamate, Quinolinate	NMDA, AMPA	(Bruyn and Stoof, 1990)
Creutzfeldt-Jakob disease	Glutamate	AMPA	(Rodríguez et al., 2005)
AIDS dementia	Glutamate, Quinolinate	NMDA	(Heyes et al., 1991; New et al., 1998)
Experimental allergic encephalomyelitis	Glutamate	NMDA	(Hardin-Pouzet et al., 1997)
Acute stress	Quinolinate, Glutamate	AMPA	(Tocco et al., 1991)
Guam-type amyo-trophic lateral sclerosis	β-N-methylaminoL-alanine	NMDA, AMPA	(Copani et al., 1991)
Olivopontocerebellar atrophy	Quinolinate	AMPA	(Feinstein and Halenda, 1988)
Schizophrenia	Glutamate	NMDA	(Olney and Farber, 1995)

infiltration and activation of inflammatory cells, death of local neurons and glial cells caused by ischemia and hypoxia along with release of excitatory amino acids, gliosis,recruitment of endothelial cells, angiogenesis, compensatory responses in the surroundings of the most severely affected area, and formation of a glial scar.

Another pathway for glutamate toxicity is transporter-mediated cell death that requires cellular expression of the cystine/glutamate antiporter system (Murphy et al., 1989). The competition between glutamate and cystine for uptake via the cystine/glutamate antiporter induces an imbalance in the homeostasis of cystine, an amino acid that is the precursor of glutathione. Thus, exposure of neural cells to high levels of glutamate results in an inability of neural cells to maintain intracellular levels of glutathione. Increases in extracellular glutamate in spinal cord trauma may also be caused by reduction in expression of glutamate transporters on glial cells and the release of glutamate by a reverse Na^+-dependent glutamate transporter. All these processes may lead to a reduced ability of neural cells to protect against free radical-mediated oxidative stress causing cell death (Murphy et al., 1989; Pereira and Resende de Oliveira, 2000). Glutamate antagonists that block glutamate-mediated inflammation and oxidative stress may be the best neuroprotective agents that can be used for the treatment of spinal cord trauma.

Table 8.3 Status of neuroinflammation, oxidative stress, and NF-κB in acute neural trauma and neurodegenerative diseases

Disorder	Neuroin-flammation	Oxidative stress	NF-κB	Reference
Ischemia	Enhanced	Enhanced	Upregulated	(Love, 1999; Farooqui et al., 2006; Stephenson et al., 2000)
Spinal cord injury	Enhanced	Enhanced	Upregulated	(Farooqui and Horrocks, 2007a; Farooqui et al., 2006; Bethea et al., 1998)
Head trauma	Enhanced	Enhanced	Upregulated	(Farooqui and Horrocks, 2007a; Farooqui et al., 2006; Hang et al., 2006)
Epilepsy	Enhanced	Enhanced	Upregulated	(Chung and Han, 2003; Albensi, 2001)
Alzheimer disease	Enhanced	Enhanced	Upregulated	(Ong and Farooqui, 2005; Phillis et al., 2006; Yoshiyama et al., 2001; Collister and Albensi, 2005)
ALS	Enhanced	Enhanced	–	(Phillis et al., 2006)
HD	Enhanced	Enhanced	Upregulated	(Calabrese et al., 2004; Chin et al., 2004; Yu et al., 2000)
AIDS dementia complex	Enhanced	Enhanced	Upregulated	(Boven et al., 1999; Ahn and Aggarwal, 2005; Álvarez et al., 2005)
CJD	Enhanced	Enhanced	–	(Phillis et al., 2006)
EAE	Enhanced	Enhanced	Upregulated	(Murphy et al., 2002)
PD	Enhanced	Enhanced	Upregulated	(Beal, 2003; Wersinger and Sidhu, 2006; Kim et al., 2004)
ALS/PDC	–	Enhanced	–	(Winton et al., 2006)
MS	Enhanced	Enhanced	No effect	(Pitt et al., 2000; Rojas et al., 2003)
DOM toxicity	–	Enhanced	–	(Walser et al., 2006)
Schizophrenia	–	Enhanced	–	(Mahadik and Scheffer, 1996)

8.3 Glutamate in Head Injury

Like spinal cord trauma, traumatic head injury consists of a primary injury, attributable to the mechanical insult itself, and a secondary injury, attributable to the series of systemic and local neurochemical changes that occur in brain after the initial traumatic insult (Klussmann and Martin-Villalba, 2005). The primary injury causes a rapid deformation of brain tissues, leading to rupture of neural cell membranes, release of intracellular contents, and disruption of blood flow and breakdown of the blood-brain barrier. In contrast, secondary injury to the brain tissue includes many neurochemical alterations such as release of cytokines, glial cell reactions involving both activated microglia and astroglia, and demyelination involving oligodendroglia (McIntosh et al., 1998). These processes associated with inflammation in head injury lead to the activation, proliferation, and hypertrophy of phagocytic cells and gliosis (Pan et al., 1999).

In traumatic head injury neurochemical changes are accompanied by widespread neuronal depolarization, accumulation of glutamate in extracellular space, and increased levels of arachidonic acid and eicosanoids as well as leukotrienes (Bullock et al., 1995; McIntosh et al., 1998). Glutamate-mediated stimulation of PLA_2 and PLC produces changes in phospholipid composition, decreased ATP, and altered ion homeostasis in traumatic as well as fluid percussion models of head injury (Dhillon et al., 1994; Homayoun et al., 1997). Among the most common mechanisms for excessive free radical generation, non-enzymic iron-catalyzed Fenton reactions as well as enzymic mechanisms involving lipoxygenases and cyclooxygenases may be the most important mechanisms associated with pathogenesis of head injury (Bazan et al., 1995; Hoffman et al., 2000).

Inflammation and oxidative stress induced by head injury is characterized, in part, by the activation of NF-κB and AP-1, synthesis of cytokines, TNF-α and IL-1β, and chemokines, induction of nitric oxide synthase (NOS) and cyclooxygenase-2 (COX-2), and the coordinated infiltration of leukocytes (McIntosh et al., 1998) (Table 8.3). Induction of NOS results in increased synthesis of NO. The reaction between NO and superoxide anion produces peroxynitrite. The generation of peroxynitrite is accompanied by a rapid increase in cortical mitochondrial 3-nitrotyrosine levels as early as 30 min following head injury (Singh et al., 2006). This suggests that the generation of peroxynitrite may be closely associated with the pathogenesis of traumatic head injury (Singh et al., 2006). Peroxynitrite acts as a signaling molecule in the activation of proto-oncogenes (Kamat, 2006) and decomposes rapidly to toxic hydroxyl radicals. Compounds that specifically scavenge peroxynitrite or peroxynitrite-derived radicals are likely to be particularly effective for the treatment of traumatic head injury.

Elevated levels of glutamate are controlled effectively by the astrocytic glutamate transporters EAAT1 and EAAT2 (Matute et al., 2006). These transporters and splice variants are down regulated shortly following the traumatic head injury (Yi and Hazell, 2006). Involvement of glutamate transporters in the pathogenesis of head injury is also supported by the observation of a transient upregulation of complexin I and complexin II (Yi et al., 2006). These two nerve terminal proteins are involved in cellular machinery associated with neurotransmitter release from synaptic vesicles. Complexins I and II are localized in inhibitory and excitatory synapses respectively (Yi et al., 2006). These proteins contribute to the modulation of neurotransmitter release and maintenance of normal synaptic function (Pabst et al., 2000). Increased complexin II expression may dysregulate glutamate release following traumatic head injury and may therefore be important in causing increased extracellular glutamate levels following injury. The lack of success with glutamate receptor antagonists, MK 801 and CGS-19755, as a potential source of clinical intervention treatment following traumatic head injury has resulted in the necessity for a better understanding of molecular mechanisms associated with head injury. These mechanisms include the regulation of glutamate transporters, glutamate-mediated oxidative stress, and neuroinflammation (Lea and Faden, 2001;

Ikonomidou and Turski, 2002; Yi and Hazell, 2006). This suggests a cocktail of glutamate receptor antagonists, glutamate transporter inhibitors, antioxidants, and anti-inflammatory drugs may have beneficial effects in traumatic head injury.

8.4 Glutamate in Epilepsy

Epilepsy is a devastating neurological disorder with disruption of normal brain function due to the occurrence of seizures. Epilepsy not only involves long-lasting neuroplastic changes in the expression of neurotransmitters, receptors, and ion channels, but also involves sprouting, reorganization of synapses, as well as reactive gliosis (Spencer, 2007). Intracerebral injections of glutamate and its analogs produce seizure activity and brain damage in rat brain. This seizure activity resembles the seizures in patients with epilepsy (Bradford and Dodd, 1975; Liu et al., 1997; Meldrum, 1993). The mechanism by which persistent seizure activity results in neurodegeneration and brain damage is not fully understood. However, epileptic seizures in rat and human brain produce markedly increased levels of glutamate and aspartate in the focal, compared to nonfocal, regions of the cerebral cortex (Nadi et al., 1987; Sherwin et al., 1988; Bazan et al., 2002). Thus alterations in glutamate and aspartate may contribute to the pathogenesis of epilepsy-induced brain damage.

Increased glutamate levels and over-stimulation of NMDA, KA, and AMPA types of glutamate receptors result in the accumulation of Ca^{2+} in rat hippocampus during seizure activity (Griffiths et al., 1983; Meldrum, 1993, 2002; Liu et al., 1997; Wasterlain et al., 1993; Bazan et al., 2002) (Table 8.2). KA-mediated epileptic seizures in rat induce TNF-α and interleukin-6 release in hippocampal slices (De Bock et al., 1996). The release of TNF-α and interleukin-6 following severe limbic seizures may be an adaptation associated with protection of brain tissue from KA-mediated toxicity. KA also mediates altered expression of many genes involved in plasticity, gliosis, and cognitive functions (Zagulska-Szymczak et al., 2001; Bazan et al., 2002). Studies on KA-mediated epileptic seizures in normal and interleukin-6 null mice indicate that reactive astrogliosis and microgliosis is accompanied by markedly increased iNOS, peroxynitrite-mediated nitration of hippocampal proteins, Mn-SOD, Cu/Zn-SOD, and metallothioneins I and II immunoreactivities (Penkowa et al., 2001). Interleukin-6 null mice show less reactive astrogliosis and microgliosis but morphological hippocampal damage, oxidative stress, and apoptotic cell death are markedly increased. Interleukin-6 deficiency increases neuronal injury and impairs inflammatory responses following KA-mediated epileptic seizures (Penkowa et al., 2001). The activation of various Ca^{2+}-regulated enzymes found in neurons may be associated with epileptogenesis. These enzymes include the non-receptor proteins tyrosine kinases Src and Fyn, a serine-threonine kinase [Ca^{2+}-calmodulin-dependent protein kinase II (CaMKII)], and the phosphatase calcineurin.

Epileptic seizures also stimulate $cPLA_2$ activity and expression with accumulation of arachidonic acid (Visioli et al., 1994; Kajiwara et al., 1996). Levels of free fatty acids and diacylglycerols in rat brain rise rapidly with the onset of epileptic seizures indicating the activation of $cPLA_2$ and PLC (Visioli et al., 1994). Studies on the pentylenetetrazol (PTZ)-induced model of epilepsy in rat brain also indicate significant elevations in $sPLA_2$ activity in cortical, hippocampal, and cerebellar regions compared to the control group. The increase in $sPLA_2$ activity is more pronounced in hippocampal and cortical regions than in the cerebellar region (Yegin et al., 2002). Collective evidence suggests that like other types of acute trauma, Ca^{2+}-mediated stimulation of protein kinases, phospholipases, and calpains may trigger a variety of biochemical processes, including the degradation of membrane phospholipids, proteolysis of cytoskeletal proteins, and protein phosphorylation that may be involved in the pathogenesis of epilepsy (Bazan et al., 2002). Understanding of molecular mechanisms associated with epileptic seizures may result in identifying molecular targets for drugs that can be used for prevention.

8.5 Glutamate in Alzheimer Disease

Alzheimer disease (AD) is a progressive neurodegenerative disorder characterized by memory impairment and severe dementia. Neuropathological changes in AD include accumulation of amyloid neuritic plaques, neurofibrillary tangles, inflammation, and loss of synapses. Although the molecular mechanism associated with the pathogenesis of AD is not clearly understood, impairment of memory and neuroinflammatory process, triggered by abnormal glutamate metabolism and β-amyloid-mediated oxidative stress plays a major role in the neurodegenerative process (Arlt et al., 2002; Farooqui and Horrocks, 2007b; Kontush, 2001; Butterfield, 2002).

Glutamatergic neurotransmission is critical to normal learning and memory. The glutamatergic system is linked to the acetylcholinergic neurotransmitter system to form the glutamatergic/aspartatergic-acetylcholinergic circuit. This circuit is associated with not only with learning and memory but also with cognition (Wenk, 2006). The cognitive abnormalities in AD may be caused by the loss of neurons in the cholinergic and glutamatergic neural systems and by irregular functioning of the surviving neurons in these two systems (Riederer and Hoyer, 2006). Glutamate and acetylcholine are ligands for the glutamatergic/aspartatergic-acetylcholinergic neurotransmitter systems.

Another mechanism of synaptic dysfunction in AD may involve amyloid β peptide (Aβ, a 40 to 42 amino acid peptide). A marked increase in Aβ levels occurs in brain tissue from AD patients. Aβ inhibits glutamatergic neurotransmission and reduces synaptic plasticity (Snyder et al., 2005). Treatment of cortical neuronal cultures with Aβ facilitates endocytosis of NMDA receptor. Aβ-mediated endocytosis of NMDA receptor requires the α-7 nicotinic receptor, protein phosphatase 2B, and the tyrosine phosphatase STEP. Dephosphorylation of the NMDA receptor subunit NR2B at Tyr1472 correlates with receptor endocytosis. The addition of a γ-secretase inhibitor not only reduces Aβ but also restores surface expression of NMDA receptors, suggesting that A plays an important role in the regulation of NMDA and AMPA receptor trafficking (Snyder et al., 2005; Morishita et al., 2005).

Two forms of synaptic plasticity, long-term potentiation (LTP) and long-term depression (LTD), are major cellular mechanisms involved in learning and memory (Bliss and Collingridge, 1993; Chen and Tonegawa, 1997). These mechanisms involve depolarization of the postsynaptic membranes and Ca^{2+} entry mediated by NMDA receptors. The induction of LTP is triggered by not only postsynaptic entry of Ca^{2+} through the NMDA receptor channel and also through the insertion of new AMPA receptors at the synapse. The maintenance of LTP is regulated by presynaptic mechanisms. NMDA-mediated stimulation of PLA_2 liberates arachidonic acid that crosses the synaptic cleft to act pre-synaptically as a retrograde messenger through activation of the γ-isoform of protein kinase C (Linden and Routtenberg, 1989; Izquierdo and Medina, 1995). This isoform of protein kinase C together with $cPLA_2$ plays an important role in the induction and maintenance of LTP (Murakami and Routtenberg, 2003; Bernard et al., 1994). Insertion of AMPA receptors at the synapse increases synaptic current as well as synaptic strength associated with maintenance of LTP (Kauer and Malenka, 2006).

Treatment of hippocampal slices with arachidonate mimics the formation of long-term depression (LTD) (Massicotte, 2000). Based on several studies, protein phosphorylation of GluR2 at Ser880 mediated by protein kinase A may modulate synaptic transmission by inducing receptor internalization, an event that may contribute to LTD formation in hippocampus as well as cerebellum (Chung et al., 2003; Seidenman et al., 2003). Collective evidence suggests that NMDA receptor-dependent LTD is associated with protein phosphatase activity and that dephosphorylation of GluR1 and GluR2 AMPA receptor subunits is a key step in its expression mechanism (Ménard et al., 2005; St-Gelais et al., 2004).

Decreased NMDA and AMPA receptors, along with glutamate transporter activity, has been reported in several brain regions from AD patients compared to brain tissue from age-matched control subjects (Greenamyre et al., 1987; Greenamyre and Young, 1989; Penney et al., 1990; Dewar et al., 1990; Masliah et al., 1996; Scheuer et al., 1996) (Table 8.2). A marked reduction in the expression of NR2A and NR2B subunit mRNA occurs in the hippocampus and entorhinal cortex in brain from AD patients (Bi and Sze, 2002). This may induce changes in glutamate homeostasis in AD that may cause a major disturbance in Ca^{2+} homeostasis (Peterson and Goldman, 1986; Peterson et al., 1985; Mattson, 2002) and the activity of Ca^{2+}-dependent enzymes including protein kinase C (Shimihama et al., 1990; Vincent and Davies, 1990) and phospholipases A_2 (Farooqui et al., 2003a,b; Farooqui and Horrocks, 2007b). Collective evidence suggests that anomalous glutamatergic activity and changes in Ca^{2+} homeostasis are associated with alterations in postsynaptic receptor function and downstream defects that not only leads to neuronal injury and death but also to cognitive deficits associated with dementia (Wenk et al., 2006).

Little is known about the relationship between glutamate receptors and the neuropathological changes observed in AD. Neurofibrillary tangles, the hallmark of AD, are composed of paired helical filaments (Terry and Katzman, 1983). Incubation of cultured human neurons with aspartate or glutamate causes the formation of paired helical filaments (Mattson, 1990). Treatment of hippocampal neurons with glutamate in the presence of Ca^{2+} not only induces their degeneration but also increases τ and ubiquitin immunostaining (Mattson, 1990). Thus collective evidence suggests that the Ca^{2+} influx caused by glutamate may lead to modifications of cytoskeleton-associated proteins similar to those seen in the neurofibrillary tangles of patients with AD. The formation of neurofibrillary tangles may also involve calpains and protein kinases (Siman et al., 1989; Mattson, 1990; Haddad, 2004). Phosphorylation/dephosphorylation mechanisms are crucial in the regulation of tau and amyloid precursor proteins. Based on many studies it is proposed that mitogen-activated protein kinases are the key regulators of not only senile plaque and tangle formation but also NF-κB mediated oxidative stress (Haddad, 2004).

The β-amyloid protein found in the neurite plaques of AD patients increases the vulnerability of cortical cultures to glutamate and its analogs (Koh et al., 1990) suggesting that excitotoxicity may play an important role in the pathogenesis of AD. The mechanism underlying the abnormal function of glutamatergic neurons in AD has yet to be elucidated. One hypothesis is that glutamate-mediated abnormalities in glycerophospholipid metabolism may induce inflammation and β-amyloid deposition, which may impair glutamatergic receptor function in such a way that the ability of these receptors to prevent the influx of Ca^{2+} in the absence of an appropriate presynaptic signal is compromised. If this hypothesis is correct, then glutamate-mediated neuroinflammation, abnormalities in glycerophospholipid metabolism, oxidative stress, and β-amyloid deposition will start long before the symptomatic onset of AD (Zafrilla et al., 2006; Farooqui and Horrocks, 2007b; Mattson, 2002). Another possibility, which does not involve β-amyloid, is age-mediated neuronal insulin receptor dysfunction. This may lead to a deficit in acetylcholine and available energy production and also may cause neuronal membrane instability due to glutamate-mediated oxidative stress (Frolich and Hoyer, 2002). According to this hypothesis the development of AD may be due to dynamic cellular biological changes in the brain that are independent of the β-amyloid deposition and genetic factors.

Early-onset familial AD may be caused by mutations in β-amyloid precursor protein (APP), presenilins 1 and 2 (PSEN1, PSEN2), and apolipoprotein E (APOE) genes. In brain tissue and lymphocytes PSEN1, PSEN2 and APP mutations induce oxidative stress and perturb Ca^{2+} signaling in ways that alter their production of cytokines and ROS that are critical for proper immune responses (Gould et al., 2002). The molecular mechanism associated with these processes is not fully understood. However, the decreased NMDA and AMPA receptors in brain tissue from AD patients suggest that decreased glutamate receptors may be related to astrocytic and microglial cell-mediated inflammatory and immune responses in AD (Greenamyre et al., 1987; Greenamyre and Young, 1989; Penney et al., 1990; Dewar et al., 1990; Kalaria et al., 1996; Lue et al., 1996). In recent years considerable attention has focused on discovering non-toxic and tolerable glutamate receptor antagonists for the treatment of AD.

8.6 Glutamate in Amyotrophic Lateral Sclerosis (ALS)

ALS is a neurological disease with damage to motor neurons in the brain and spinal cord. It is characterized by progressive muscle weakness, atrophy, and spasticity. Characteristic neuropathological features include the loss of anterior horn motor neurons as well as degeneration of the corticospinal tracts (Shaw and Ince, 1997). Motor neurons have a high density of glutamate receptors. These neurons differ from other neural cells by their large size, their high ratio of axonal length to cell soma diameter, their high metabolic rate, and their high content of neurofilament proteins and antioxidant enzymes (Shaw and Ince, 1997). The primary cause of ALS remains elusive but excitotoxicity and oxidative stress may be closely associated (Rao and Weiss, 2004). Involvement of excitotoxicity in ALS is strongly supported by recent reports on the deficient RNA editing of AMPA receptor subunit GluR2 at the Q/R site in motor neurons in ALS (Kawahara et al., 2004, 2006; Kwak and Kawahara, 2005). The under-editing of the GluR2 Q/R site in AMPA markedly enhances the Ca^{2+} permeability of AMPA receptors (Hume et al., 1991; Burnashev et al., 1992), which may produce neuronal cell demise due to increased Ca^{2+} influx through the receptor channel. Hence, mice with RNA editing abnormalities at the GluR2 Q/R site die young (Kwak and Kawahara, 2005) and mice transgenic with abnormal Ca^{2+} permeability develop motor neuron disease 12 months after birth (Kuner et al., 2005; Kawahara et al., 2006). Furthermore, a specific functional defect in the Na^{+}-dependent glutamate uptake system in synaptosomal preparations from spinal cord and motor cortex of ALS patient was reported (Rothstein et al., 1992). This may lead to inefficient clearance of glutamate from the synaptic cleft, thus promoting excitotoxicity. This suggestion is supported by studies on the expression of glial glutamate transporters in brain tissue from sporadic ALS patients in which glial glutamate transporter GLT-1 is markedly decreased in motor cortex from ALS (Rothstein et al., 1995). This may result in accumulation of glutamate in the synaptic cleft and induce neuronal death through excitotoxicity (Reynolds et al., 1996). Numerous reports demonstrate mitochondrial dysfunction in ALS (Bacman et al., 2006). Based on studies in a mouse model of ALS, glutamate may mediate the intrinsic mitochondrial dysfunction programs that lead to mitochondrial swelling, release of cytochrome c, and apoptotic cell death in ALS (Bacman et al., 2006). Upregulation of p38 mitogen-activated protein kinase (p38MAPK) has been reported in motor neurons of spinal cord from familial ALS transgenic mice (SOD1G93A) but no changes have been reported in its mRNA levels (Tortarolo et al., 2003). As the ALS progressed, activated p38MARK also accumulates in hypertrophic astrocytes and reactive microglia (Tortarolo et al., 2003). These studies indicate that the activation of p38MARK in motor neurons and then in reactive glial cells may not only contribute to phosphorylation of cytoskeletal proteins and activation of cytokine release but also to the upregulation of nitric oxide synthase. All these may facilitate the development and progression of motor neuron pathology in SOD1G93A mice (Tortarolo et al., 2003).

Glutamate generates ROS that inhibit high affinity-glutamate uptake in cell culture, thus increasing the synaptic glutamate concentration (Volterra et al., 1994). Continuous exposure of neurons to glutamate not only causes excitotoxicity but also retards cystine uptake via the glutamate/cystine anti-porter (Murphy et al., 1989; Ye et al., 1999), inducing oxidative stress. Thus, a cross talk exists between excitotoxicity and oxidative stress. Most cases of ALS are sporadic, but 10% are familial. The sporadic form of this disease is characterized by a prominent neuroinflammatory component, upregulation of COX-2 mRNA, and oxidative stress (Yasojima et al., 2001). Mutations in the Cu/Zn superoxide dismutase (SOD1) gene occur in the familial form of ALS (Beal, 1998a). Upregulation of COX-2 mRNA also occurs in SOD1 transgenic mice at the onset of ALS (Almer et al., 2001). It is not known why and how mutations of SOD1 are related to ALS. In an organotypic cell culture model of ALS, the addition of a selective COX-2 inhibitor, SC236, blocked the destruction of motor neurons (Drachman and Rothstein, 2000), suggesting that COX-2 may play an important role in inflammatory processes in ALS. Similarly, treatment with celecoxib, another COX-2 inhibitor, prolongs the survival of neurons in the SOD1 mouse model of ALS (Drachman et al., 2002). In the spinal cord of a transgenic mouse model of ALS (G93A mice) (Okuno et al., 2004), the expression of CD40 is increased in both reactive microglia and astrocytes. In contrast, COX-2 activity is especially upregulated in astrocytes. In vitro studies also indicate that CD40-mediated stimulation of microglia and astrocytes results in upregulation of COX-2 and COX-2 inhibitors that protect motor neurons from neurodegeneration (Okuno et al., 2004). Collectively these studies suggest that CD40, which is upregulated in reactive glial cells in ALS, participates in motor neuron loss via induction of COX-2. Although upregulation of COX-2 and PGE_2 levels may not be the root cause of ALS, alterations in these parameters may be responsible for the induction and maintenance of inflammation and oxidative stress during the progression of ALS.

8.7 Glutamate in Huntington Disease

Huntington disease (HD) is a rare and progressive movement disorder caused by an autosomal dominant mutation in a gene located on chromosome 4 (Cepeda et al., 2001). This gene encodes a novel protein called huntingtin. This protein is distributed in both the peripheral body tissues as well as brain. HD is characterized by an expansion of a CAG repeat that encodes a polyglutamine tract in huntingtin. This leads to a dramatic loss of striatal medium-sized spiny GABAergic projection neurons (MSN) and gliosis (reactive astrocytosis) in the striatum and cerebral cortex (Cepeda et al., 2001). Despite identification of the gene mutation more than a decade ago, the normal function of ubiquitously expressed huntingtin is still unknown and the mechanisms associated with selective neurodegeneration in HD remain poorly understood (Cepeda et al., 2001; Bonilla, 2000). Huntingtin is associated with synaptic vesicles and/or microtubules and plays an important role in vesicular transport and/or the binding to the cytoskeleton (Bonilla, 2000).

Huntingtin is important for embryogenesis. Its mutant form alters the function of the mitochondrial respiratory chain. Detailed investigations on ATP generation and functioning of the respiratory complexes in clonal striatal cells, from Hdh(Q7) (wild-type) and Hdh(Q111) (mutant huntingtin knock-in) mouse embryos indicate that mitochondrial respiration and ATP synthesis are significantly decreased in the mutant striatal cells compared with the wild-type cells when either glutamate/malate or succinate is used as the substrate (Milakovic et al., 2006). However, there were no differences in mitochondrial respiration in the two cell lines when the artificial electron donor TMPD/ascorbate, which feeds into complex IV, is used as the substrate. Attenuation of mitochondrial respiration and ATP generation when either glutamate/malate or succinate is used as a substrate may not be due to impairment of the respiratory complexes, because their activities are equivalent in both cell lines. These studies demonstrate that the presence of mutant huntingtin impairs mitochondrial ATP production through one or more mechanisms that do not directly affect the function of the respiration complexes (Milakovic et al., 2006).

Huntingtin inhibits glyceraldehyde-3-phosphate dehydrogenase through its interactions with polyglutamine repeats (Burke et al., 1996). Interactions of huntingtin with other proteins may promote a cellular dysfunction or its own polymerization to form insoluble aggregates. The intraneuronal aggregates of huntingtin may modulate gene transcription, protein interactions, protein transport inside the nucleus and cytoplasm, and vesicular transport. However, since a dissociation between the aggregation of huntingtin and the selective pattern of striatal neuronal loss has been demonstrated, other properties of the mutant huntingtin, like proteolysis and the interactions with other proteins that affect vesicular trafficking and nuclear transport, may be associated with neurodegeneration (Bonilla, 2000).

Over-stimulation of NMDA-type of glutamate receptors may contribute to the selective degeneration of medium spinal neurons in HD (Zeron et al., 2004) (Table 8.2). Injections of NMDAR agonists into the striatum of rodents or non-human primates recapitulate the pattern of neuronal damage observed in HD. Altered NMDA receptor function has also been reported in corticostriatal synapses in a mouse model of HD. NMDA receptor-mediated current and/or toxicity is potentiated in striatal neurons from several HD mouse models as well as in heterologous cells expressing the mutant huntingtin protein. Changes in NMDA receptor activity correlate with altered Ca^{2+} homeostasis, mitochondrial membrane depolarization, and caspase activation. NMDA receptor stimulation is also closely linked to mitochondrial function, as treatment with mitochondrial toxins produces striatal damage that can be reversed by the addition of NMDA receptor antagonists. Recent efforts have focused on the elucidation of molecular pathways linking huntingtin to NMDA receptors, as well as the mechanisms which underlie the enhancement of NMDA receptor activity by mutant huntingtin (Fan and Raymond, 2007). Enhanced excitotoxicity mediated through a NMDA receptor is mediated by the mitochondrial-associated apoptotic pathway in cultured MSN from YAC transgenic mice expressing full-length huntingtin with a polyglutamine expansion of 46 or 72 (YAC46 or YAC72) (Zeron et al., 2004; Cepeda et al., 2001; Li et al., 2004). The Ca^{2+} transients and mitochondrial membrane depolarization mediated through a NMDA receptor are significantly increased in YAC transgenic mice compared to wild-type mice MSN. Inhibitors of the mitochondrial permeability transition (mPT), cyclosporin A and bongkrekic acid, and coenzyme Q10, an anti-oxidant involved in bioenergetic metabolism, dramatically diminished cell death mediated through a NMDA receptor. In YAC46 MSN, NMDA stimulates caspase-3 and caspase-9 but not caspase-8. Cyclosporin can block activation of caspase-3 and -9 mediated through a NMDA receptor, suggesting a link among NMDA receptors, mitochondrial dysfunction, and apoptotic cell death.

NMDA receptor-mediated oxidative damage also occurs in HD. The caudate nucleus of brain from HD patients has increased levels of 8-hydroxy-2-deoxyguanosine in nuclear DNA (Browne et al., 1997). This observation is supported by studies on a yeast model of HD. This model is constructed by expressing a human huntingtin fragment containing a mutant polyglutamine tract of 103Q and fusing it with green fluorescent protein (GFP) (Solans et al., 2006). ROS generation is significantly enhanced in cells expressing 103Q. Quenching of ROS with resveratrol partially prevents the cell respiration defect. It is proposed that oxidative stress and mitochondrial dysfunction in this yeast model of HD leads to increased ROS production. The oxidative damage may preferentially affect the stability and function of enzymes containing iron-sulfur clusters such as complexes II and III (Solans et al., 2006).

8.8 Glutamate in AIDS Dementia Complex

Levels of glutamate are elevated in blood and cerebrospinal fluid from subjects with AIDS Dementia Complex (Famularo et al., 1999; Ferrarese et al., 2001) (Table 8.2). Chronic over-activation of glutamate receptors in brains of patients infected with HIV-1 may contribute to the pathogenesis of the HIV-1-associated dementia complex. HIV-1 enters the brain very soon after peripheral infection and can induce severe and debilitating neurological problems that include behavioral abnormalities, motor dysfunction, and frank dementia (Kaul and Lipton, 2006). Neuronal damage in HIV infection results mainly from microglial activation and involves glutamate-mediated neurotoxicity, oxidative stress, and apoptosis (Gras et al., 2003). The microglial cells then serve as a reservoir for viral replication, resulting in increased expression of pro-inflammatory factors. Microglial cells are activated by interaction with viral proteins, such as Tat and gp120 (see below).

Glutamate toxicity commences through two distinct mechanisms: an excitotoxic mechanism in which glutamate receptors are over-stimulated, and an oxidative mechanism in which cystine uptake is inhibited, resulting in glutathione depletion and oxidative stress (Table 8.3). HIV-1 infection also initiates activation of chemokine receptors, inflammatory mediators, and extracellular matrix-degrading enzymes. All these processes can activate numerous downstream signaling pathways and disturb neuronal and glial function (Gras et al., 2003; Porcheray et al., 2006). These include generation of nitric oxide, quinolinic acid, and platelet-activating factor (Lipton, 1998). Astrocytes normally take up glutamate, keeping extracellular glutamate concentration low in the brain and preventing excitotoxicity. This action is inhibited in HIV infection, probably due to the effects of inflammatory mediators and viral proteins (Gras et al., 2003).

In brain tissue, HIV-1 sheds a coat glycoprotein called gp120 (Fig. 8.1). This protein, like glutamate, increases the intracellular Ca^{2+} concentration and produces neurotoxicity in neuronal cultures at picomolar concentrations (Sucher et al., 1991). The molecular mechanisms through which glial cells affect neuronal viability after exposure to gp120 protein are not fully understood. However, it is proposed that interaction of gp120 with the chemokine receptors in the CNS causes the release of arachidonic acid, probably through the stimulation of PLA_2 activity, which then interferes with the high-affinity uptake system of astrocytes for glutamate,

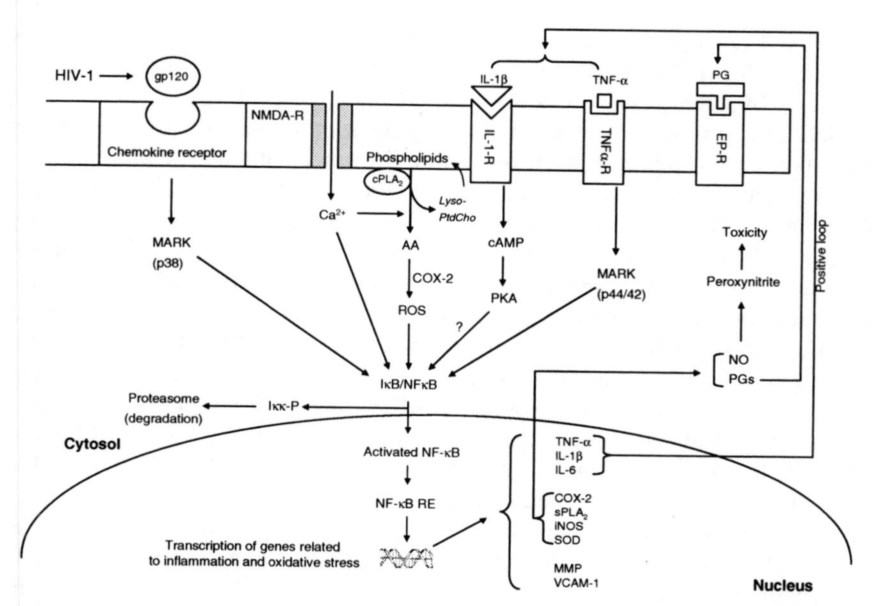

Fig. 8.1 A schematic diagram illustrating the involvement of NF-κB in gp120, ROS, NO, PG, IL-1β and TNF-α-mediated neurotoxicity. NMDA-R, N-Methyl-D-aspartate receptor; cPLA2, cytosolic phospholipase A_2; lyso-PtdCho, lysophosphatidylcholine; AA, arachidonic acid; cAMP, cyclic adenosine monophosphate; PKA, protein kinase A; TNF-α, tumor necrosis factor-α; TNF-α-R, TNF-α-receptor; IL-1β, interleukin-1β; IL-1β-R, IL-1β-receptor; IL-6, interleukin-6; MARK, mitogen-activated protein kinase; NO, nitric oxide; PG, prostaglandins; EP-R, prostaglandin receptors; NF-κB, nuclear factor-κB; NF-κB-RE, nuclear factor-κB-response element; IκB, inhibitory subunit of NF-κB; HIV-1, human immunodeficiency virus type 1; gp120, HIV-1 coat glycoprotein; COX-2, cyclooxygenase-2; iNOS, inducible nitric oxide synthase; $sPLA_2$, secretory phospholipase A_2; SOD, superoxide dismutase; MMP, matrix metalloproteinase; and VCAM-1, vascular adhesion molecule-1

resulting in an elevation of the extracellular glutamate concentration that causes Ca^{2+}-mediated excitotoxic brain damage in patients with AIDS Dementia Complex (Dreyer and Lipton, 1995).

In several human neuroblastoma cell lines that express CXCR4 and CCR5 but not CD4, gp120 up-regulates COX-2 mRNA and protein (Álvarez et al., 2005). Moreover, the gp120 also up-regulates the transcription of COX-2 promoter in neuronal cultures. Sequential deletions of the promoter show that deletion of a distal NF-κB site down-regulates gp120-dependent transcription. More importantly, the overexpression of the NF-κB inhibitory subunit, I-κBα, completely blocks the gp120-mediated increase in COX-2 activity (Álvarez et al., 2005). However, transfection of p65/relA NF-κB is not enough to induce COX-2 transcription, suggesting that NF-κB is necessary but not sufficient to control gp120-mediated COX-2 transcription. In addition to NF-κB, activating protein-1 (AP-1) but not nuclear factor of activated T cells (NFAT)-dependent transcription is also induced by gp120. Transfection of a dominant negative mutant c-Jun protein, TAM-67, efficiently inhibits the upregulation of COX-2 promoter by gp120, confirming the AP-1 requirement. Moreover, gp120 rapidly activates the c-Jun amino-terminal kinase (JNK) and p38 mitogen-activated protein kinase phosphorylation. The importance of NF-κB and AP-1 in COX-2 promoter and protein induction is strongly supported by NF-κB, p38, and JNK inhibitors (Álvarez et al., 2005).

Also, gp120 protein induces the rapid generation of reactive oxygen species and expression of interleukin-1β (Fig. 8.1). Treatment of the co-culture with an antibody against interleukin-1β prevents increased [Ca^{2+}]i-mediated neuronal cell death but not the production of reactive oxygen species in glia. Prior incubation of glial cells with Trolox, an antioxidant analog of vitamin E, down-regulated interleukin-1β expression and completely prevents neuron cell death (Viviani et al., 2001). Recent studies from this group indicate that exposure of neurons to gp120 increases phosphorylation of a NMDA receptor subunit, NR2B at Tyr-1472, and the interleukin-1 receptor antagonist (IL-1ra) can block this phosphorylation (Viviani et al., 2006). Phosphorylation of NR2B mediated by gp120 produces a sustained increase in intracellular Ca^{2+} in neurons and a significant increase in NR2B binding to PSD95. Ifenprodil, an antagonist that selectively inhibits NR2B receptors, IL-1ra, and ca-pYEEIE, a Src family SH2 inhibitor peptide, can block the increase in Ca^{2+}. Collectively these studies suggest that gp120-mediated IL-1β from glia up-regulates tyrosine phosphorylation of NMDA receptors and this may be relevant for neurodegeneration in AIDS dementia complex (Viviani et al., 2006). Furthermore, the degradation of endogenous glutamate by glutamate-pyruvate transaminase protects neurons from gp120-mediated neurodegeneration suggesting that gp120 and glutamate are necessary for neurodegeneration as synergistic effectors (Lipton, 1998).

Learning and memory impairments are frequently observed in patients suffering from AIDS dementia complex. These effects have been linked to the presence of gp120. The amnestic effect of gp120 can be overcome with glutamine or with precursors of glutamate synthesis, but only weakly by glutamate. The amnestic effect of gp120 may be due to an impaired uptake of glutamate by astrocytes and a subsequent interruption of glutamine supply to neurons (Fernandes et al., 2007).

Collectively these studies suggest that in AIDS dementia complex, HIV-infected macrophages, astrocytes, and microglia, (activated by the shed HIV-1 envelope protein, gp120, or other viral proteins and cytokines), secrete excitants and neurotoxins. These substances may include arachidonic acid, platelet-activating factor, free radicals ($NO^{\cdot}$ and $^{\cdot}O_2^{-}$), glutamate, quinolinate, cysteine, amines, and as yet unidentified factors emanating from stimulated macrophages and reactive astrocytes (Lipton, 1998). However, it still remains unknown whether the glutamate-mediated neurochemical changes in AIDS dementia complex are the primary event or a secondary effect of an abnormal signal transduction process related to inflammation and oxidative stress in degenerating neurons.

8.9 Glutamate in Creutzfeldt-Jakob Disease (CJD)

CJD is the most prevalent form of prion disease in humans. CJD is characterized by neuronal loss, astrocytic gliosis, and accumulation of an abnormal extracellular β-helix rich prion protein (PrP^{sc}). This abnormal protein is an isoform of a normally occurring α-helix rich prion protein (PrP^c). PrP^c is a glycolipid-anchored sialoglycoprotein with a molecular mass of 25–35 kDa. PrP^c and its mRNA are found in all neural cells, namely neurons, astrocytes, oligodendrocyte, and microglial cells (Siso et al., 2002). The function of PrP^c has not been clearly elucidated (Prusiner, 2001; Brown, 1999). However, several observations indicate that PrP^c may function as a copper chaperone. It regulates Cu^{2+}/Zn^{2+}-superoxide dismutase, redox changes, and oxidative stress. These processes are closely associated with the pathogenesis of CJD (Capellari et al., 1999; Brown, 2005). Thus unlike other neurodegenerative diseases, CJD results in a remarkable loss of antioxidant defense. This includes not only the loss of antioxidant activity of prion protein but also an increase in oxidative substances (Brown, 2005). These processes create a paradox as the loss of activity is not accompanied by the down-regulation of protein expression. It is stated that continued expression of non-functional prion protein is essential for the progression of prion diseases (Brown, 2005).

There are similarities between brain pathological changes caused by glutamate toxicity and by prion infection. Like glutamate-mediated neurotoxicity, prion infection leads to oxidative stress (Brown, 2005). Thus, the end-stage neuropathological changes caused by prion infection have many of the characteristics of glutamate-mediated neuronal necrosis. The NMDA receptor-channel antagonists such as memantine (1-amino-3,5- dimethyladamantane; MEM), MDADA (1-*N*-methylamino-3,5-dimethyladamantane) or dizocilpine (MK-801, (+)-5-methyl-10,11-dihydro-5H-dibenzo(a,d)cyclohepten-5,10-imine malate) produce cytoprotective effects in prion diseases (Scallet et al., 2003).

Up-regulation of arachidonic acid metabolism also occurs in a cell culture model of prion diseases. Thus, the treatment of primary cerebellar granule neuronal cultures with prion peptide (PrP106-126) stimulates PLA_2 (Bate and Williams, 2004) and results in the release and upregulation of arachidonic acid metabolism (Stewart et al., 2001; Farooqui et al., 2006). COX-1, COX-2, and 5-LOX activities are also increased in brain and CSF samples from CJD patients compared to age-matched controls, implicating both enzymes in the pathogenesis of CJD (Deininger et al., 2003; Minghetti et al., 2000).

The antimalarial drug quinacrine and some phenothiazine derivatives, acepromazine, chlorpromazine, and promazine, have been used for the treatment of prion diseases (Doh-ura et al., 2000; Korth et al., 2001; May et al., 2003). The molecular mechanism associated with the inhibition of PrP^{sc} formation by quinacrine remains unknown. However, it is proposed that quinacrine binds with human prion protein at the Tyr-225, Tyr-226, and Gln-227 residues of helix $\alpha 3$ (Vogtherr et al., 2003) and provides neuroprotection. Quinacrine may also act as an antioxidant and reduce the toxicity of prP^{sc} (Turnbull et al., 2003).

In a cell culture model of CJD, the 5-LOX inhibitors, nordihydroguaiaretic acid and caffeic acid, block PrP106-126-mediated neurodegeneration. These inhibitors also prevent the PrP106-126-induced caspase-3 activation and annexin V binding involved in 5-LOX-mediated apoptosis (Stewart et al., 2001). In contrast, indomethacin, a COX-1 and COX-2 inhibitor, and baicalein, a 12-LOX inhibitor, do not affect PrP106-126 induced neurotoxicity in cerebellar granule neuronal cultures (Stewart et al., 2001). Collectively these studies suggest that glutamate-mediated enhanced arachidonic acid metabolism along with neuroinflammation and oxidative stress may be closely associated with the pathogenesis of CJD. However, it still remains unknown whether PLA_2/COX/LOX-mediated enhancement of arachidonic acid metabolism in CJD is the primary event or a secondary effect of an abnormal signal transduction process related to inflammation and oxidative stress in degenerating neurons in CJD.

8.10 Glutamate in Parkinson Disease

The gradual and selective loss of dopaminergic neurons in the substantia nigra pars compacta characterizes PD. These neurons regulate body movement. Degeneration of these neurons in PD results in resting tremor, rigidity, bradykinesia, and gait disturbance in the patients with PD. The molecular mechanisms associated with the pathogenesis of PD have not yet been elucidated. However, dopaminergic neurons may be particularly vulnerable to oxidative stress because monoamine oxidase-mediated dopamine metabolism produces hydrogen peroxide. Thus increased turnover of dopamine neurons in the nigrostriatal tract, generation of peroxide with oxidative stress, impairment of energy metabolism with mitochondrial dysfunction, neuroinflammation, proteasomal dysfunction, and environmental and/or endogenous neurotoxins promote the progressive loss of nigral neurons (Beal, 1998b; Jenner and Olanow, 2006) (Table 8.3).

Loss of dopamine with formation of deaminated metabolites is accompanied by a significant rise in the level of oxidized glutathione (GSSG) in brain, strongly supporting the view that oxidative stress and energy impairment play an important role in pathogenesis of PD (Zeevalk et al., 1998). Six different genes associated with the pathogenesis of familial PD have been identified recently. These genes cause specific mutations in proteins such as α-synuclein, parkin, UCHL1, DJ1, PINK1, and LRRK2. DJ1 and PINK1 are mitochondrial proteins. The overexpression of α-synuclein and parkin results in mitochondrial dysfunction. It is interesting to note that these same proteins participate in the response of neural cells to oxidative stress and proteasomal dysfunction (Schapira, 2006; McNaught et al., 2006). It still remains to be seen whether mitochondrial and proteasomal dysfunctions play a primary or a secondary role in the initiation or progression of the neurodegeneration in PD.

The conversion of 1-methyl-4-phenyl-1,2,3,6-tetrahydropyridine (MPTP) to 1-methyl-4-phenylpyridinium ion (MPP^+) has been used to develop animal models for the study of cell death in nigrostriatal dopaminergic neurons. MPP^+ causes neuronal cell death through mitochondrial dysfunction at the level of the complex I rotenone binding site (Chiba et al., 1984; Ramsay et al., 1991). Marked increases in extracellular glutamate and aspartate occur following striatal administration of MPP^+. The injection of ionotropic glutamate receptor antagonists in rats can prevent this increase (Carboni et al., 1990; Turski et al., 1991). These observations implicate glutamate in the pathogenesis of experimental PD. However, studies on this topic have been controversial because the use of glutamate antagonists has provided mixed results. Some investigators report full protection

(Srivastava et al., 1993), whilst others found partial (Brouillet and Beal, 1993) or no protection (Finiels-Marlier et al., 1993). Recently, (2S, 2'R, 3'R)-2-(2'3'-dicarboxycyclopropyl)glycine, a selective agonist for group II metabotropic glutamate receptors protects dopaminergic nerve terminals against MPP^+ toxicity (Matarredona et al., 2001). This protective effect is dependent on protein synthesis because cycloheximide abolishes the neuroprotective effect mediated by (2S, 2'R, 3'R)-2-(2'3'-dicarboxycyclopropyl)glycine. Based on these studies, in a MPTP-induced animal model of PD, mitochondrial dysfunction may lead to secondary excitotoxic neuronal damage (Blandini et al., 1996).

It is well known that $cPLA_2$ and COX-2 activities are closely associated with neuroinflammation and oxidative stress in neurological disorders and $cPLA_2$ and COX-2-deficient mice are resistant to MPTP neurotoxicity (Klivenyi et al., 1998; Feng et al., 2003; Teismann et al., 2003). MPP^+-mediated toxicity stimulates $cPLA_2$ activity with release of arachidonic acid in the GH3 cell culture model of PD (Yoshinaga et al., 2000). Arachidonoyl trifluoromethylketone blocks this release of arachidonic acid supporting the involvement of $cPLA_2$ in this PD model. Similarly, MPTP stimulates COX-2 activity and COX-2 inhibitors prevent the neurotoxicity mediated by MPTP (Feng et al., 2003; Patrignani et al., 2005).

In the glutathione (GSH) depletion cell culture model of PD (Bharath et al., 2002), a decrease in GSH triggers the activation of neuronal 12-LOX, resulting in the production of 12-LOX generated products and peroxides (Li et al., 1997). 12-LOX inhibitors block neuronal cell death induced by GSH depletion in cell cultures. This suggests that a LOX pathway may also promote neuronal degeneration in PD. Collective evidence suggests that changes in PLA_2, COX, and LOX, activities, loss of glutathione, and impairment of energy metabolism may contribute to the oxidative stress observed in PD. These changes may not be the primary cause, but secondary processes associated with PD pathogenesis.

8.11 Guam-Type Amyotrophic Lateral Sclerosis/Parkinsonism-Dementia

Amyotrophic lateral sclerosis/parkinsonism-dementia complex (ALS/PDC) is a progressive neurodegenerative disease affecting the indigenous Chamorro population of Guam. Neuropathologically, ALS/PDC is characterized by neuronal loss in the substantia nigra pars compacta with severe widespread neurofibrillary tangles (NFTs) similar to those observed in AD, and is thus considered a tauopathy (Winton et al., 2006; Sebeo et al., 2004). In sharp contrast to AD, levels of the lipid peroxidation product 8, 12-iso-$iPF_{2\alpha}$-VI isoprostane are not elevated in ALS/PDC brains relative to controls (Winton et al., 2006). Collectively these studies indicate that tau pathologies of ALS/ PDC share similarities with AD in certain characteristics but the composite ALS/PDC neuropathology profile of tau, α-synuclein, and 8, 12-iso-$iPF_{2\alpha}$-VI isoprostane that occur in ALS/PDC closely resembles other tauopathies including frontotemporal dementias. It is proposed that ALS/PDC and frontotemporal dementias tauopathies may share common mechanisms of neurodegeneration (Winton et al., 2006).

Important diagnostic indicators of ALS/PDC include rigido-akinetic type Parkinsonism with severe dementia. In ALS/PDC, rigidity is so marked that postural deformities such as a generally flexed posture become prominent along with gait disturbances, hyperreflexia, and spinal muscular atrophy, developing mainly in the distal extremities. The cycad hypothesis for ALS/PDC has recent strong evidence, based on a new understanding of Chamorro food practices, a cyanobacterial origin of beta-methylaminoalanine (BMAA) in cycad tissue, and a possible mechanism of biomagnification of this neurotoxic amino acid in the food chain. BMAA is one of two cycad chemicals with known neurotoxic properties, the other is cycasin, a proven developmental neurotoxin, among the many substances that exist in these highly poisonous plants, the seeds of which are used by Chamorros for food and medicine (Murch et al., 2004; Ince and Codd, 2005). BMAA, a non-protein amino acid toxin, induces increased intracellular Ca^{2+} levels through a glutamate receptor channel in a concentration-dependent manner (Brownson et al., 2002). The BMAA-mediated increase in Ca^{2+} depends not only on extracellular Ca^{2+} concentrations, but also on the presence of bicarbonate ion. Increasing concentrations of sodium bicarbonate potentiate the BMAA-mediated $[Ca^{2+}]i$ elevation. The bicarbonate dependence does not produce an increase in Na^{+} concentration or alkalinization of the buffer. Thus, it's tempting to speculate that the neurotoxicity of BMAA is due to an excitotoxic mechanism, involving elevated intracellular Ca^{2+} levels and HCO_3^- (Brownson et al., 2002). Since BMAA alone does not produce an increase in Ca^{2+} levels, these results suggest the involvement of a product of BMAA and CO_2, namely a beta-carbamate, which has a structure similar to other excitatory amino acids (EAA) such as glutamate; thus, the causative agent for ALS-PDC on Guam and elsewhere may be the beta-carbamate of BMAA (Brownson et al., 2002).

Patients with ALS/PDC also show a moderate loss of choline acetyl transferase activity in the midfrontal and inferior parietal cortex and a severe loss in the superior temporal cortex (Masliah et al., 2001). This deficit is similar to that seen in Alzheimer disease and less severe than in Lewy body disease. Thus, cholinergic deficits in the neocortex may contribute to some of the cognitive alterations in ALS/PDC.

8.12 Glutamate in Multiple Sclerosis (MS)

MS is the most prevalent inflammation-mediated demyelinating disease. Initial symptoms include vertigo, fatigue, optic neuritis, and weakness in the limb extremities. In MS, the immune system attacks the white matter of the brain and spinal cord, leading to disability and/or paralysis. Myelin, oligodendrocytes, and neurons are lost due to the release by immune cells of cytotoxic cytokines, autoantibodies, and toxic amounts of the excitatory neurotransmitter, glutamate. Oligodendroglial cells are highly vulnerable to glutamate-mediated neurotoxicity at various stages, preoligodendroglial cells and mature oligodendroglial cells, of their development in cell culture (McDonald et al., 1998). Glutamate treatment increases $[Ca^{2+}]i$, and this effect can be totally abolished by the non-NMDA receptor antagonist CNQX or by removal of Ca^{2+} from the culture medium (Alberdi et al., 2002). NBQX treatment can prevent oligodendroglial cell death, suggesting that oligodendrocytic cell death is a glutamate receptor-mediated process. Treatment of mice with NBQX sensitizes them to experimental autoimmune encephalomyelitis (EAE), a demyelinating model that mimics many of characteristics of MS. NBQX treatment causes substantial amelioration of the disease, promotes oligodendroglial cell survival, and down-regulates dephosphorylation of neurofilament H, an indicator of axonal damage (Pitt et al., 2000). Despite the clinical differences, treatment with NBQX has no effect on lesion size and does not down-regulate the degree of neuroinflammation. In addition, NBQX does not alter the proliferative activity of antigen-primed T cells in vitro, further indicating a lack of effect on the immune system.

Furthermore, the incubation of oligodendroglial cells with glutamate, followed by exposure to complement, is lethal to oligodendrocytes in vitro and in freshly isolated optic nerves (Alberdi et al., 2006). Complement toxicity is induced by activation of KA receptors but not of AMPA receptors and can be retarded by removing Ca^{2+} from the medium during glutamate priming. Dose-response experiments indicate that sensitization to complement attack is induced by two distinct KA receptor populations. One has a high and the other a low affinity for glutamate (Alberdi et al., 2006). Oligodendrocyte death by complement requires the formation of the membrane attack complex, which in turn increases membrane conductance and induces Ca^{2+} overload and mitochondrial depolarization as well as a rise in the level of ROS. Treatment with Trolox antioxidant and inhibition of poly(ADP-ribose) polymerase-1 protects oligodendrocytes against complement-mediated damage. Glutamate sensitization of oligodendrocytes to complement attack may contribute to white matter damage in acute and chronic neurological disorders (Alberdi et al., 2006). Collective evidence suggests that glutamate-mediated neuroinflammation and oxidative stress may be important mechanisms associated with autoimmune demyelination in MS, and neuroinflammation and oxidative stress can be treated with AMPA/KA antagonists and antioxidants (Pitt et al., 2000; Alberdi et al., 2006).

MRI and PET studies indicate that microglial cell activity in markedly increased around the site of the MS lesions (Banati et al., 2000). Once at the site of the lesion in MS, microglia increase COX-2 (Rose et al., 2004) and iNOS expression (Hill et al., 2004), which are necessary for the generation of PGE_2 and nitric oxide respectively. The production of nitric oxide through iNOS results in the generation of peroxynitrite which has been implicated in the pathogenesis of white matter injury that occurs in MS (Zhang et al., 2006). Treatment of mature oligodendrocyte cultures with 3-morpholinosydnonimine, a peroxynitrite generator, results in toxicity. N, N, N', N'-tetrakis (2-pyridylmethyl) ethylenediamine, a zinc ion chelator (Zhang et al., 2006), completely blocks this toxicity suggesting that peroxynitrite toxicity may be mediated through Zn^{2+}. The peroxynitrite-mediated activation of extracellular signal-regulated kinase 42/44 (ERK42/44), 12-lipoxygenase, and generation of ROS depend upon the intracellular release of Zn^{2+} (Zhang et al., 2006), which is closely involved with excitotoxic injury in cerebral ischemia, epilepsy, and brain trauma (Capasso et al., 2005) (Frederickson et al., 2004). Intracellular Zn^{2+}homeostasis is sensitive to pathophysiological changes such as inflammation and oxidative stress through redox-sensitive transcription factor signaling involving NF- κB, p53, and AP1 (Capasso et al., 2005; Ong and Farooqui, 2005; Hao and Maret, 2005). It is likely that the above mechanism may be associated with oligodendroglial cell injury and death in MS.

8.13 Domoic Acid Neurotoxicity

The phytoplankton-derived neurotoxin, domoic acid (DOM), causes poisoning of marine animals and poses an increasing threat to public health through contamination of seafood (Larm et al., 1997; Berman and Murray, 1997). This toxin characteristically damages the hippocampus in exposed humans, rodents, and marine mammals. Thus at 1 mM DOM produces both ipsilateral and contralateral neuronal cell death in CA1, CA3, as well as dentate gyrus subfields. Animal behavioral changes after microinjection are similar to those observed in previous studies through systemic DOM injection. Neurodegeneration is paralleled by reduced glutamate receptor NR1, GluR1, and GluR6/7 immunolabeling throughout the whole hippocampal formation. Pre-injections of the AMPA/KA receptor antagonist, NBQX, retard DOM-mediated neurodegeneration as well as behavioral symptoms (Qiu et al., 2006; Qiu and Curras-Collazo, 2006). At low DOM concentration neither histopathological changes are observed microscopically, nor are there differences in the level of immunostaining of NR1, GluR1, GluR6/7. However, increased immunolabeling of autophosphorylated Ca^{2+}-calmodulin-dependent kinase II (CaMKII, p-Thr286) and phosphorylated cAMP response element binding protein (CREB, p-Ser133) were observed at 24 h post-injection, suggesting that the altered intracellular signal transduction mediated by GluRs may be an adaptive cellular protective mechanism against DOM-induced neurotoxicity (Qiu et al., 2006; Qiu and Curras-Collazo, 2006). In vitro, 24-h exposure of neuronal cultures to DOM causes substantial concentration-dependent neuronal cell death. The CNQX and AMPA receptor-selective antagonist LY293558 ((3S,4aR,6R,8aR)-6-[2-(1H-tetrazol-5-yl)-ethyl]-1,2,3,4,4a,5,6,7,8,8a-decahydroisoquinoline-3-carboxylic acid) can attenuate this cell death, but it is not affected by NS-102 (5-nitro-6,7,8,9-tetrahydrobenzo[g]indole-2, 3-dione-3-oxime), a low-affinity kainate receptor antagonist (Larm et al., 1997). Collective evidence suggests that DOM acts via high-affinity AMPA- and kainate-sensitive glutamate receptors to produce excitotoxic cell death (Larm et al., 1997; Berman and Murray, 1997; Qiu et al., 2006; Qiu and Curras-Collazo, 2006). DOM-mediated neurodegeneration can also be prevented by melatonin, which has potential for the treatment of pathologies associated with DOM-mediated brain damage (Chandrasekaran et al., 2004).

Although the mechanism underlying the neuronal damage induced by DOM is not yet fully understood, recent reverse transcription-polymerase chain reaction analyses show the upregulation of nNOS expression and down-regulation of NMDAR1 expression at 5 days after DOM administration (Ananth et al., 2003a,c; Chandrasekaran et al., 2004). Although nNOS mRNA expression is rapidly induced at 5 days after DOM administration, in situ hybridization analysis reveals the complete loss of nNOS mRNA expression in the region of neuronal degeneration in the hippocampus at 24 h and 5 days after DOM administration. This suggests that nNOS may be involved in DOM-mediated neurotoxicity.

In neuronal cultures, DOM also produces oxidative stress through its affects on cellular glutathione metabolism. Thus, cerebellar granule neurons (CGN) from mice lacking the modifier subunit of glutamate-cysteine ligase [Gclm (-/-)] have very low levels of GSH and are 10-fold more sensitive to DOM-mediated neurotoxicity than CGN from Gclm (+/+) mice. GSH ethyl ester down-regulates, whereas the glutamate-cysteine ligase (Gcl) inhibitor buthionine sulfoximine up-regulates the DOM-mediated neurotoxicity. Antagonists of AMPA/KA receptors and NMDA receptors retard DOM-mediated neurotoxicity (Walser et al., 2006). Ca^{2+} chelators and antioxidants prevent DOM-mediated neurotoxicity.

DOM treatment also rapidly decreases cellular GSH, which precedes neurotoxicity. This decrease is primarily due to DOM-mediated GSH efflux. DOM also induces an increase in oxidative stress as indicated by increases in ROS and lipid peroxidation products, which follow GSH efflux. Astrocytes from both genotypes are resistant to DOM-mediated neurotoxicity and present a diminished Ca^{2+} response to DOM-mediated toxicity (Walser et al., 2006). Exposure of neonatal rat microglia to DOM triggers the release of TNF-α and matrix metalloproteinase-9 (MMP-9) (Mayer et al., 2001). These molecules are involved in the modulation of neuroinflammation in brain (Farooqui et al., 2007). Collective evidence suggests that DOM-mediated neurodegeneration involves changes in cellular redox, oxidative stress, and increased expression of cytokines, nitric oxide synthase, NADPH diaphorase, and matrix metalloproteinase-9 (Walser et al., 2006; Chandrasekaran et al., 2004; Ananth et al., 2003a,b; Mayer et al., 2001).

8.14 Glutamate in Schizophrenia

Schizophrenia is a complex neuropsychiatric disorder with a prevalence of nearly 1% of the world population. Clinically, schizophrenia is characterized by chronic psychotic symptoms and psychosocial impairment. At least two core syndromes of schizophrenia, the positive and the negative, are recognized (Liddle, 1987). Those in whom positive symptoms predominate have hallucinations, delusions, and paranoid ideation. Those with negative symptoms have apathy and anledonia. Besides impairment in the cognitive domain, individuals with schizophrenia show a deficit in emotion processing, as indicated by a markedly reduced ability to perceive, process, and express facial emotions (Aleman et al., 1999).

Although schizophrenia is a disorder of dopaminergic neurotransmission, modulation of the dopaminergic system by glutamatergic neurotransmission plays a key role in the pathogenesis of schizophrenia (Muller and Schwarz, 2006). Recent studies on the interaction between dopaminergic and glutamatergic neurotransmitter systems in prefrontal cortex indicate that application of the D4 receptor agonist, PD168077, produces a reversible decrease in the NMDA receptor-mediated current and this effect can be prevented by selective D4 receptor antagonists (Wang et al., 2003). Modulation of NMDA receptors in prefrontal cortex by dopamine D4 receptors involves the inhibition of protein kinase A, activation of protein phosphatase 1, and the ensuing inhibition of active Ca^{2+}-calmodulin-dependent kinase II (CaMKII). Moreover, PD168077 not only down-regulates the surface expression of NMDAR, but also triggers the internalization of NMDAR in a manner dependent on CaMKII activity. These results identify a mechanistic link between D4 dopamine receptors and NMDA glutamate receptors in prefrontal cortex pyramidal neurons (Wang et al., 2003; Tseng and O'Donnell, 2004).

These results provide a strong support for the view that interactions between dopaminergic and glutamatergic systems in prefrontal cortex (PFC) play an important role in normal mental function and may be involved in the pathophysiology of schizophrenia (Wang et al., 2003). Suggested modulation of the dopaminergic system by glutamatergic neurotransmission in the pathophysiology of schizophrenia is also supported by genetic findings on the neuregulin and dysbindin genes. These genes have a functional impact on the glutamatergic system. Hypofunction of the glutamatergic system is induced by NMDA receptor antagonism (MacDonald and Chafee, 2006). Kynurenic acid, the endogenous antagonist, inhibits NMDA receptors and also blocks the nicotinergic acetylcholine receptor (Muller and Schwarz, 2006). Thus an increase in kynurenic acid causes cognitive deterioration due to the inhibition of NMDA receptor and also produces psychotic symptoms because the nicotinergic acetylcholine receptor is blocked. Levels of kynurenic acid are higher in CSF preparations of schizophrenic patients compared to CSF from age-matched control subjects (Muller and Schwarz, 2006; MacDonald and Chafee, 2006).

The association of glutamate receptors with the pathophysiology of schizophrenia is also supported by the findings that phencyclidine (PCP), the non-competitive NMDA receptor antagonist, exacerbates psychotomimetic symptoms in humans that resemble schizophrenia, whilst treatment of experimental animals with PCP or MK-801 mimics the action of the dopaminergic agent amphetamine (Kerwin et al., 1990; Ulas and Cotman, 1993; Olney and Farber, 1995). Although pharmacological data implicate the NMDA glutamate receptors in schizophrenia, abnormalities in AMPA and KA receptor expression were also reported (Akbarian et al., 1995; Meador-Woodruff et al., 2001). Alterations in AMPA receptor editing of GluR2 mRNA occur in the prefrontal cortex of schizophrenic patients (Akbarian et al., 1995). Altered GluR2 editing may produce excessive Ca^{2+} influx and inhibition of neurite outgrowth associated with the pathophysiology of schizophrenia. Decreased kainate receptor binding is observed in brain tissue from subjects with schizophrenia, but is restricted to infragranular laminae of the prefrontal cortex (Meador-Woodruff et al., 2001). No differences in kainate receptor binding or subunit mRNA levels are found in striatum or occipital cortex, suggesting that these findings may be restricted to prefrontal cortex. Abnormalities in AMPA and KA receptors may potentially lead to NMDA receptor hypoactivity in schizophrenic patients. Collective evidence suggests that in schizophrenia, dopaminergic neurotransmission and many glutamate-regulated processes are perturbed (Konradi and Heckers, 2003). These events may be closely associated with the pathophysiology of schizophrenia.

It is well known that ROS cause cell injury when they are generated in excess or the antioxidant defense is impaired. Both processes are affected in schizophrenia

(Mahadik and Mukherjee, 1996; Yao et al., 2001). Excess ROS and other lipid peroxidation products were observed in blood and CSF from schizophrenic patients (Mahadik and Scheffer, 1996). ROS within cell membranes can destabilize neural structure, alter membrane fluidity and permeability, and impair signal transduction. Recent findings suggest that multiple neurotransmitter systems become faulty due to a high level of catecholamine-mediated oxidative stress. Collectively these studies suggest that ROS-mediated impairment of neuroplasticity and abnormal glutamate and dopamine receptor function can lead to impaired dopamine-glutamate balance. These processes may be associated with the pathophysiology of schizophrenia (Fendri et al., 2006; Muller and Schwarz, 2006).

8.15 Mechanism of Glutamate-Mediated Neural Cell Injury in Neurological Disorders

Ca^{2+} influx mediates excitotoxicity. A massive increase in intracellular Ca^{2+} has numerous consequences in addition to ATP depletion. They include: activation of Ca^{2+}-dependent enzymes, activation of second messenger and changes in gene expression, and increased generation of reactive oxygen species (ROS), which in turn interfere with the generation of ATP (Love, 1999). Potential targets of glutamate and Ca^{2+}-mediated toxicity include PLA_2, PLC, PLD, proteases, protein kinases, protein phosphatases, nitric oxide synthases, endonucleases (Farooqui and Horrocks, 1991, 1994; Farooqui et al., 2001; Wang et al., 2005), cyclooxygenase-2 (COX-2), lipoxygenases (LOX), epoxygenases (EPOX) (Phillis et al., 2006) and mitochondrial dysfunction. Accumulation of oxygenated arachidonic acid metabolites along with abnormal homeostasis and lack of energy generation is associated with neural cell injury and cell death in acute neural trauma, ischemia, epilepsy, head injury, and spinal cord trauma, and in neurodegenerative diseases such as Alzheimer disease (AD), Parkinson disease (PD), multiple sclerosis (MS), and prion diseases like Creutzfeldt-Jakob disease (CJD).

As stated earlier, glutamate can damage glial cells by mechanisms that do not involve glutamate receptor activation, but rather glutamate uptake through glutamate transporters (Oka et al., 1993; Matute et al., 2006). Five distinct glutamate transporters have been cloned. All glutamate transporters except EAAT5 are distributed in the molecular layer in the cerebellum, especially near the excitatory synapses. Exposure of astroglial, oligodendroglial, and microglial cell cultures to glutamate causes glial cell demise by a transporter-related mechanism involving the inhibition of cystine uptake. This decreases glutathione and makes glial cells vulnerable to toxic free radicals (Oka et al., 1993; Matute et al., 2006).

Prostaglandins and leukotrienes, the oxygenated products of arachidonic acid generated through COX and LOX action (Phillis et al., 2006), modulate glutamate receptors in the hippocampus (Chabot et al., 1998) through cross talk among glutamate, prostaglandin, leukotriene, and thromboxane receptors. Under normal conditions, this cross talk refines their communication, but under pathological situations, it promotes neuronal injury that depends on the magnitude of PLA_2, COX, and LOX expression, production of arachidonic acid metabolites, and generation of ROS. Increased activities of PLA_2 isoforms and high levels of their reaction products are likely involved in extending excitotoxicity and oxidative stress during neurodegeneration (Farooqui and Horrocks, 2006). In addition, the indirect mechanism of arachidonic acid release involving the PLC/diacylglycerol pathway may also participate in prolonging excitotoxicity and oxidative stress in glutamate-mediated neurotoxicity (Farooqui et al., 1993).

8.16 Conclusion

Glutamate is a major excitatory transmitter in mammalian CNS. It acts on NMDA, KA, AMPA, and metabotropic types of glutamate receptors. Abnormalities in glutamate neurotransmitter systems are not only involved in acute neural trauma such as ischemia, spinal cord injury, head trauma, and epilepsy, but also in neurodegenerative disorders such as AD, PD, HD, ALS, AIDS complex, and domoic acid neurotoxicity. Glutamate toxicity is accompanied by a rapid and sustained increase in intracellular Ca^{2+} that stimulates and regulates phospholipases, proteases, protein kinases, and endonucleases. Under normal conditions of intermittent trans-synaptic stimulation of glutamate receptors, the resulting increased intracellular free Ca^{2+} is rapidly reversed by the intervention of several mechanisms that regulate intracellular Ca^{2+} homeostasis. However, under pathological situations, the resulting increased extracellular glutamate levels elicit a constant stimulation of the glutamate receptors. This results in a persistent increase in intracellular free Ca^{2+}.This may cause abnormal metabolism of neural membrane glycerophospholipids, with generation of high levels of free fatty acids, eicosanoids, lipid peroxides, and ROS.

These metabolites, along with abnormal ion homeostasis, altered redox status, and lack of energy generation, may be responsible for neural cell death in acute neural trauma and neurodegenerative diseases. Glutamate-induced alterations in Ca^{2+} and enhanced neural membrane glycerophospholipid metabolism may be a common mechanism involved in the cellular response in acute neural trauma and neurodegenerative disease. It remains controversial whether these processes are the cause or consequence of neuroinflammation and oxidative stress-mediated neurodegeneration (Andersen, 2004; Juranek and Bezek, 2005). The fact that high levels of glutamate promote neural injury in a variety of neurological disorders suggests that excitotoxicity, neuroinflammation and oxidative stress can be used to monitor neurodegenerative processes in human subjects. This information can be valuable for developing diagnostic tests and measuring therapeutic responses.

Our emphasis on the interplay among excitotoxicity, oxidative stress, and neuroinflammation does not rule out the participation of other mechanisms involved in neural cell injury. However, it is timely and appropriate to apply the concept of interplay among excitotoxicity, oxidative stress, and neuroinflammation to neural cell injury in acute neural trauma and neurodegenerative diseases. Lack of coordination among the above parameters may control the time taken by neural cells to die. For example, in ischemic, traumatic brain, and spinal cord injuries, neuronal cell death occurs within hours to days, whereas in neurodegenerative diseases, neuronal damage takes years to develop. For ischemia and traumatic brain and spinal cord injuries, there is a therapeutic window of 4–6 h (Leker and Shohami, 2002) for partial restoration of many body functions. In contrast, there is no therapeutic window for neurodegenerative diseases and drug intervention may be appropriate as soon as the first symptoms of the disease appear.

References

Ahn K. S. and Aggarwal B. B. (2005). Transcription factor NF-κB: a sensor for smoke and stress signals. Ann. N. Y Acad. Sci. 1056:218–233.

Ahn M. J., Sherwood E. R., Prough D. S., Lin C. Y., and DeWitt D. S. (2004). The effects of traumatic brain injury on cerebral blood flow and brain tissue nitric oxide levels and cytokine expression. J. Neurotrauma 21:1431–1442.

Akbarian S., Smith M. A., and Jones E. G. (1995). Editing for an AMPA receptor subunit RNA in prefrontal cortex and striatum in Alzheimer's disease, Huntington's disease and schizophrenia. Brain Res. 699:297–304.

Albensi B. C. (2001). Potential roles for tumor necrosis factor and nuclear factor-κB in seizure activity. J. Neurosci. Res. 66:151–154.

Alberdi E., Sánchez-Gómez M. V., Marino A., and Matute C. (2002). Ca^{2+} influx through AMPA or kainate receptors alone is sufficient to initiate excitotoxicity in cultured oligodendrocytes. Neurobiol. Dis. 9:234–243.

Alberdi E., Sánchez-Gómez M. V., Torre I., Domercq M., Pérez-Samartín A., Pérez-Cerdá F., and Matute C. (2006). Activation of kainate receptors sensitizes oligodendrocytes to complement attack. J. Neurosci. 26:3220–3228.

Aleman A., Hijman R., de Haan E. H., and Kahn R. S. (1999). Memory impairment in schizophrenia: a meta-analysis. Am. J. Psychiatry 156:1358–1366.

Almer G., Guegan C., Teismann P., Naini A., Rosoklija G., Hays A. P., Chen C. P., and Przedborski S. (2001). Increased expression of the pro-inflammatory enzyme cyclooxygenase-2 in amyotrophic lateral sclerosis. Ann. Neurol. 49:176–185.

Álvarez S., Serramía M. J., Fresno M., and Muñoz-Fernández M. A. (2005). Human immunodeficiency virus type 1 envelope glycoprotein 120 induces cyclooxygenase-2 expression in neuroblastoma cells through a nuclear factor-κB and activating protein-1 mediated mechanism. J. Neurochem. 94:850–861.

Ananth C., Dheen S. T., Gopalakrishnakone P., and Kaur C. (2003a). Distribution of NADPH-diaphorase and expression of nNOS, N-methyl-D-aspartate receptor (NMDAR1) and non-NMDA glutamate receptor (GlutR2) genes in the neurons of the hippocampus after domoic acid-induced lesions in adult rats. Hippocampus 13:260–272.

Ananth C., Gopalakrishnakone P., and Kaur C. (2003b). Induction of inducible nitric oxide synthase expression in activated microglia following domoic acid (DA)-induced neurotoxicity in the rat hippocampus. Neurosci. Lett. 338:49–52.

Ananth C., Gopalakrishnakone P., and Kaur C. (2003c). Protective role of melatonin in domoic acid-induced neuronal damage in the hippocampus of adult rats. Hippocampus 13: 375–387.

Andersen J. K. (2004). Oxidative stress in neurodegeneration: cause or consequence? Nature Rev. Neurosci.S18-S25.

Arlt S., Beisiegel U., and Kontush A. (2002). Lipid peroxidation in neurodegeneration: new insights into Alzheimer's disease. Curr. Opin. Lipidol. 13:289–294.

Arundine M. and Tymianski M. (2004). Molecular mechanisms of glutamate-dependent neurodegeneration in ischemia and traumatic brain injury. Cell Mol. Life Sci. 61:657–668.

Auer R. N. and Siesjö B. K. (1988). Biological differences between ischemia, hypoglycemia, and epilepsy. Ann. Neurol. 24:699–707.

Auger C. and Attwell D. (2000). Fast removal of synaptic glutamate by postsynaptic transporters. Neuron 28:547–558.

Bacman S. R., Bradley W. G., and Moraes C. T. (2006). Mitochondrial involvement in amyotrophic lateral sclerosis - Trigger or target? Mol. Neurobiol. 33:113–131.

Banati R. B., Newcombe J., Gunn R. N., Cagnin A., Turkheimer F., Heppner F., Price G., Wegner F., Giovannoni G., Miller D. H., Perkin G. D., Smith T., Hewson A. K., Bydder G., Kreutzberg G. W., Jones T., Cuzner M. L., and Myers R. (2000). The peripheral benzodiazepine binding site in the brain in multiple sclerosis: quantitative *in vivo* imaging of microglia as a measure of disease activity. Brain 123 (Pt 11):2321–2337.

Bate C. and Williams A. (2004). Role of glycosylphosphatidylinositols in the activation of phospholipase A_2 and the neurotoxicity of prions. J. Gen. Virol. 85:3797–3804.

Bazan N. G., Rodriguez de Turco E. B., and Allan G. (1995). Mediators of injury in neurotrauma: intracellular signal transduction and gene expression. J. Neurotrauma 12:791–814.

Bazan N. G., Tu B., and Rodriguez de Turco E. B. (2002). What synaptic lipid signaling tells us about seizure-induced damage and epileptogenesis. In: Sutula T. and Pitkanen A. (eds.), *Do Seizures Damage the Brain.* Perspectives in Analytical Philosophy Elsevier Science BV, Amsterdam, pp. 175–185.

Beal M. F. (1998a). Excitotoxicity and nitric oxide in Parkinson's disease pathogenesis. Ann. Neurol. 44:S110-S114.

Beal M. F. (1998b). Mitochondrial dysfunction in neurodegenerative diseases. Biochim. Biophys. Acta Bioenergetics 1366:211–223.

Beal M. F. (2003). Mitochondria, oxidative damage, and inflammation in Parkinson's disease. Ann. N. Y Acad. Sci. 991:120–131.

Beattie M. S., Farooqui A. A., and Bresnahan J. C. (2000). Review of current evidence for apoptosis after spinal cord injury. J. Neurotrauma 17:915–925.

Benveniste H., Jorgensen M. B., Sandberg M., Christensen T., Hagberg H., and Diemer N. H. (1989). Ischemic damage in hippocampal CA1 is dependent on glutamate release and intact innervation from CA3. J. Cereb. Blood Flow Metab. 9:629–639.

Berman F. W. and Murray T. F. (1997). Domoic acid neurotoxicity in cultured cerebellar granule neurons is mediated predominantly by NMDA receptors that are activated as a consequence of excitatory amino acid release. J. Neurochem. 69:693–703.

Bernard J., Lahsaini A., and Massicotte G. (1994). Potassium-induced long-term potentiation in area CA1 of the hippocampus involves phospholipase activation. Hippocampus 4:447–453.

Bethea J. R., Castro M., Keane R. W., Lee T. T., Dietrich W. D., and Yezierski R. P. (1998). Traumatic spinal cord injury induces nuclear factor-κB activation. J. Neurosci. 18: 3251–3260.

Bharath S., Hsu M., Kaur D., Rajagopalan S., and Andersen J. K. (2002). Glutathione, iron and Parkinson's disease. Biochem. Pharmacol. 64:1037–1048.

Bi H. and Sze C. I. (2002). *N*-methyl-D-aspartate receptor subunit NR2A and NR2B messenger RNA levels are altered in the hippocampus and entorhinal cortex in Alzheimer's disease. J. Neurol. Sci. 200:11–18.

Blandini F., Porter R. H., and Greenamyre J. T. (1996). Glutamate and Parkinson's disease. Mol. Neurobiol. 12:73–94.

Bliss T. V. P. and Collingridge G. L. (1993). A synaptic model of memory: long-term potentiation in the hippocampus. Nature 361:31–39.

Bonilla E. (2000). [Huntington disease. A review]. Invest. Clin. 41:117–141.

Boven L. A., Gomes L., Hery C., Gray F., Verhoef J., Portegies P., Tardieu M., and Nottet H. S. L. M. (1999). Increased peroxynitrite activity in AIDS dementia complex: implications for the neuropathogenesis of HIV-1 infection. J. Immunol. 162:4319–4327.

Bradford H. F. and Dodd P. R. (1975). Convulsions and activation of epileptic foci induced by monosodium glutamate and related compounds. Biochem. Pharmacol. 26:253–254.

Bramlett H. M. and Dietrich W. D. (2004). Pathophysiology of cerebral ischemia and brain trauma: Similarities and differences. J. Cereb. Blood Flow Metab. 24:133–150.

Brouillet E. and Beal M. F. (1993). NMDA antagonists partially protect against MPTP induced neurotoxicity in mice. NeuroReport 4:387–390.

Brown D. R. (1999). Prion protein peptide neurotoxicity can be mediated by astrocytes. J. Neurochem. 73:1105–1113.

Brown D. R. (2005). Neurodegeneration and oxidative stress: prion disease results from loss of antioxidant defence. Folia Neuropathol. 43:229–243.

Browne S. E., Bowling A. C., MacGarvey U., Baik M. J., Berger S. C., Muqit M. M., Bird E. D., and Beal M. F. (1997). Oxidative damage and metabolic dysfunction in Huntington's disease: selective vulnerability of the basal ganglia. Ann. Neurol. 41:646–653.

Brownson D. M., Mabry T. J., and Leslie S. W. (2002). The cycad neurotoxic amino acid, *β-N*-methylamino-l-alanine (BMAA), elevates intracellular calcium levels in dissociated rat brain cells. J. Ethnopharmacol. 82:159–167.

Bruyn R. P. M. and Stoof J. C. (1990). The quinolinic acid hypothesis in Huntington's chorea. J. Neurol. Sci. 95:29–38.

Bullock R., Zauner A., Myseros J. S., Marmarou A., Woodward J. J., and Young H. F. (1995). Evidence for prolonged release of excitatory amino acids in severe human head trauma-Relationship to clinical events. Ann. N. Y. Acad. Sci. 765:290–297.

Burke J. R., Enghild J. J., Martin M. E., Jou Y. S., Myers R. M., Roses A. D., Vance J. M., and Strittmatter W. J. (1996). Huntingtin and DRPLA proteins selectively interact with the enzyme GAPDH. Nat. Med. 2:347–350.

Burnashev N., Monyer H., Seeburg P. H., and Sakmann B. (1992). Divalent ion permeability of AMPA receptor channels is dominated by the edited form of a single subunit. Neuron 8: 189–198.

Butterfield D. A. (2002). Amyloid beta-peptide (1-42)-induced oxidative stress and neurotoxicity: Implications for neurodegeneration in Alzheimer's disease brain. A review. Free Radical Res. 36:1307–1313.

Calabrese V., Boyd-Kimball D., Scapagnini G., and Butterfield D. A. (2004). Nitric oxide and cellular stress response in brain aging and neurodegenerative disorders: the role of vitagenes. *In Vivo* 18:245–267.

Capasso M., Jeng J. M., Malavolta M., Mocchegiani E., and Sensi S. L. (2005). Zinc dyshomeostasis: A key modulator of neuronal injury. J. Alzheimer's Dis. 8:93–108.

Capellari S., Zaidi S. I. A., Urig C. B., Perry G., Smith M. A., and Petersen R. B. (1999). Prion protein glycosylation is sensitive to redox change. J. Biol. Chem. 274:34846–34850.

Carboni S., Melis F., Pani L., Hadjiconstantinou M., and Rossetti Z. L. (1990). The non-competitive NMDA-receptor antagonist MK-801 prevents the massive release of glutamate and aspartate from rat striatum induced by 1-methyl-4-phenylpyridinium (MPP^+). Neurosci. Lett. 117: 129–133.

Cepeda C., Ariano M. A., Calvert C. R., Flores-Hernandez J., Chandler S. H., Leavitt B. R., Hayden M. R., and Levine M. S. (2001). NMDA receptor function in mouse models of Huntington disease. J. Neurosci. Res. 66:525–539.

Chabot C., Gagné J., Giguère C., Bernard J., Baudry M., and Massicotte G. (1998). Bidirectional modulation of AMPA receptor properties by exogenous phospholipase A_2 in the hippocampus. Hippocampus 8:299–309.

Chandrasekaran A., Ponnambalam G., and Kaur C. (2004). Domoic acid-induced neurotoxicity in the hippocampus of adult rats. Neurotox. Res. 6:105–117.

Chen C. and Tonegawa S. (1997). Molecular genetic analysis of synaptic plasticity, activity-dependent neural development, learning, and memory in the mammalian brain. Annu. Rev. Neurosci. 20:157–184.

Chiba K., Trevor A., and Castagnoli N., Jr. (1984). Metabolism of the neurotoxic tertiary amine, MPTP, by brain monoamine oxidase. Biochem. Biophys. Res. Commun. 120: 574–578.

Chin J. H., Buckholz T. M., and DeLorenzo R. J. (1985). Calmodulin and protein phosphorylation: implications in brain ischemia. Prog. Brain Res. 63:169–184.

Chin P. C., Liu L., Morrison B. E., Siddiq A., Ratan R. R., Bottiglieri T., and D'Mello S. R. (2004). The c-Raf inhibitor GW5074 provides neuroprotection in vitro and in an animal model of

neurodegeneration through a MEK-ERK and Akt- independent mechanism. J. Neurochem. 90:595–608.
Choi D. W. (1988). Glutamate neurotoxicity and diseases of the nervous system. Neuron 1: 628–634.
Choi D. W. (1990). Cerebral hypoxia: Some new approaches and unanswered questions. J. Neurosci. 10:2493–2501.
Chung S. Y. and Han S. H. (2003). Melatonin attenuates kainic acid-induced hippocampal neurodegeneration and oxidative stress through microglial inhibition. J. Pineal Res. 34:95–102.
Chung H. J., Steinberg J. P., Huganir R. L., and Linden D. J. (2003). Requirement of AMPA receptor GluR2 phosphorylation for cerebellar long-term depression. Science 300:1751–1755.
Collister K. A. and Albensi B. C. (2005). Potential therapeutic targets in the NF-κB pathway for Alzheimer's disease. Drug News Perspect. 18:623–629.
Copani A., Canonico P. L., Catania M. V., Aronica E., Bruno V., Ratti E., Van Amsterdam F. T. M., Gaviraghi G., and Nicoletti F. (1991). Interaction between β-N-methylamino-L-alanine and excitatory amino acid receptors in brain slices and neuronal cultures. Brain Res. 558:79–86.
Crocker J. F., Lee S. H., Love J. A., Malatjalian D. A., Renton K. W., Rozee K. R., and Murphy M. G. (1991). Surfactant-potentiated increases in intracranial pressure in a mouse model of Reye's syndrome. Exp. Neurol. 111:95–97.
De Bock F., Dornand J., and Rondouin G. (1996). Release of TNFalpha in the rat hippocampus following epileptic seizures and excitotoxic neuronal damage. NeuroReport 7:1125–1129.
Deininger M. H., Bekure-Nemariam K., Trautmann K., Morgalla M., Meyermann R., and Schluesener H. J. (2003). Cyclooxygenase-1 and -2 in brains of patients who died with sporadic Creutzfeldt-Jakob disease. J. Mol. Neurosci. 20:25–30.
Demediuk P., Daly M. P., and Faden A. I. (1988). Free amino acid levels in laminectomized and traumatized rat spinal cord. Trans. Am. Soc. Neurochem. 19:176.
Dewar D., Chalmers D. T., Shand A., Graham D. I., and McCulloch J. (1990). Selective reduction of quisqualate (AMPA) receptors in Alzheimer cerebellum. Ann. Neurol. 28:805–810.
Dewar D., Chalmers D. T., Graham D. I., and McCulloch J. (1991). Glutamate metabotropic and AMPA binding sites are reduced in Alzheimer's disease: an autoradiographic study of the hippocampus. Brain Res. 553:58–64.
Dhillon H. S., Donaldson D., Dempsey R. J., and Prasad M. R. (1994). Regional levels of free fatty acids and Evans blue extravasation after experimental brain injury. J. Neurotrauma 11: 405–415.
Doble A. (1999). The role of excitotoxicity in neurodegenerative disease: implications for therapy. Pharmacol. Ther. 81:163–221.
Doh-ura K., Iwaki T., and Caughey B. (2000). Lysosomotropic agents and cysteine protease inhibitors inhibit scrapie-associated prion protein accumulation. J. Virol. 74:4894–4897.
Drachman D. B. and Rothstein J. D. (2000). Inhibition of cyclooxygenase-2 protects motor neurons in an organotypic model of amyotrophic lateral sclerosis. Ann. Neurol. 48:792–795.
Drachman D. B., Frank K., Dykes-Hoberg M., Teismann P., Almer G., Przedborski S., and Rothstein J. D. (2002). Cyclooxygenase 2 inhibition protects motor neurons and prolongs survival in a transgenic mouse model of ALS. Ann. Neurol. 52:771–778.
Dreyer E. B. and Lipton S. A. (1995). The coat protein gp120 of HIV-1 inhibits astrocyte uptake of excitatory amino acids via macrophage arachidonic acid. Eur. J. Neurosci. 7: 2502–2507.
Edgar A. D., Strosznajder J., and Horrocks L. A. (1982). Activation of ethanolamine phospholipase A_2 in brain during ischemia. J. Neurochem. 39:1111–1116.
Ellis R. C., Earnhardt J. N., Hayes R. L., Wang K. K. W., and Anderson D. K. (2004). Cathepsin B mRNA and protein expression following contusion spinal cord injury in rats. J. Neurochem. 88:689–697.
Faden A. I. and Simon R. P. (1988). A potential role for excitotoxins in the pathophysiology of spinal cord injury. Ann. Neurol. 23:623–626.
Famularo G., Moretti S., Alesse E., Trinchieri V., Angelucci A., Santini G., Cifone G., and De Simone C. (1999). Reduction of glutamate levels in HIV-infected subjects treated with acetylcarnitine. J. NeuroAIDS 2:65–73.

Fan M. M. and Raymond L. A. (2007). N-Methyl-d-aspartate (NMDA) receptor function and excitotoxicity in Huntington's disease. Prog. Neurobiol. 81:272–293.

Farooqui A. A. and Horrocks L. A. (1991). Excitatory amino acid receptors, neural membrane phospholipid metabolism and neurological disorders. Brain Res. Rev. 16:171–191.

Farooqui A. A. and Horrocks L. A. (1994). Excitotoxicity and neurological disorders: involvement of membrane phospholipids. Int. Rev. Neurobiol. 36:267–323.

Farooqui A. A. and Horrocks L. A. (2006). Phospholipase A_2-generated lipid mediators in the brain: the good, the bad, and the ugly. Neuroscientist 12:245–260.

Farooqui A. A. and Horrocks L. A. (2007a). Glutamate and cytokine-mediated alterations of phospholipids in head injury and spinal cord trauma. In: Banik N. (ed.), *Brain and Spinal Cord Trauma.* Handbook of Neurochemistry. Springer, New York, in press. Lajtha, A. (ed.).

Farooqui A. A. and Horrocks L. A. (2007b). *Glycerophospholipids in the Brain: Phospholipases A_2 in Neurological Disorders*, pp. 1–394. Springer, New York.

Farooqui A. A., Anderson D. K., and Horrocks L. A. (1993). Effect of glutamate and its analogs on diacylglycerol and monoacylglycerol lipase activities of neuron-enriched cultures. Brain Res. 604:180–184.

Farooqui A. A., Yang H.-C., and Horrocks L. A. (1997). Involvement of phospholipase A_2 in neurodegeneration. Neurochem. Int. 30:517–522.

Farooqui A. A., Ong W. Y., Horrocks L. A., and Farooqui T. (2000). Brain cytosolic phospholipase A_2: Localization, role, and involvement in neurological diseases. Neuroscientist 6: 169–180.

Farooqui A. A., Ong W. Y., Lu X. R., Halliwell B., and Horrocks L. A. (2001). Neurochemical consequences of kainate-induced toxicity in brain: involvement of arachidonic acid release and prevention of toxicity by phospholipase A_2 inhibitors. Brain Res. Rev. 38:61–78.

Farooqui A. A., Ong W. Y., Lu X. R., and Horrocks L. A. (2002). Cytosolic phospholipase A_2 inhibitors as therapeutic agents for neural cell injury. Curr. Med. Chem.-Anti-Inflammatory & Anti-Allergy Agents 1:193–204.

Farooqui A. A., Ong W. Y., and Horrocks L. A. (2003a). Plasmalogens, docosahexaenoic acid, and neurological disorders. In: Roels F., Baes M., and de Bies S. (eds.), *Peroxisomal Disorders and Regulation of Genes.* Kluwer Academic/Plenum Publishers, London, pp. 335–354.

Farooqui A. A., Ong W. Y., and Horrocks L. A. (2003b). Stimulation of lipases and phospholipases in Alzheimer disease. In: Szuhaj B. and van Nieuwenhuyzen W. (eds.), *Nutrition and Biochemistry of Phospholipids.* AOCS Press, Champaign, pp. 14–29.

Farooqui A. A., Ong W. Y., and Horrocks L. A. (2004). Biochemical aspects of neurodegeneration in human brain: involvement of neural membrane phospholipids and phospholipases A_2. Neurochem. Res. 29:1961–1977.

Farooqui A. A., Ong W. Y., and Horrocks L. A. (2006). Inhibitors of brain phospholipase A_2 activity: Their neuropharmacological effects and therapeutic importance for the treatment of neurologic disorders. Pharmacol. Rev. 58:591–620.

Farooqui A. A., Horrocks L. A., and Farooqui T. (2007). Modulation of inflammation in brain: a matter of fat. J. Neurochem. 101:577–599.

Feinstein M. B. and Halenda S. P. (1988). Arachidonic acid mobilization in platelets: The possible role of protein kinase C and G-proteins. Experientia 44:101–104.

Fendri C., Mechri A., Khiari G., Othman A., Kerkeni A., and Gaha L. (2006). Implication du stress oxydant dans la physiopathologie de la schizophrénie: revue de la littérature [Oxidative stress involvement in schizophrenia pathophysiology: a review]. L'Encéphale 32:244–252.

Feng Z., Li D., Fung P. C., Pei Z., Ramsden D. B., and Ho S. L. (2003). COX-2-deficient mice are less prone to MPTP-neurotoxicity than wild-type mice. NeuroReport 14: 1927–1929.

Fernandes S. P., Edwards T. M., Ng K. T., and Robinson S. R. (2007). HIV-1 protein gp120 rapidly impairs memory in chicks by interrupting the glutamate-glutamine cycle. Neurobiol. Learn. Mem. 87:1–8.

Ferrarese C., Aliprandi A., Tremolizzo L., Stanzani L., De Micheli A., Dolara A., and Frattola L. (2001). Increased glutamate in CSF and plasma of patients with HIV dementia. Neurology 57:671–675.

Finiels-Marlier F., Marini A. M., Williams P., and Paul S. M. (1993). The N-methyl-D-aspartate antagonist MK-801 fails to protect dopaminergic neurons from 1-methyl-4-phenylpyridinium toxicity in vitro. J. Neurochem. 60:1968–1971.

Fiskum G., Murphy A. N., and Beal M. F. (1999). Mitochondria in neurodegeneration: Acute ischemia and chronic neurodegenerative diseases. J. Cereb. Blood Flow Metab 19: 351–369.

Frederickson C. J., Maret W., and Cuajungco M. P. (2004). Zinc and excitotoxic brain injury: a new model. Neuroscientist 10:18–25.

Frolich L. and Hoyer S. (2002). The etiological and pathogenetic heterogeneity of Alzheimer's disease. Nervenarzt 73:422–427.

Gabriel C., Justicia C., Camins A., and Planas A. M. (1999). Activation of nuclear factor-κB in the rat brain after transient focal ischemia. Brain Res. Mol. Brain Res. 65:61–69.

Gally J. A., Montague P. R., Reeke G. N., Jr., and Edelman G. M. (1990). The NO hypothesis: Possible effects of a short-lived, rapidly diffusible signal in the development and function of the nervous system. Proc. Natl. Acad. Sci. USA 87:3547–3551.

Gould T. D., Chen G., and Manji H. K. (2002). Mood stabilizer psychopharmacology. Clin. Neurosci. Res. 2:193–212.

Gras G., Chrétien F., Vallat-Decouvelaere A. V., Le Pavec G., Porcheray F., Bossuet C., Léone C., Mialocq P., Dereuddre-Bosquet N., Clayette P., Le Grand R., Créminon C., Dormont D., Rimaniol A. C., and Gray F. (2003). Regulated expression of sodium-dependent glutamate transporters and synthetase: a neuroprotective role for activated microglia and macrophages in HIV infection? Brain Pathol. 13:211–222.

Greenamyre J. T. and Young A. B. (1989). Excitatory amino acids and Alzheimer's disease. Neurobiol. Aging 10:593–602.

Greenamyre J. T., Penney J. B., D'Amato C. J., and Young A. B. (1987). Dementia of the Alzheimer's type: Changes in hippocampal L-[3H]glutamate binding. J. Neurochem. 48: 543–551.

Griffiths T., Evans M. C., and Meldrum B. S. (1983). Temporal lobe epilepsy, excitotoxins and the mechanism of selective neuronal loss. In: Fuxe K., Roberts P., and Schwarcz R. (eds.), *Excitotoxins*. Macmillan Publ. Co. Inc., New York, pp. 331–342.

Haba K., Ogawa N., Mizukawa K., and Mori A. (1991). Time course of changes in lipid peroxidation, pre- and postsynaptic cholinergic indices, NMDA receptor binding and neuronal death in the gerbil hippocampus following transient ischemia. Brain Res. 540:116–122.

Haddad J. J. (2004). Mitogen-activated protein kinases and the evolution of Alzheimer's: a revolutionary neurogenetic axis for therapeutic intervention? Prog. Neurobiol. 73:359–377.

Halpain S., Girault J.-A., and Greengard P. (1990). Activation of NMDA receptors induces dephosphorylation of DARPP-32 in rat striatal slices. Nature 343:369–372.

Hang C. H., Chen G., Shi J. X., Zhang X., and Li J. S. (2006). Cortical expression of nuclear factor κB after human brain contusion. Brain Res. 1109:14–21.

Hao Q. and Maret W. (2005). Imbalance between pro-oxidant and pro-antioxidant functions of zinc in disease. J. Alzheimer's Dis. 8:161–170.

Hardin-Pouzet H., Krakowski M., Bourbonnière L., Didier-Bazes M., Tran E., and Owens T. (1997). Glutamate metabolism is down-regulated in astrocytes during experimental allergic encephalomyelitis. Glia 20:79–85.

Hattori H. and Wasterlain C. G. (1990). Excitatory amino acids in the developing brain: ontogeny, plasticity, and excitotoxicity. Pediatr. Neurol. 6:219–228.

Hayes R. L., Jenkins L. W., and Lyeth B. G. (1992). Neurotransmitter-mediated mechanisms of traumatic brain injury: Acetylcholine and excitatory amino acids. J. Neurotrauma 9: S173-S187.

Hayes K. C., Hull T. C., Delaney G. A., Potter P. J., Sequeira K. A., Campbell K., and Popovich P. G. (2002). Elevated serum titers of proinflammatory cytokines and CNS autoantibodies in patients with chronic spinal cord injury. J. Neurotrauma 19:753–761.

Heyes M. P., Swartz K. J., Markey S. P., and Beal M. F. (1991). Regional brain and cerebrospinal fluid quinolinic acid concentrations in Huntington's disease. Neurosci. Lett. 122: 265–269.

Hill I. E., MacManus J. P., Rasquinha I., and Tuor U. I. (1995). DNA fragmentation indicative of apoptosis following unilateral cerebral hypoxia-ischemia in the neonatal rat. Brain Res. 676:398–403.

Hill K. E., Zollinger L. V., Watt H. E., Carlson N. G., and Rose J. W. (2004). Inducible nitric oxide synthase in chronic active multiple sclerosis plaques: distribution, cellular expression and association with myelin damage. J. Neuroimmunol. 151:171–179.

Hoffman S. W., Rzigalinski B. A., Willoughby K. A., and Ellis E. F. (2000). Astrocytes generate isoprostanes in response to trauma or oxygen radicals. J. Neurotrauma 17:415–420.

Homayoun P., Rodriguez de Turco E. B., Parkins N. E., Lane D. C., Soblosky J., Carey M. E., and Bazan N. G. (1997). Delayed phospholipid degradation in rat brain after traumatic brain injury. J. Neurochem. 69:199–205.

Hume R. I., Dingledine R., and Heinemann S. F. (1991). Identification of a site in glutamate receptor subunits that controls calcium permeability. Science 253:1028–1031.

Ikonomidou C. and Turski L. (2002). Why did NMDA receptor antagonists fail clinical trials for stroke and traumatic brain injury? Lancet Neurol. 1:383–386.

Ince P. G. and Codd G. A. (2005). Return of the cycad hypothesis-does the amyotrophic lateral sclerosis/parkinsonism dementia complex (ALS/PDC) of Guam have new implications for global health? Neuropathol. Appl. Neurobiol. 31:345–353.

Izquierdo I. and Medina J. H. (1995). Correlation between the pharmacology of long-term potentiation and the pharmacology of memory. Neurobiol. Learn. Mem. 63:19–32.

Jenner P. and Olanow C. W. (2006). The pathogenesis of cell death in Parkinson's disease. Neurology 66:S24-S36.

Juranek I. and Bezek S. (2005). Controversy of free radical hypothesis: Reactive oxygen species-Cause or consequence of tissue injury? Gen. Physiol. Biophys. 24:263–278.

Kajiwara K., Nagawawa H., Shimizu-Nishikawa S., Ookuri T., Kimura M., and Sugaya E. (1996). Molecular characterization of seizure-related genes isolated by differential screening. Biochem. Biophys. Res. Commun. 219:795–799.

Kalaria R. N., Harshbarger-Kelly M., Cohen D. L., and Premkumar D. R. D. (1996). Molecular aspects of inflammatory and immune responses in Alzheimer's disease. Neurobiol. Aging 17:687–693.

Kamat J. P. (2006). Peroxynitrite: A potent oxidizing and nitrating agent. Indian J. Exp. Biol. 44:436–447.

Katayama Y., Shimizu J., Suzuki S., Memezawa H., Kashiwagi F., Kamiya T., and Terashi A. (1990). Role of arachidonic acid metabolism on ischemic brain edema and metabolism. Adv. Neurol. 52:105–108.

Katsuki H., Tomita M., Takenaka C., Shirakawa H., Shimazu S., Ibi M., Kume T., Kaneko S., and Akaike A. (2001). Superoxide dismutase activity in organotypic midbrain-striatum co-cultures is associated with resistance of dopaminergic neurons to excitotoxicity. J. Neurochem. 76:1336–1345.

Kauer J. A. and Malenka R. C. (2006). LTP: AMPA receptors trading places. Nat. Neurosci. 9: 593–594.

Kaul M. and Lipton S. A. (2006). Mechanisms of neuronal injury and death in HIV-1 associated dementia. Curr. HIV. Res. 4:307–318.

Kawahara Y., Ito K., Sun H., Ito M., Kanazawa I., and Kwak S. (2004). GluR4c, an alternative splicing isoform of GluR4, is abundantly expressed in the adult human brain. Brain Res. Mol. Brain Res. 127:150–155.

Kawahara Y., Sun H., Ito K., Hideyama T., Aoki M., Sobue G., Tsuji S., and Kwak S. (2006). Underediting of GluR2 mRNA, a neuronal death inducing molecular change in sporadic ALS, does not occur in motor neurons in ALS1 or SBMA. Neurosci. Res. 54:11–14.

Kerwin R., Patel S., and Meldrum B. (1990). Quantitative autoradiographic analysis of glutamate binding sites in the hippocampal formation in normal and schizophrenic brain post mortem. Neuroscience 39:25–32.

Kim S. H., Engelhardt J. I., Henkel J. S., Siklos L., Soos J., Goodman C., and Appel S. H. (2004). Widespread increased expression of the DNA repair enzyme PARP in brain in ALS. Neurology 62:319–322.

Klivenyi P., Beal M. F., Ferrante R. J., Andreassen O. A., Wermer M., Chin M. R., and Bonventre J. V. (1998). Mice deficient in group IV cytosolic phospholipase A_2 are resistant to MPTP neurotoxicity. J. Neurochem. 71:2634–2637.

Klussmann S. and Martin-Villalba A. (2005). Molecular targets in spinal cord injury. J. Mol. Med. 83:657–671.

Koh J.-Y., Yang L. L., and Cotman C. W. (1990). β-Amyloid protein increases the vulnerability of cultured cortical neurons to excitotoxic damage. Brain Res. 533:315–320.

Konradi C. and Heckers S. (2003). Molecular aspects of glutamate dysregulation: implications for schizophrenia and its treatment. Pharmacol. Ther. 97:153–179.

Kontush A. (2001). Amyloid-beta: An antioxidant that becomes a pro-oxidant and critically contributes to Alzheimer's disease. Free Radical Biol. Med. 31:1120–1131.

Korth C., May B. C., Cohen F. E., and Prusiner S. B. (2001). Acridine and phenothiazine derivatives as pharmacotherapeutics for prion disease. Proc. Natl. Acad. Sci. USA 98: 9836–9841.

Kuner R., Groom A. J., Bresink I., Kornau H. C., Stefovska V., Müller G., Hartmann B., Tschauner K., Waibel S., Ludolph A. C., Ikonomidou C., Seeburg P. H., and Turski L. (2005). Late-onset motoneuron disease caused by a functionally modified AMPA receptor subunit. Proc. Natl. Acad. Sci. U. S. A 102:5826–5831.

Kwak S. and Kawahara Y. (2005). Deficient RNA editing of GluR2 and neuronal death in amyotropic lateral sclerosis. J. Mol. Med. 83:110–120.

Larm J. A., Beart P. M., and Cheung N. S. (1997). Neurotoxin domoic acid produces cytotoxicity via kainate- and AMPA-sensitive receptors in cultured cortical neurones. Neurochem. Int. 31:677–682.

Lea P. M. and Faden A. I. (2001). Traumatic brain injury: developmental differences in glutamate receptor response and the impact on treatment. Ment. Retard. Dev. Disabil. Res. Rev. 7: 235–248.

Lee H. G., Casadesus G., Zhu X. W., Takeda A., Perry G., and Smith M. A. (2004). Challenging the amyloid cascade hypothesis - Senile plaques and amyloid-beta as protective adaptations to Alzheimer disease. In: DeGrey A. D. N. (ed.), *Strategies for Engineered Negligible Senescence: Why Genuine Control of Aging May Be Foreseeable.* Annals of the New York Academy of Sciences New York Acad Sciences, pp. 1–4.

Leker R. R. and Shohami E. (2002). Cerebral ischemia and trauma - different etiologies yet similar mechanisms: neuroprotective opportunities. Brain Res. Rev. 39:55–73.

Li S. and Stys P. K. (2000). Mechanisms of ionotropic glutamate receptor-mediated excitotoxicity in isolated spinal cord white matter. J. Neurosci. 20:1190–1198.

Li Y., Maher P., and Schubert D. (1997). A role for 12-lipoxygenase in nerve cell death caused by glutathione depletion. Neuron 19:453–463.

Li L., Murphy T. H., Hayden M. R., and Raymond L. A. (2004). Enhanced striatal NR2B-containing *N*-methyl-D-aspartate receptor-mediated synaptic currents in a mouse model of Huntington disease. J. Neurophysiol. 92:2738–2746.

Liddle P. F. (1987). The symptoms of chronic schizophrenia. A re-examination of the positive-negative dichotomy. Br. J. Psychiatry 151:145–151.

Linden D. J. and Routtenberg A. (1989). The role of protein kinase C in long-term potentiation: a testable model. Brain Res. Rev. 14:279–296.

Lipton S. A. (1998). Neuronal injury associated with HIV-1: approaches to treatment. Annu. Rev. Pharmacol. Toxicol. 38:159–177.

Liu Z., Stafstrom C. E., Sarkisian M. R., Yang Y., Hori A., Tandon P., and Holmes G. L. (1997). Seizure-induced glutamate release in mature and immature animals: an *in vivo* microdialysis study. NeuroReport 8:2019–2023.

Love S. (1999). Oxidative stress in brain ischemia. Brain Pathol. 9:119–131.

Lue L. F., Brachova L., Civin W. H., and Rogers J. (1996). Inflammation, Abeta deposition, and neurofibrillary tangle formation as correlates of Alzheimer's disease neurodegeneration. J. Neuropathol. Exp. Neurol. 55:1083–1088.

MacDonald A. W. I. and Chafee M. V. (2006). Translational and developmental perspective on N-methyl-D-aspartate synaptic deficits in schizophrenia. Dev. Psychopathol. 18:853–876.

Mahadik S. P. and Mukherjee S. (1996). Free radical pathology and antioxidant defense in schizophrenia: a review. Schizophr. Res. 19:1–17.

Mahadik S. P. and Scheffer R. E. (1996). Oxidative injury and potential use of antioxidants in schizophrenia. Prostaglandins Leukot. Essent. Fatty Acids 55:45–54.

Masliah E., Alford M., DeTeresa R., Mallory M., and Hansen L. (1996). Deficient glutamate transport is associated with neurodegeneration in Alzheimer's disease. Ann. Neurol. 40: 759–766.

Masliah E., Alford M., Galasko D., Salmon D., Hansen L. A., Good P. F., Perl D. P., and Thal L. (2001). Cholinergic deficits in the brains of patients with parkinsonism-dementia complex of Guam. NeuroReport 12:3901–3903.

Massicotte G. (2000). Modification of glutamate receptors by phospholipase A2: its role in adaptive neural plasticity. Cell Mol. Life Sci. 57:1542–1550.

Matarredona E. R., Santiago M., Venero J. L., Cano J., and Machado A. (2001). Group II metabotropic glutamate receptor activation protects striatal dopaminergic nerve terminals against MPP^+-induced neurotoxicity along with brain-derived neurotrophic factor induction. J. Neurochem. 76:351–360.

Mattson M. P. (1990). Antigenic changes similar to those seen in neurofibrillary tangles are elicited by glutamate and Ca^{2+} influx in cultured hippocampal neurons. Neuron 2:105–117.

Mattson M. P. (2002). Oxidative stress, perturbed calcium homeostasis, and immune dysfunction in Alzheimer's disease. J. Neurovirol. 8:539–550.

Matute C., Domercq M., and Sánchez-Gómez M. V. (2006). Glutamate-mediated glial injury: Mechanisms and clinical importance. Glia 53:212–224.

May B. C. H., Fafarman A. T., Hong S. B., Rogers M., Deady L. W., Prusiner S. B., and Cohen F. E. (2003). Potent inhibition of scrapie prion replication in cultured cells by bis-acridines. Proc. Natl. Acad. Sci. USA 100:3416–3421.

Mayer A. M. S., Hall M., Fay M. J., Lamar P., Pearson C., Prozialeck W. C., Lehmann V. K. B., Jacobson P. B., Romanic A. M., Uz T., and Manev H. (2001). Effect of a short-term in vitro exposure to the marine toxin domoic acid on viability, tumor necrosis factor-alpha, matrix metalloproteinase-9 and superoxide anion release by rat neonatal microglia. BMC Pharmacol. 1:7.

McDonald J. W., Garofalo E. A., Hood T., Sackellares C., Gilman S., McKeever P. E., Troncoso J. C., and Johnston M. V. (1991). Altered excitatory and inhibitory amino acid receptor binding in hippocampus of patients with temporal lobe epilepsy. Ann. Neurol. 29:529–541.

McDonald J. W., Levine J. M., and Qu Y. (1998). Multiple classes of the oligodendrocyte lineage are highly vulnerable to excitotoxicity. NeuroReport 9:2757–2762.

McGeer E. G. and McGeer P. L. (1998). The importance of inflammatory mechanisms in Alzheimer disease. Exp. Gerontol. 33:371–378.

McIntosh T. K., Saatman K. E., Raghupathi R., Graham D. I., Smith D. H., Lee V. M., and Trojanowski J. Q. (1998). The molecular and cellular sequelae of experimental traumatic brain injury: pathogenetic mechanisms. Neuropathol. Appl. Neurobiol. 24:251–267.

McNaught K. S. P., Jackson T., Jnobaptiste R., Kapustin A., and Olanow C. W. (2006). Proteasomal dysfunction in sporadic Parkinson's disease. Neurology 66:S37-S49.

Meador-Woodruff J. H., Davis K. L., and Haroutunian V. (2001). Abnormal kainate receptor expression in prefrontal cortex in schizophrenia. Neuropsychopharmacology 24:545–552.

Meldrum B. S. (1993). Excitotoxicity and selective neuronal loss in epilepsy. Brain Pathol. 3: 405–412.

Meldrum B. S. (2002). Implications for neuroprotective treatments. In: Sutula T. and Pitkanen A. (eds.), *Do Seizures Damage the Brain. Perspectives in Analytical Philosophy*, Elsevier Science BV, Amsterdam, pp. 487–495.

Ménard C., Valastro B., Martel M. A., Chartier T., Marineau A., Baudry M., and Massicotte G. (2005). AMPA receptor phosphorylation is selectively regulated by constitutive phospholipase A_2 and 5-lipoxygenase activities. Hippocampus 15:370–380.

Milakovic T., Quintanilla R. A., and Johnson G. V. (2006). Mutant huntingtin expression induces mitochondrial calcium handling defects in clonal striatal cells: functional consequences. J. Biol. Chem. 281:34785–34795.

Ming Y., Zhang H., Long L., Wang F., Chen J., and Zhen X. (2006). Modulation of Ca^{2+} signals by phosphatidylinositol-linked novel D1 dopamine receptor in hippocampal neurons. J. Neurochem. 98:1316–1323.

Minghetti L., Greco A., Cardone F., Puopolo M., Ladogana A., Almonti S., Cunningham C., Perry V. H., Pocchiari M., and Levi G. (2000). Increased brain synthesis of prostaglandin E_2 and F_2-isoprostane in human and experimental transmissible spongiform encephalopathies. J. Neuropathol. Exp. Neurol. 59:866–871.

Moreira P. I., Oliveira C. R., Santos M. S., Nunomura A., Honda K., Zhu X. W., Smith M. A., and Perry G. (2005a). A second look into the oxidant mechanisms in Alzheimer's disease. Curr. Neurovasc. Res. 2:179–184.

Moreira P. L., Smith M. A., Zhu X. W., Honda K., Lee H. G., Aliev G., and Perry G. (2005b). Oxidative damage and Alzheimer's disease: Are antioxidant therapies useful? Drug News Perspect. 18:13–19.

Morishita W., Marie H., and Malenka R. C. (2005). Distinct triggering and expression mechanisms underlie LTD of AMPA and NMDA synaptic responses. Nat. Neurosci. 8:1043–1050.

Muller N. and Schwarz M. (2006). Schizophrenia as an inflammation-mediated dysbalance of glutamatergic neurotransmission. Neurotox. Res. 10:131–148.

Murakami K. and Routtenberg A. (2003). The role of fatty acids in synaptic growth and plasticity. In: Peet M., Glen L., and Horrobin D. F. (eds.), *Phospholipid Spectrum Disorders in Psychiatry and Neurology*. Marius Press, Carnforth, Lancashire, pp. 77–92.

Murch S. J., Cox P. A., and Banack S. A. (2004). A mechanism for slow release of biomagnified cyanobacterial neurotoxins and neurodegenerative disease in Guam. Proc. Natl. Acad. Sci. USA 101:12228–12231.

Murphy T. H., Miyamoto M., Sastre A., Schnaar R. L., and Coyle J. T. (1989). Glutamate toxicity in a neuronal cell line involves inhibition of cystine transport leading to oxidative stress. Neuron 2:1547–1558.

Murphy P., Sharp A., Shin J., Gavrilyuk V., Dello R. C., Weinberg G., Sharp F. R., Lu A., Heneka M. T., and Feinstein D. L. (2002). Suppressive effects of ansamycins on inducible nitric oxide synthase expression and the development of experimental autoimmune encephalomyelitis. J. Neurosci. Res. 67:461–470.

Nadi N. S., Wyler A. R., and Porter R. J. (1987). Amino acids and catecholamines in the epileptic focus from the human brain. Neurology 37:106.

Nakagawa Y., Kurihara K., Sugiura T., and Waku K. (1985). Heterogeneity in the metabolism of the arachidonoyl molecular species of glycerophospholipids of rabbit alveolar macrophages. The relationship between metabolic activities and chemical structures of the arachidonoyl molecular species. Eur. J. Biochem. 153:263–268.

New D. R., Maggirwar S. B., Epstein L. G., Dewhurst S., and Gelbard H. A. (1998). HIV-1 Tat induces neuronal death via tumor necrosis factor-α and activation of non-*N*-methyl-D-aspartate receptors by a NFκB-independent mechanism. J. Biol. Chem. 273:17852–17858.

Novelli A., Nicoletti F., Wroblewski J. T., Alho H., Costa A. E., and Guidotti A. (1987). Excitatory amino acid receptors coupled with guanylate cyclase in primary cultures of cerebellar granule cells. J. Neurosci. 7:40–47.

Numazawa S., Ishikawa M., Yoshida A., Tanaka S., and Yoshida T. (2003). Atypical protein kinase C mediates activation of NF-E2-related factor 2 in response to oxidative stress. Am. J. Physiol. Cell Physiol. 285:C334-C342.

Oka A., Belliveau M. J., Rosenberg P. A., and Volpe J. J. (1993). Vulnerability of oligodendroglia to glutamate: pharmacology, mechanisms, and prevention. J. Neurosci. 13: 1441–1453.

Okuno T., Nakatsuji Y., Kumanogoh A., Koguchi K., Moriya M., Fujimura H., Kikutani H., and Sakoda S. (2004). Induction of cyclooxygenase-2 in reactive glial cells by the CD40 pathway: relevance to amyotrophic lateral sclerosis. J. Neurochem. 91:404–412.

Olney J. W. and Farber N. B. (1995). Glutamate receptor dysfunction and schizophrenia. Arch. Gen. Psychiatr. 52:998–1007.

Olney J. W., Wozniak D. F., and Farber N. B. (1997). Excitotoxic neurodegeneration in Alzheimer disease. New hypothesis and new therapeutic strategies. Arch. Neurol. 54:1234–1240.

Ong W. Y. and Farooqui A. A. (2005). Iron, neuroinflammation, and Alzheimer's disease. J. Alzheimer's Dis. 8:183–200.

Pabst S., Hazzard J. W., Antonin W., Sudhof T. C., Jahn R., Rizo J., and Fasshauer D. (2000). Selective interaction of complexin with the neuronal SNARE complex. Determination of the binding regions. J. Biol. Chem. 275:19808–19818.

Pan W., Kastin A. J., Bell R. L., and Olson R. D. (1999). Upregulation of tumor necrosis factor α transport across the blood-brain barrier after acute compressive spinal cord injury. J. Neurosci. 19:3649–3655.

Panter S. S., Yum S. W., and Faden A. I. (1990). Alteration in extracellular amino acids after traumatic spinal cord injury. Ann. Neurol. 27:96–99.

Park E., Velumian A. A., and Fehlings M. G. (2004). The role of excitotoxicity in secondary mechanisms of spinal cord injury: a review with an emphasis on the implications for white matter degeneration. J. Neurotrauma 21:754–774.

Patrignani P., Tacconelli S., Sciulli M. G., and Capone M. L. (2005). New insights into COX-2 biology and inhibition. Brain Res. Rev. 48:352–359.

Pavel J., Lukácová N., Mar²ala J., and Mar²ala M. (2001). The regional changes of the catalytic NOS activity in the spinal cord of the rabbit after repeated sublethal ischemia. Neurochem. Res. 26:833–839.

Penkowa M., Molinero A., Carrasco J., and Hidalgo J. (2001). Interleukin-6 deficiency reduces the brain inflammatory response and increases oxidative stress and neurodegeneration after kainic acid-induced seizures. Neuroscience 102:805–818.

Penney J. B., Maragos W. F., Greenamyre J. T., Debowey D. L., Hollingsworth Z., and Young A. B. (1990). Excitatory amino acid binding sites in the hippocampal region of Alzheimer's disease and other dementias. J. Neurol. Neurosurg. Psychiatr. 53:314–320.

Pereira C. F. M. and Resende de Oliveira C. (2000). Oxidative glutamate toxicity involves mitochondrial dysfunction and perturbation of intracellular Ca^{2+} homeostasis. Neurosci. Res. 37:227–236.

Peterson C. and Goldman J. E. (1986). Alterations in calcium content and biochemical processes in cultured skin fibroblasts from aged and Alzheimer donors. Proc. Natl. Acad. Sci. USA 83:2758–2762.

Peterson C., Gibson G. E., and Blass J. P. (1985). Altered calcium uptake in cultured skin fibroblasts from patients with Alzheimer's disease. New Eng. J. Med. 312:1063–1069.

Phillis J. W. and O'Regan M. H. (1996). Mechanisms of glutamate and aspartate release in the ischemic rat cerebral cortex. Brain Res. 730:150–164.

Phillis J. W., Horrocks L. A., and Farooqui A. A. (2006). Cyclooxygenases, lipoxygenases, and epoxygenases in CNS: Their role and involvement in neurological disorders. Brain Res. Rev. 52:201–243.

Pitt D., Werner P., and Raine C. S. (2000). Glutamate excitotoxicity in a model of multiple sclerosis. Nat. Med. 6:67–70.

Porcheray F., Léone C., Samah B., Rimaniol A. C., Dereuddre-Bosquet N., and Gras G. (2006). Glutamate metabolism in HIV-infected macrophages: implications for the CNS. Am. J. Physiol. Cell Physiol. 291:C618-C626.

Prusiner S. B. (2001). Shattuck lecture - Neurodegenerative diseases and prions. New Eng. J. Med. 344:1516–1526.

Qiu S. and Curras-Collazo M. C. (2006). Histopathological and molecular changes produced by hippocampal microinjection of domoic acid. Neurotoxicol. Teratol. 28:354–362.

Qiu S., Pak C. W., and Curras-Collazo M. C. (2006). Sequential involvement of distinct glutamate receptors in domoic acid-induced neurotoxicity in rat mixed cortical cultures: effect of multiple dose/duration paradigms, chronological age, and repeated exposure. Toxicol. Sci. 89: 243–256.

Ramsay R. R., Krueger M. J., Youngster S. K., Gluck M. R., Casida J. E., and Singer T. P. (1991). Interaction of 1-methyl-4-phenylpyridinium ion (MPP^+) and its analogs with the rotenone/piericidin binding site of NADH dehydrogenase. J. Neurochem. 56:1184–1190.

Rao S. D. and Weiss J. H. (2004). Excitotoxic and oxidative cross-talk between motor neurons and glia in ALS pathogenesis. Trends Neurosci. 27:17–23.

Ray S. K., Hogan E. L., and Banik N. L. (2003). Calpain in the pathophysiology of spinal cord injury: neuroprotection with calpain inhibitors. Brain Res. Rev. 42:169–185.

Reynolds I. J., Hoyt K. R., White J., and Stout A. K. (1996). Intracellular signalling in glutamate excitotoxicity. In: Fiskum G. (ed.), *Neurodegenerative Diseases*. Plenum Press, New York, pp. 1–7.

Riederer P. and Hoyer S. (2006). From benefit to damage. Glutamate and advanced glycation end products in Alzheimer brain. J. Neural Transm. 113:1671–1677.

Rodríguez A., Freixes M., Dalfó E., Martín M., Puig B., and Ferrer I. (2005). Metabotropic glutamate receptor phospholipase C pathway: A vulnerable target to Creutzfeldt-Jakob disease in the cerebral cortex. Neuroscience 131:825–832.

Rojas C. V., Martinez J. I., Flores I., Hoffman D. R., and Uauy R. (2003). Gene expression analysis in human fetal retinal explants treated with docosahexaenoic acid. Invest. Ophthalmol. Vis. Sci. 44:3170–3177.

Rose J. W., Hill K. E., Watt H. E., and Carlson N. G. (2004). Inflammatory cell expression of cyclooxygenase-2 in the multiple sclerosis lesion. J. Neuroimmunol. 149:40–49.

Rothstein J. D., Martin L. J., and Kuncl R. W. (1992). Decreased glutamate transport by the brain and spinal cord in amyotrophic lateral sclerosis. N. Engl. J. Med. 326:1464–1468.

Rothstein J. D., Van Kammen M., Levey A. I., Martin L. J., and Kuncl R. W. (1995). Selective loss of glial glutamate transporter GLT-1 in amyotrophic lateral sclerosis. Ann. Neurol. 38: 73–84.

Rothstein J. D., Dykes-Hoberg M., Pardo C. A., Bristol L. A., Jin L., Kuncl R. W., Kanai Y., Hediger M. A., Wang Y., Schielke J. P., and Welty D. F. (1996). Knockout of glutamate transporters reveals a major role for astroglial transport in excitotoxicity and clearance of glutamate. Neuron 16:675–686.

Saluja I., O'Regan M. H., Song D. K., and Phillis J. W. (1999). Activation of $cPLA_2$, PKC, and ERKs in the rat cerebral cortex during ischemia/reperfusion. Neurochem. Res. 24:669–677.

Samdani A. F., Dawson T. M., and Dawson V. L. (1997). Nitric oxide synthase in models of focal ischemia. Stroke 28:1283–1288.

Saunders D. E., Howe F. A., van den Boogaart A., McLean M. A., Griffiths J. R., and Brown M. M. (1995). Continuing ischemic damage after acute middle cerebral artery infarction in humans demonstrated by short-echo proton spectroscopy. Stroke 26:1007–1013.

Scallet A. C., Carp R. I., and Ye X. (2003). Pathophysiology of transmissible spongiform encephalopathies. Curr. Med. Chem. - Immunol. Endo. Metab. Agents 3:171–184.

Schapira A. H. V. (2006). Etiology of Parkinson's disease. Neurology 66:S10-S23.

Scheuer K., Maras A., Gattaz W. F., Cairns N., Förstl H., and Müller W. E. (1996). Cortical NMDA receptor properties and membrane fluidity are altered in Alzheimer's disease. Dementia. 7:210–214.

Schinder A. F., Olson E. C., Spitzer N. C., and Montal M. (1996). Mitochondrial dysfunction is a primary event in glutamate neurotoxicity. J. Neurosci. 16:6125–6133.

Schneider A., Martin-Villalba A., Weih F., Vogel J., Wirth T., and Schwaninger M. (1999). NF-kappaB is activated and promotes cell death in focal cerebral ischemia. Nature Med. 5: 554–559.

Sebeo J., Hof P. R., and Perl D. P. (2004). Occurrence of α-synuclein pathology in the cerebellum of Guamanian patients with parkinsonism-dementia complex. Acta Neuropathol. (Berl) 107: 497–503.

Seidenman K. J., Steinberg J. P., Huganir R., and Malinow R. (2003). Glutamate receptor subunit 2 Serine 880 phosphorylation modulates synaptic transmission and mediates plasticity in CA1 pyramidal cells. J. Neurosci. 23:9220–9228.

Shaw P. J. and Ince P. G. (1997). Glutamate, excitotoxicity and amyotrophic lateral sclerosis. J. Neurol. 244 Suppl 2:S3-S14.

Sherwin A., Robitaille Y., and Quesney F. (1988). Excitatory amino acids are elevated in human epileptic cerebral cortex. Neurology 38:920–923.

Shimihama S., Ninomiya H., Saitoh T., Terry R. D., Fukunaga R., Taniguchi T., Fujiwara M., Kimura J., and Kameyama M. (1990). Changes in signal transduction in Alzheimer's disease. J. Neural Transm. 30:69–78.

Siesjö B. K. (1990). Calcium in the brain under physiological and pathological conditions. Eur. Neurol. 30:3–9.

Siman R. and Noszek J. C. (1988). Excitatory amino acids activate calpain I and induce structural protein breakdown *in vivo*. Neuron 1:279–287.

Siman R., Noszek J. C., and Kegerise C. (1989). Calpain I activation is specifically related to excitatory amino acid induction of hippocampal damage. J. Neurosci. 9:1579–1590.

Singh I. N., Sullivan P. G., Deng Y., Mbye L. H., and Hall E. D. (2006). Time course of post-traumatic mitochondrial oxidative damage and dysfunction in a mouse model of focal traumatic brain injury: implications for neuroprotective therapy. J. Cereb. Blood Flow Metab. 26:1407–1418.

Siso S., Puig B., Varea R., Vidal E., Acin C., Prinz M., Montrasio F., Badiola J., Aguzzi A., Pumarola M., and Ferrer I. (2002). Abnormal synaptic protein expression and cell death in murine scrapie. Acta Neuropathol. (Berl) 103:615–626.

Snyder E. M., Nong Y., Almeida C. G., Paul S., Moran T., Choi E. Y., Nairn A. C., Salter M. W., Lombroso P. J., Gouras G. K., and Greengard P. (2005). Regulation of NMDA receptor trafficking by amyloid-β. Nat. Neurosci. 8:1051–1058.

Solans A., Zambrano A., Rodríguez M., and Barrientos A. (2006). Cytotoxicity of a mutant huntingtin fragment in yeast involves early alterations in mitochondrial OXPHOS complexes II and III. Hum. Mol. Genet. 15:3063–3081.

Spencer S. (2007). Epilepsy: clinical observations and novel mechanisms. Lancet Neurol. 6:14–16.

Srivastava R., Brouillet E., Beal M. F., Storey E., and Hyman B. T. (1993). Blockade of 1-methyl-4-phenylpyridinium ion (MPP^+) nigral toxicity in the rat by prior decortication or MK-801 treatment: a stereological estimate of neuronal loss. Neurobiol. Aging 14:295–301.

St-Gelais F., Ménard C., Congar P., Trudeau L. E., and Massicotte G. (2004). Postsynaptic injection of calcium-independent phospholipase A2 inhibitors selectively increases AMPA receptor-mediated synaptic transmission. Hippocampus 14:319–325.

Stephenson D., Yin T., Smalstig E. B., Hsu M. A., Panetta J., Little S., and Clemens J. (2000). Transcription factor nuclear factor-kappa B is activated in neurons after focal cerebral ischemia. J. Cereb. Blood Flow Metab 20:592–603.

Stewart L. R., White A. R., Jobling M. F., Needham B. E., Maher F., Thyer J., Beyreuther K., Masters C. L., Collins S. J., and Cappai R. (2001). Involvement of the 5-lipoxygenase pathway in the neurotoxicity of the prion peptide PrP106-126. J. Neurosci. Res. 65:565–572.

Sucher N. J., Lei S. Z., and Lipton S. A. (1991). Calcium channel antagonists attenuate NMDA receptor-mediated neurotoxicity of retinal ganglion cells in culture. Brain Res. 297:297–302.

Sullivan P. G., Springer J. E., Hall E. D., and Scheff S. W. (2004). Mitochondrial uncoupling as a therapeutic target following neuronal injury. J. Bioenerg. Biomembr. 36:353–356.

Sundström E. and Mo L. L. (2002). Mechanisms of glutamate release in the rat spinal cord slices during metabolic inhibition. J. Neurotrauma 19:257–266.

Szatkowski M. and Attwell D. (1994). Triggering and execution of neuronal death in brain ischaemia: two phases of glutamate release by different mechanisms. Trends Neurosci. 17:359–365.

Teismann P., Vila M., Choi D. K., Tieu K., Wu D. C., Jackson-Lewis V., and Przedborski S. (2003). COX-2 and neurodegeneration in Parkinson's disease. Ann. N. Y. Acad. Sci. 991:272–277.

Terry R. D. and Katzman R. (1983). Senile dementia of the Alzheimer type. Ann. Neurol. 14: 497–506.

Tocco G., Shors T. J., Baudry M., and Thompson R. F. (1991). Selective increase of AMPA binding to the AMPA/quisqualate receptor in the hippocampus in response to acute stress. Brain Res. 559:168–171.

Tortarolo M., Veglianese P., Calvaresi N., Botturi A., Rossi C., Giorgini A., Migheli A., and Bendotti C. (2003). Persistent activation of p38 mitogen-activated protein kinase in a mouse model of familial amyotrophic lateral sclerosis correlates with disease progression. Mol. Cell. Neurosci. 23:180–192.

Tseng K. Y. and O'Donnell P. (2004). Dopamine-glutamate interactions controlling prefrontal cortical pyramidal cell excitability involve multiple signaling mechanisms. J. Neurosci. 24: 5131–5139.

Turnbull S., Tabner B. J., Brown D. R., and Allsop D. (2003). Quinacrine acts as an antioxidant and reduces the toxicity of the prion peptide PrP106-126. NeuroReport 14:1743–1745.

Turski L., Bressler K., Rettig K.-J., Loschmann P.-A., and Wachtel H. (1991). Protection of substantia nigra from MPP^+ neurotoxicity by N-methyl-D-aspartate antagonists. Nature 349: 414–417.

Ulas J. and Cotman C. W. (1993). Excitatory amino acid receptors in schizophrenia. Schizophr. Bull. 19:105–117.

Vincent I. J. and Davies P. (1990). Phosphorylation characteristics of the A68 protein in Alzheimer's disease. Brain Res. 531:127–135.

Visioli F., Rodriguez de Turco E. B., Kreisman N. R., and Bazan N. G. (1994). Membrane lipid degradation is related to interictal cortical activity in a series of seizures. Metab Brain Dis. 9:161–170.

Viviani B., Corsini E., Binaglia M., Galli C. L., and Marinovich M. (2001). Reactive oxygen species generated by glia are responsible for neuron death induced by human immunodeficiency virus-glycoprotein 120 in vitro. Neuroscience 107:51–58.

Viviani B., Gardoni F., Bartesaghi S., Corsini E., Facchi A., Galli C. L., Di Luca M., and Marinovich M. (2006). Interleukin-1β released by gp120 drives neural death through tyrosine phosphorylation and trafficking of NMDA receptors. J. Biol. Chem. 281: 30212–30222.

Vogtherr M., Grimme S., Elshorst B., Jacobs D. M., Fiebig K., Griesinger C., and Zahn R. (2003). Antimalarial drug quinacrine binds to C-terminal helix of cellular prion protein. J. Med. Chem. 46:3563–3564.

Volterra A., Trotti D., Floridi S., and Racagni G. (1994). Reactive oxygen species inhibit high-affinity glutamate uptake: molecular mechanism and neuropathological implications. Ann. N.Y. Acad. Sci. 738:153–162:153–162.

Walser B., Giordano R. M., and Stebbins C. L. (2006). Supplementation with omega-3 polyunsaturated fatty acids augments brachial artery dilation and blood flow during forearm contraction. Eur. J. Appl. Physiol. 97:347–354.

Wang X., Zhong P., Gu Z., and Yan Z. (2003). Regulation of NMDA receptors by dopamine D4 signaling in prefrontal cortex. J. Neurosci. 23:9852–9861.

Wang Q., Yu S., Simonyi A., Sun G. Y., and Sun A. Y. (2005). Kainic acid-mediated excitotoxicity as a model for neurodegeneration. Mol. Neurobiol. 31:3–16.

Wasterlain C. G., Fujikawa D. G., Penix L., and Sankar R. (1993). Pathophysiological mechanisms of brain damage from status epilepticus. Epilepsia 34:S37-S53.

Wells J. E. A., Rice T. K., Nuttall R. K., Edwards D. R., Zekki H., Rivest S., and Yong V. W. (2003). An adverse role for matrix metalloproteinase 12 after spinal cord injury in mice. J. Neurosci. 23:10107–10115.

Wenk G. L. (2006). Neuropathologic changes in Alzheimer's disease: potential targets for treatment. J. Clin. Psychiatry 67 Suppl 3:3–7.

Wenk G. L., Parsons C. G., and Danysz W. (2006). Potential role of N-methyl-D-aspartate receptors as executors of neurodegeneration resulting from diverse insults: focus on memantine. Behav. Pharmacol. 17:411–424.

Wersinger C. and Sidhu A. (2006). An inflammatory pathomechanism for Parkinson's disease? Curr. Medicinal Chem. 13:591–602.

Winton M. J., Joyce S., Zhukareva V., Practico D., Perl D. P., Galasko D., Craig U., Trojanowski J. Q., and Lee V. M. (2006). Characterization of tau pathologies in gray and white matter of Guam parkinsonism-dementia complex. Acta Neuropathol. (Berl) 111: 401–412.

Xu Z., Wang B. R., Wang X., Kuang F., Duan X. L., Jiao X. Y., and Ju G. (2006). ERK1/2 and p38 mitogen-activated protein kinase mediate iNOS-induced spinal neuron degeneration after acute traumatic spinal cord injury. Life Sci. 79:1895–1905.

Yager J. Y. and Thornhill J. A. (1997). The effect of age on susceptibility to hypoxic-ischemic brain damage. Neurosci. Biobehav. Rev. 21:167–174.

Yager J. Y., Shuaib A., and Thornhill J. (1996). The effect of age on susceptibility to brain damage in a model of global hemispheric hypoxia-ischemia. Brain Res. Dev. Brain Res. 93:143–154.

Yao J. K., Reddy R. D., and Van Kammen D. P. (2001). Oxidative damage and schizophrenia - An overview of the evidence and its therapeutic implications. CNS Drugs 15:287–310.

Yasojima K., Tourtellotte W. W., McGeer E. G., and McGeer P. L. (2001). Marked increase in cyclooxygenase-2 in ALS spinal cord: implications for therapy. Neurology 57:952–956.

Ye Z. C., Rothstein J. D., and Sontheimer H. (1999). Compromised glutamate transport in human glioma cells: reduction-mislocalization of sodium-dependent glutamate transporters and enhanced activity of cystine-glutamate exchange. J. Neurosci. 19:10767–10777.

Yegin A., Akbas S. H., Ozben T., and Korgun D. K. (2002). Secretory phospholipase A_2 and phospholipids in neural membranes in an experimental epilepsy model. Acta Neurol. Scand. 106:258–262.

Yi J. H. and Hazell A. S. (2006). Excitotoxic mechanisms and the role of astrocytic glutamate transporters in traumatic brain injury. Neurochem. Int. 48:394–403.

Yi J. H., Hoover R., McIntosh T. K., and Hazell A. S. (2006). Early, transient increase in complexin I and complexin II in the cerebral cortex following traumatic brain injury is attenuated by N-acetylcysteine. J. Neurotrauma 23:86–96.

Yoshinaga N., Yasuda Y., Murayama T., and Nomura Y. (2000). Possible involvement of cytosolic phospholipase A_2 in cell death induced by 1-methyl-4-phenylpyridinium ion, a dopaminergic neurotoxin, in GH3 cells. Brain Res. 855:244–251.

Yoshiyama Y., Arai K., and Hattori T. (2001). Enhanced expression of I-κB with neurofibrillary pathology in Alzheimer's disease. NeuroReport 12:2641–2645.

Yu Z., Zhou D., Cheng G., and Mattson M. P. (2000). Neuroprotective role for the p50 subunit of NF-κB in an experimental model of Huntington's disease. J. Mol. Neurosci. 15:31–44.

Zafrilla P., Mulero J., Xandri J. M., Santo E., Caravaca G., and Morillas J. M. (2006). Oxidative stress in Alzheimer patients in different stages of the disease. Curr. Medicinal Chem. 13: 1075–1083.

Zagulska-Szymczak S., Filipkowski R. K., and Kaczmarek L. (2001). Kainate-induced genes in the hippocampus: lessons from expression patterns. Neurochem. Int. 38:485–501.

Zeevalk G. D., Bernard L. P., and Nicklas W. J. (1998). Role of oxidative stress and the glutathione system in loss of dopamine neurons due to impairment of energy metabolism. J. Neurochem. 70:1421–1430.

Zeron M. M., Fernandes H. B., Krebs C., Shehadeh J., Wellington C. L., Leavitt B. R., Baimbridge K. G., Hayden M. R., and Raymond L. A. (2004). Potentiation of NMDA receptor-mediated excitotoxicity linked with intrinsic apoptotic pathway in YAC transgenic mouse model of Huntington's disease. Mol. Cell Neurosci. 25:469–479.

Zhang Y. M., Wang H., Li J. R., Dong L., Xu P., Chen W. Z., Neve R. L., Volpe J. J., and Rosenberg P. A. (2006). Intracellular zinc release and ERK phosphorylation are required upstream of 12-lipoxygenase activation in peroxynitrite toxicity to mature rat oligodendrocytes. J. Biol. Chem. 281:9460–9470.

Chapter 9
Endogenous Antioxidant Mechanisms and Glutamate Neurotoxicity

9.1 Introduction

A major theory of neurodegeneration in acute neural trauma and neurodegenerative diseases is glutamate receptor-mediated neurotoxicity. Two pathways may be involved in glutamate-mediated toxicity. Both result in production of reactive oxygen species (ROS). One pathway utilizes excitotoxicity (Olney et al., 1979; Choi, 1988; Salinska et al., 2005) and the other pathway involves oxidative glutamate toxicity through the cystine/glutamate antiporter system. Details of the neural cell death mediated by excitotoxicity are discussed in Chapters 7 and 8. The cystine/glutamate antiporter system mediates Na^+-independent exchange of cystine and glutamate (McBean, 2002).

Oxidative glutamate toxicity involves the competition between glutamate and cystine for the cystine/glutamate antiporter. This produces an imbalance in cystine homeostasis in neural cells. Because this amino acid is the precursor of glutathione (GSH) (Dubovsky et al., 1989; Pereira and Resende de Oliveira, 2000), inhibition of cystine uptake by the high concentration of glutamate inhibits GSH synthesis and exposes neural cells to oxidative stress. Antioxidants such as vitamin E, idebenone, and selegiline can block this pathway (Miyamoto et al., 1989; Schiller et al., 1993; Davis and Maher, 1994; Ortiz et al., 2001; Ayasolla et al., 2002; Pereira and Resende de Oliveira, 1997). Brain is rich in both metal ions and unsaturated fatty acids that are prone to oxidation. Growing evidence suggests that excitotoxicity and oxidative stress play a fundamental role in inducing and maintaining inflammation and oxidative stress through the activation of phospholipases, PLA_2 and PLC, cyclooxygenases, COX-1 and COX-2, stress kinases, JNK, MAPK, p38, and redox-sensitive transcription factors such as NF-κB and AP-1. The transcription factors differentially regulate the genes for proinflammatory mediators and protective antioxidant genes such as γ-glutamylcysteine synthetase, Mn-superoxide dismutase, and heme oxygenase-1 (Rahman and MacNee, 2000; Farooqui et al., 2007b). The critical balance between the induction of proinflammatory mediators and antioxidant genes and the regulation of the levels of GSH in response to oxidative stress at the site of inflammation is not fully understood. However, studies on molecular mechanisms of redox GSH regulation and gene transcription in inflammation and oxidative stress may lead to better understanding of relationship between oxidative stress and neuroinflammation (Rahman and MacNee, 2000).

Evidence supporting the interplay among oxidative stress, evolution of inflammation, and inflammatory-related responses is growing not only from an immunologic perspective (Haddad and Harb, 2005), but also through studies on molecular mechanisms associated with ROS-dependent inflammation (Gilgun-Sherki et al., 2002; Farooqui and Horrocks, 2007b). Cytokines are key players when it comes to defining an intimate relationship among reduction-oxidation (redox) signals, inflammation, and excitotoxicity (Farooqui and Horrocks, 2007a; Farooqui et al., 2007b). Identification of the molecular basis of this intricate association among excitotoxicity, oxidative stress, and inflammation comes from studies on transcription factors (Haddad, 2004; Farooqui et al., 2007b; Sun et al., 2007). Emerging information on glutamate-mediated changes in glutathione and cytokines is refining the role of oxidative signaling in terms of regulating the milieu of inflammatory mediators, ostensibly via the modulation (expression/repression) of oxygen- and redox-responsive transcription factors (Yoneda et al., 2001; Haddad and Harb, 2005). Oxygen sensing, oxidative stress and inflammation, and nuclear factor-kappaB (NF-κB) are key players that determine antioxidant/ prooxidant responses to oxidative and inflammatory challenges (Haddad, 2004). Thus antioxidant/ prooxidant mechanisms that modulate interplay among excitotoxicity, neuroinflammation, and oxidative stress are cornerstones of cell death associated with neurological disorders (Wullner et al., 1999; Gilgun-Sherki et al., 2002; Margaill et al., 2005; Farooqui et al., 2007b). These studies support the view that excitotoxicity, inflammation, and oxidative stress cooperate in the pathogenesis of brain damage in neurological disorders (Farooqui and Horrocks, 1994, 2007b).

9.2 Effects of Oxidative Stress on Neural Cell Membrane Components

In spite of its low weight, brain utilizes about 20% of basal oxygen consumption. Most of the oxygen, 95%, is utilized by mitochondria during oxidative phosphorylation. ATP production is required to maintain the ionic gradient, high intracellular K^+, low Na^+ and very low free Ca^{2+}, and to maintain the membranes (Purdon et al., 2002). Under normal conditions, 1–2% of the consumed oxygen is converted to ROS through mitochondria (Emerit et al., 2004). In brain tissue, oxidative stress is characterized by a major increase in the amount of oxidatively damaged cellular components and mediated through ROS, which include superoxide, singlet oxygen, hydrogen peroxide, and the hydroxyl radical.

Monoamine oxidase, tyrosine hydroxylase, and L-amino acid oxidase generate hydrogen peroxide as their reaction product. Hydrogen peroxide is also produced by auto-oxidation of catecholamines in the presence of vitamin C. Moreover, phospholipase A_2 (PLA_2), cyclooxygenase (COX), and lipoxygenase (LOX), the enzymes associated with arachidonic acid release and the arachidonic acid cascade, generate superoxide radicals that enhance oxidative stress and eicosanoids that promote neuroinflammation (Farooqui et al., 2007b). The chemical reactivity of ROS varies considerably. Hydroxyl radicals, which are formed *in vivo* in the presence of iron and copper from hydrogen peroxide by Fenton or Haber-Weiss reactions, are highly reactive, whereas the superoxide radical is minimally reactive. Neural cells generate superoxide as a byproduct of respiration in mitochondria and from enzymic reactions in the plasma membrane and endoplasmic reticulum. Among neural cells, neurons are most susceptible to direct oxidative injury by ROS. ROS can also indirectly contribute to brain damage by activating a number of cellular pathways inducing the expression of stress-sensitive genes and proteins to produce oxidative injury (Wang et al., 2006). In brain tissue oxidative damage also initiates glial cell-mediated inflammatory processes through the generation of proinflammatory cytokines and chemokines. This process results in secondary neuronal damage (Farooqui et al., 2007b).

ROS play a critical role in initiation of apoptosis through changes in mitochondrial permeability, and poly(ADP-ribose) polymerase (PARP) activation. These processes provide additional mechanisms for oxidative damage in acute neural trauma and neurodegenerative diseases (Warner et al., 2004). PARP activation is accompanied by the depletion of nicotinamide adenine dinucleotide, NAD. Depletion of NAD leads to depletion of ATP, which in turn promotes neuronal cell death (Zhang et al., 1994; Ishikawa et al., 1999).

The superoxide oxide radical interacts with nitric oxide to produce peroxynitrite at a rate which three times faster than the rate at which superoxide dismutase utilizes superoxide (Beckman, 1994). Peroxynitrite is capable of diffusing to distant places in neural cells where it induces lipid peroxidation and may be involved in synaptosomal and myelin damage (Van der Veen and Roberts, 1999). After protonation and decomposition, peroxynitrite produces more hydroxyl radicals. This mechanism of hydroxyl radical generation is not dependent on redox active metal ions and may be involved in initiating lipid and protein peroxidation *in vivo* (Warner et al., 2004).

It is well known that prostaglandin H_2 synthase-1 (PGHS-1) has two catalytic centers. The cyclooxygenase center adds 2 oxygen atoms to arachidonic acid to form the hydroxy endoperoxide called PGG_2. The peroxidase active center reduces PGG_2 to PGH_2. Peroxynitrite activates the peroxidase site of PGHS-1 independently of the cyclooxygenase site (Deeb et al., 2006). After peroxynitrite exposure, holoPGHS-1 cannot metabolize arachidonic acid. In contrast, apoPGHS-1 retains full activity once reconstituted with heme. Incubation of holoPGHS-1 with peroxynitrite results in a decrease in heme absorbance, but to a lesser extent than the loss in enzymic function, suggesting the contribution of more than one process during enzymic inactivation (Deeb et al., 2006). Hydroperoxide scavengers increase the enzymic activity, whereas hydroxyl radical scavengers do not protect from the effects of peroxynitrite (Deeb et al., 2006). Mass spectral analyses indicate that tyrosine 385 (Tyr 385) is a target for nitration by peroxynitrite only when heme is present. The heme may play a role in catalyzing Tyr 385 nitration by peroxynitrite and inactivation of PGHS-1 (Deeb et al., 2006).

During non-enzymic lipid peroxidation, arachidonic acid is converted to prostaglandin-like F_2-isoprostanes (IsoP) and docosahexaenoic acid is transformed into neuroprostanes (NeuroP) (Fam and Morrow, 2003; Roberts, II et al., 1998; Tanaka et al., 2006). The addition of submicromolar concentrations of these metabolites to neuronal cultures results in apoptotic cell death, initial depletion of reduced glutathione, and increased ROS production (Tanaka et al., 2006). IsoP-mediated cell death also involves 12-lipoxygenase activation and phosphorylation of extracellular signal-regulated kinase 1/2 and the redox sensitive adaptor protein p66 (shc), which is associated with caspase-3 cleavage. IsoP also potentiate oxidative glutamate toxicity. IsoP are the most reliable markers of lipid peroxidation *in vivo* (Musiek et al., 2005b; Tanaka et al., 2006). Information on the molecular mechanism of their generation may provide valuable knowledge about the pathogenesis of human neurodegenerative diseases (Fam and Morrow, 2003; Montine and Morrow, 2005; Musiek et al., 2005a). As stated in Chapters 6 and 7, ROS can cause cellular damage through their attack on neural membrane phospholipids, proteins, and DNA.

Enzymic oxidation of arachidonic acid generates 4-hydroxynonenal (4-HNE), oxidative damage to proteins increases protein carbonyl content and induces the formation of nitrotyrosine by peroxynitrite. DNA oxidation produces 8-hydroxy-2'-deoxyguanosine (8-OHdG) (Dalle-Donne et al., 2003; Farooqui and Horrocks, 2006; Soule et al., 2006). Another index of oxidative stress is the activation of the redox sensitive transcription factor, NF-κB, which modulates the expression of genes associated with inflammation and oxidative stress (Chapters 6 and 7). Included are genes for secretory phospholipase A_2 ($sPLA_2$), cyclooxygenase-2 (COX-2), superoxide dismutase (SOD), inducible nitric oxide synthase (iNOS), and cytokines (TNF-α, IL-1β, IL-6). The expression of these proinflammatory enzymes and cytokines intensifies neuroinflammation and oxidative stress (Farooqui et al., 2007a,b). Collective evidence suggests that the pathophysiology of neurological disorders may involve multiple pathways associated with excitotoxicity and oxidative stress (Farooqui and Horrocks, 2007b).

Although the involvement of oxidative stress in the pathogenesis of ischemia, spinal cord trauma, and head injury can be explained through the concept of exhaustion of antioxidants by oxygen radicals, it does not fit with the situation in chronic neurodegenerative diseases (Perry et al., 2002). Oxidative stress in AD may serve as a homeostatic response to stress resulting in a shift of neuronal priority from normal function to basic survival (Perry et al., 2002). Evidence that antioxidants can affect the clinical course of neurodegenerative diseases is limited. Antioxidants such as α-tocopherol, γ-tocopherol, ascorbic acid, GSH ethyl ester, and a combination of ascorbate and α-tocopherol have different capacities to inhibit IsoP and NeuroP generation. Among the above, the most potent effects on IsoP formation are observed for ascorbate, GSH ethyl ester, and the α-tocopherol-ascorbate combination (Montine et al., 2003). α-Tocopherol has no effect at NeuroP formation, even at 100 mM. The dose-dependent effect of antioxidants was less potent for NeuroP than IsoP. Collectively these studies suggest that α-tocopherol alone may not be an effective antioxidant to block oxidant-mediated stress in the neural cells (Montine et al., 2003). A mixture of antioxidants with a range of redox potentials is required.

In addition to vitamin E, intracellular levels of ROS are controlled by a number of free radical scavenging enzymes including superoxide dismutase, catalase, peroxidase, heme oxygenase, proteins including ferritin and apolipoprotein D,

Table 9.1 Antioxidant enzymes and endogenous antioxidant in brain tissue

Antioxidant enzyme/antioxidant	Reference
Superoxide dismutase	(Loomis et al., 1976; Brooksbank and Balazs, 1983, 1984)
Catalase	(Ambani and Van Woert, 1973; Houdou et al., 1991)
Glutathione peroxidase	(Brooksbank and Balazs, 1984; Kaplan and Ansari, 1984; Kish et al., 1985)
Heme oxidase	(Schipper et al., 1995; Hara et al., 1996; Takahashi et al., 1996)
Glutathione	(Sofic et al., 1987; Slivka et al., 1987; Perry et al., 1988)
Lipoic acid	(Barbiroli et al., 1995; Schwedhelm et al., 2004)
Vitamin C	(Mefford et al., 1981; Terpstra and Gruetter, 2004)
Vitamin E	(Metcalfe et al., 1989; Adams, Jr. et al., 1991)
Melatonin	(Kopp et al., 1980)
Coenzyme Q_{10}	(Barbiroli et al., 1997; Matthews et al., 1998)
Plasmalogens	(Zoeller et al., 1999; Engelmann et al., 1992; Farooqui et al., 2000)
Gangliosides	(Ferrari et al., 1993; Favaron et al., 1988)

and low molecular weight endogenous antioxidants such as glutathione, reduced coenzyme Q_{10}, vitamin C, melatonin, uric acid, lipoic acid, plasmalogens, and ganglioside GM1 (Table 9.1). Among the low molecular endogenous antioxidants, glutathione and reduced coenzyme Q_{10}, plasmalogens, and gangliosides are synthesized endogenously and are neural cell components, whereas vitamin E, vitamin C, lipoic acid, lycopene, and docosahexaenoic acid are derived from the diet.

Antioxidant strategies have been proposed for the treatment of neurodegenerative diseases (Willcox et al., 2004; Gilgun-Sherki et al., 2001, 2006; Farooqui et al., 2007b; Hsiao et al., 2004; Contestabile, 2001; Zhao, 2005). These strategies include: inhibition of ROS production, scavenging of ROS, and increase of ROS degradation by using agents mimicking the enzymic activity of endogenous antioxidants. None of these strategies has been successful for the treatment of neurological disorders. A cocktail of multiple antioxidants with anti-inflammatory agents may be a better neuroprotectant in neurodegenerative diseases than antioxidants alone (Wang et al., 2006).

9.3 Brain Oxidant and Antioxidant Proteins in Glutamate-Mediated Neurotoxicity

9.3.1 Multiple Forms of PLA_2, COX, LOX in Glutamate Neurotoxicity

The treatment of neuronal cultures with glutamate and its analogs results in a dose-dependent increase in $cPLA_2$, COX, and LOX activities and arachidonic acid, eicosanoids and ROS release (Dumuis et al., 1988; Lazarewicz et al., 1990; Farooqui et al., 2003b; Adams et al., 1996; Manev et al., 1998). Glutamate receptor antagonists (Sanfeliu et al., 1990; Kim et al., 1995; Farooqui et al., 2003b), $cPLA_2$, COX, and LOX inhibitors, and antioxidants (Farooqui et al., 2003a, 2005; Koistinaho et al., 1999; Pepicelli et al., 2002, 2005) can block this. These studies strongly suggest that the stimulation of $cPLA_2$, COX, LOX, and ROS generation is a process mediated by glutamate receptors associated with neuroinflammation and oxidative stress. Thus $cPLA_2$, COX, and LOX can be regarded as oxidative enzymes because they specifically release arachidonic acid, which generates eicosanoids, 4-hydroxynonenal (4-HNE), and ROS (Sun et al., 2007). The activity of group IIA secretory PLA_2 is stimulated by KA in neuronal cultures and rat brain (Thwin et al., 2003). It can be regarded as an inflammatory and oxidative enzyme because its action along with $cPLA_2$ potentiates the release of arachidonic acid from neural membrane glycerophospholipids and arachidonic acid is metabolized to eicosanoids and ROS (Balboa et al., 2003). Finally, kainic acid also up-regulates plasmalogen-selective PLA_2 activity in neuronal cultures. Bromoenol lactone, a specific inhibitor of PlsEtn-PLA_2, can retard this up-regulation in enzymic activity (Farooqui et al., 2003a). This enzyme can be considered as an anti-inflammatory and antioxidant enzyme because it liberates docosahexaenoic acid (DHA) in response to physiologic, pharmacologic, and pathologic stimuli.

DHA acts as an anti-excitotoxic agent (Högyes et al., 2003) and antioxidant (Hossain et al., 1998, 1999) and is metabolized to resolvins and neuroprotectins (Bazan, 2005a; Serhan, 2005). Docosahexaenoic acid inhibits neuronal apoptosis mediated by soluble β-amyloid oligomers and significantly increases neuronal survival by preventing cytoskeleton perturbations, caspase activation, and promoting extra signal-related kinase pathway (Park et al., 2004). DHA attenuates endothelial cyclooxygenase-2 induction through NADPH oxidase and protein kinase Cε inhibition (Massaro et al., 2006). DHA enhances the antioxidant response of human fibroblasts by up-regulating γ-glutamyl-cysteinyl ligase and glutathione reductase activities (Arab et al., 2006).

Docosahexaenoic acid, resolvins, and neuroprotectins are endogenous neuroprotectants (Serhan, 2005; Mukherjee et al., 2004; Farooqui and Horrocks, 2007b). They retard inflammation and oxidative stress by inhibiting p65 subunit transcription factor NF-κB, preventing cytokine secretion, blocking the synthesis of eicosanoids, blocking toll like receptor-mediated activation of macrophages, and reducing leukocyte trafficking (Farooqui and Horrocks, 2006; Farooqui et al., 2006; Duffield et al., 2006; Massaro et al., 2006; Li et al., 2005). DHA has been used for the treatment of neurological disorders (Table 9.2). The dietary depletion of DHA activates caspases and decreases NMDA receptors in Tg2576 mice brains. Dietary supplementation of DHA partially protects from NMDA receptor subunit loss (Calon et al., 2005). DHA promotes neuronal survival by facilitating neural membrane translocation/activation of Akt through its capacity to increase PtdSer in neural membranes (Akbar et al., 2005). DHA also stimulates extracellular signal regulated kinase in photoreceptors (German et al., 2006). Collectively these studies suggest that glutamate-mediated alterations in activities of multiple forms of PLA_2 and changes in levels of PLA_2-derived lipid mediators are closely associated with oxidative, inflammatory, and anti-inflammatory responses in neurodegenerative diseases. Elucidation of the signaling pathways linking multiple forms of PLA_2 with

Table 9.2 Effects of endogenous neuroprotectants (ganglioside, lipoic acid, and vitamin E) in neurological disorders

Endogenous neuroprotectant	Neurological disorder	Therapeutic effect	Reference
GM1 ganglioside	Stroke	Beneficial	(Hernandez et al., 1994)
Vitamin E		Beneficial	(Zhang et al., 2004)
GM1 ganglioside	Spinal cord injury	Beneficial	(Geisler et al., 1991)
Vitamin E		Beneficial	(Hall et al., 1992)
GM1 ganglioside	Alzheimer disease	Beneficial	(Svennerholm, 1994; Matsuoka et al., 2003)
Vitamin E		Beneficial	(Sano et al., 1997; Morris et al., 2002; Ayasolla et al., 2002)
GM1 ganglioside	Parkinson disease	Beneficial	(Schneider, 1992)
Vitamin E		Beneficial	(Ebadi et al., 1996)
GM1 or GM4 ganglioside	EAE EAE	Beneficial	(Mullin et al., 1986)
Vitamin E		Beneficial	(Gilgun-Sherki et al., 2005)
Lipoic acid	R6/2 transgenic mice	Beneficial	(Surai and Sparks, 2001)

downstream inflammatory and anti-inflammatory signal transduction processes as well as oxidative stress may be important in understanding the complex mechanisms associated with neurodegeneration in various neurological disorders (Farooqui et al., 2006; Phillis et al., 2006; Farooqui and Horrocks, 2007b).

9.3.2 Superoxide Dismutase in Glutamate Neurotoxicity

Superoxide dismutase (SOD) catalyzes the reaction between two molecules of superoxide radical with two protons to generate one molecule of hydrogen peroxide and one molecule of oxygen (ground state). Two SOD exist, copper zinc SOD, and manganese SOD. Most CuZnSOD is located in the cytosol, but some appears to be associated with the nucleus and the space between inner and outer mitochondrial membranes, whereas MnSOD is largely located in the mitochondria. Mice defective in CuZnSOD appear normal, although as they age, neurological damage and cancers have been claimed to develop at an accelerated rate. Those lacking in MnSOD die within the first 10 days of life with cardiac abnormalities accompanied by severe mitochondrial abnormalities in the heart and other tissues. Those that survive the first 10 days quickly succumb to a variety of abnormalities including anemia and neurodegeneration.

Hippocampal CuZnSOD mRNA levels are increased after systemic injection of kainate at 7 days after injection. The increased mRNA levels correlated with increased specific activities of the enzyme. The CA1 and CA3 neurons lost their CuZnSOD–like immunoreactivity whereas glial cells showed intense immunoreactivity at 3 and 7 days after KA (Kim et al., 2000a,b). Overexpression of CuZnSOD did not protect against OGD- or kainic-acid-induced toxicity *in vivo*. On the other hand, it significantly worsened toxicity, hydrogen peroxide accumulation, and lipid peroxidation induced by kainic acid (Zemlyak et al., 2006). Cells may be unable to detoxify the excess hydrogen peroxide produced as a result of CuZnSOD overexpression.

As with CuZnSOD, CA1 and CA3 had decreased immunoreactivity to MnSOD, but intense immunoreactivity was observed in astrocytes at 3 and 7 days after kainate injection. (Kim et al., 2000a,b). The early loss of both CuZu SOD and Mn SOD are possible causes of increased oxidative stress after kainate-induced excitotoxic injury.

9.3.3 Catalase and Glutathione Peroxidase in Glutamate Neurotoxicity

Catalases convert two molecules of hydrogen peroxide to two molecules of water and one molecule of oxygen (ground state). Hydrogen peroxide can also be removed by the action of peroxidases. The catalase activity of animal and plant tissues is largely or completely located in subcellular organelles known as peroxisomes. Catalase activity was significantly increased at 5 days after kainate administration, and returned to the pre-injection level by 3 weeks. Lipid peroxidation, as indicated by malondialdehyde (MDA) concentration, exhibited an early significant increase at 8 and 16 h, followed at 48 h and 5 days by a significant decrease (Bruce and Baudry, 1995).

Glutathione peroxidases (GPX) remove hydrogen peroxide by coupling its reduction to water with oxidation of reduced glutathione (GSH). Glutathione peroxidase enzymes degrading hydrogen peroxide are widely distributed in animal tissues and are specific for GSH as a hydrogen donor. However, they can act on peroxides other than hydrogen peroxide, and can catalyze GSH-dependent reduction of fatty acid hydroperoxides. Oxidized glutathione (GSSG) is reduced back to glutathione by the enzyme glutathione reductase.

$$\mathrm{GSSG} + \mathrm{NADPH} + \mathrm{H}^+ - - > 2\mathrm{GSH} + \mathrm{NADP}^+$$

The NADPH required is produced by several enzyme systems but the best known is the oxidative pentose phosphate pathway. Glutathione peroxidase and catalase thus co-operate to remove hydrogen peroxide *in vivo*.

Glutathione peroxidase activity increased significantly at 5 days post-injection, and remained elevated at 3 weeks after kainate injury (Bruce and Baudry, 1995). In cultured neurons, over expression of genes for the antioxidant enzymes catalase (CAT) or glutathione peroxidase (GPX) resulted in increased enzymic activity of the cognate protein. These enzymes also decreased the neurotoxicity induced by kainic acid, glutamate, sodium cyanide, and oxygen/glucose deprivation (Wang et al., 2003). Studies *in vivo* however, showed that overexpression of GPX in mice resulted in increased rather than decreased KA susceptibility including increased seizure activity and neuronal hippocampal damage (Boonplueang et al., 2005). This could be due to alterations in the redox state of the glutathione system resulting in elevated glutathione disulfide (GSSG) levels that, in turn, may directly activate NMDA receptors. Conversely, mice with knock-out of the glutathione peroxidase gene are resistant to kainate-induced seizures and neurodegeneration (Jiang et al., 2000). A possible explanation for these observations is that at low micromolar concentrations, glutathione depolarizes neurons by binding to its own receptors and modulates glutamatergic excitatory neurotransmission by displacing glutamate from its ionotropic receptors, whereas at higher concentrations, reduced glutathione may increase N-methyl-D-aspartate receptor responses by interacting with its redox sites (Regan and Guo, 1999).

9.3.4 Heme Oxygenase in Glutamate Neurotoxicity

Heme oxygenases catalyze the breakdown of biliverdin, carbon monoxide, and oxygen, with release of one atom of iron, Fe^{3+}. The biliverdin is quickly converted to bilirubin. Bilirubin is a powerful scavenger of peroxyl radicals and singlet oxygen. However, in excess, bilirubin can also sensitize the formation of singlet oxygen in the presence of light. High levels in premature babies are toxic in the brain. Fe^{3+} is a pro-oxidant. It is therefore not quite clear whether heme oxygenase is a pro- or antioxidant enzyme. ICV injection of kainate resulted in strong induction of HO-1 protein, but with no change in HO-2 protein levels. One day after treatment, the level of HO-1 reached a maximum and then gradually decreased over a period of 3–7 days (Matsuoka et al., 1998). Induction of HO-1 was observed in neurons at short time intervals after kainate lesions, and in astrocytes at later time intervals. The increased HO-1 is accompanied by increased bilirubin immunoreactivity in these cells, indicating that HO-1 is enzymically active (Huang et al., 2005) (Fig. 9.1). Experiments using an inhibitor to heme oxygenase (tin protoporphyrin) permeable to the blood-brain barrier showed that inhibition of heme oxygenase resulted in increased neuronal death after kainate injury (Huang et al., 2005). These results suggest that HO-1 has a net protective effect on neurons. HO-1 expression in the hippocampus decreased from approximately 2 week after kainate lesion, whereas Fe^{3+} accumulation continued in the hippocampus after that time (Huang et al., 2005). These results indicate that the Fe^{3+} increase in the hippocampus is not likely due to a mere redistribution of the Fe^{3+}, but also involved increased Fe^{3+} uptake across the blood-brain barrier.

9.3.5 Ferritin in Glutamate Neurotoxicity

One form of antioxidant defense may be the binding of excess Fe^{3+} and other transition metal ions, preventing Fe^{3+}, and other transition metal pro-oxidants from catalyzing free radical reactions. Most intracellular Fe^{3+} is stored in ferritin. Mammalian ferritins consist of a hollow protein shell 12–13 nm outside diameter

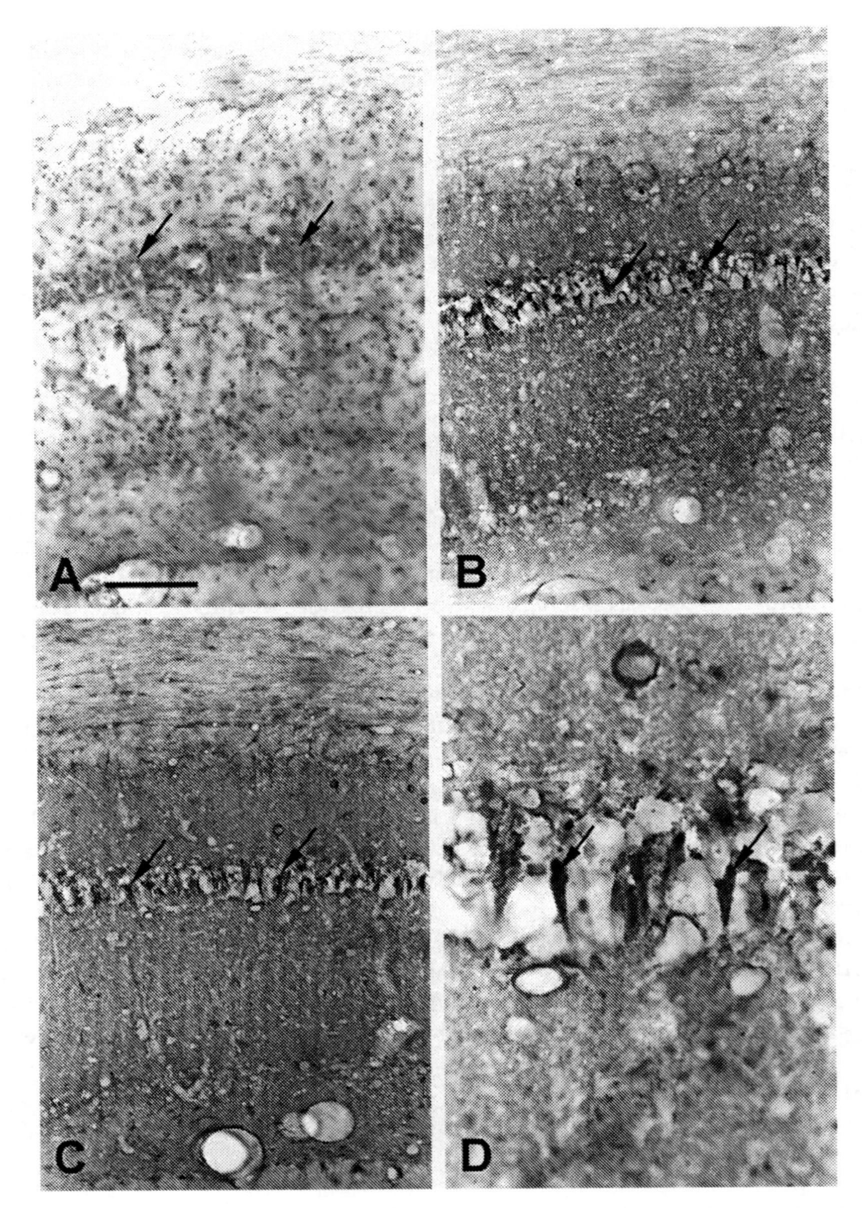

Fig. 9.1 Light micrographs of Nissl-stained and HO-1- and bilirubin-immunolabeled sections from the 3 day post-kainate lesioned hippocampus. A: Nissl-stained section, showing loss of pyramidal neurons in field CA1 (arrows). B, C: An increase in HO-1 (B) and bilirubin (C) staining is also observed in pyramidal neurons (arrows) in field CA1. D: Higher magnification, showing dense bilirubin immunoreactivity in field CA1 pyramidal neurons (arrows). The staining intensity is greater than that of reactive astrocytes. Scale bar = 150 μm for A–C, 40 μm for D. Reproduced with kind permission from Huang et al., 2005, Journal of Neuroscience Research, 80:268–278, Wiley-Liss

composed of 24 subunits. The shells surround a Fe^{3+} core that can hold up to 4500 Fe^{3+} per molecule. Iron enters ferritin as Fe^{2+}, which is oxidized to Fe^{3+} and deposited in the core as an insoluble hydrated ferric oxide. In vitro, iron can be released as Fe^{2+} by several reducing agents including ascorbate. The expression of ferritin has been studied after kainate lesions (Huang and Ong, 2005). An early induction of ferritin expression was observed in microglia and oligodendrocytes after kainate lesions, but expression peaked at 1 month after kainate injection. At 2 months after lesion, there was decreased ferritin expression concomitant with a rise in the level of ferrous ion (Fig. 9.2). The decrease in ferritin expression could have been a cause for the increase in unbound ferrous ions at late time intervals after kainate injury. This could result in a mismatch between increased iron ion accumulation as a result of increased divalent metal transporter activity in the hippocampus after kainate lesions (Ong et al., 1999, 2006; Wang et al., 2002a,b), and ability of ferritin to bind the Fe^{3+}, resulting in increased amounts of Fe^{3+} that is present in the unbound form capable of catalyzing free radical reactions.

9.3.6 Apolipoprotein D in Glutamate Neurotoxicity

Another possible form of antioxidant defense against lipid peroxidation may be the induction of proteins that bind excess lipids, thus preventing their oxidation and possible propagation of free radical damage. ApoD is an atypical apolipoprotein, and belongs to the lipocalin superfamily of proteins with the ability to bind small hydrophobic ligands. This protein increases after various forms of CNS injury, including kainate treatment (Ong et al., 1997) (Fig. 9.3), Niemann Pick C disease (Ong et al., 2002), and stroke (Rickhag et al., 2006). Clozapine increases ApoD expression in mouse brain striatum (Thomas et al., 2001). In control animals ApoD is predominantly expressed in astrocytes. In contrast, 2 weeks after clozapine injection, mice show substantially increased ApoD expression in neurons of the striatum, globus pallidus, and thalamus. Clozapine-mediated increase in ApoD expression may be associated with psychoses (Thomas et al., 2001).

Fruit flies with increased expression of their homologue of ApoD have increased longevity in a high oxygen environment (Walker et al., 2006). Conversely, flies with a knock-out showed decreased ability to handle oxidative stress and reduced longevity in the high O_2 environment. Likewise, ApoD knock out mice showed increased neuronal damage after intrastriatal nitropropionic acid (NPA) injections. Treatment of ApoD knockout fibroblasts with kainate also gave increased levels of F_2 isoprostanes, indicating increased oxidative stress, whereas external application of purified ApoD protein to hippocampal cultures reduced the levels of F_2 isoprostanes and neuronal injury (Ong et al., 2006). One possibility is that ApoD could bind to arachidonic acid preventing its oxidation to F_2 isoprostane. Another possibility, because increased cholesterol and cholesterol oxidation products were detected in the kainate-lesioned rat hippocampus (Ong et al., 2003; He et al., 2006), is that ApoD could bind to cholesterol thus reducing its oxidation to toxic oxidation products of cholesterol. It is thus likely that it represents an inducible form of

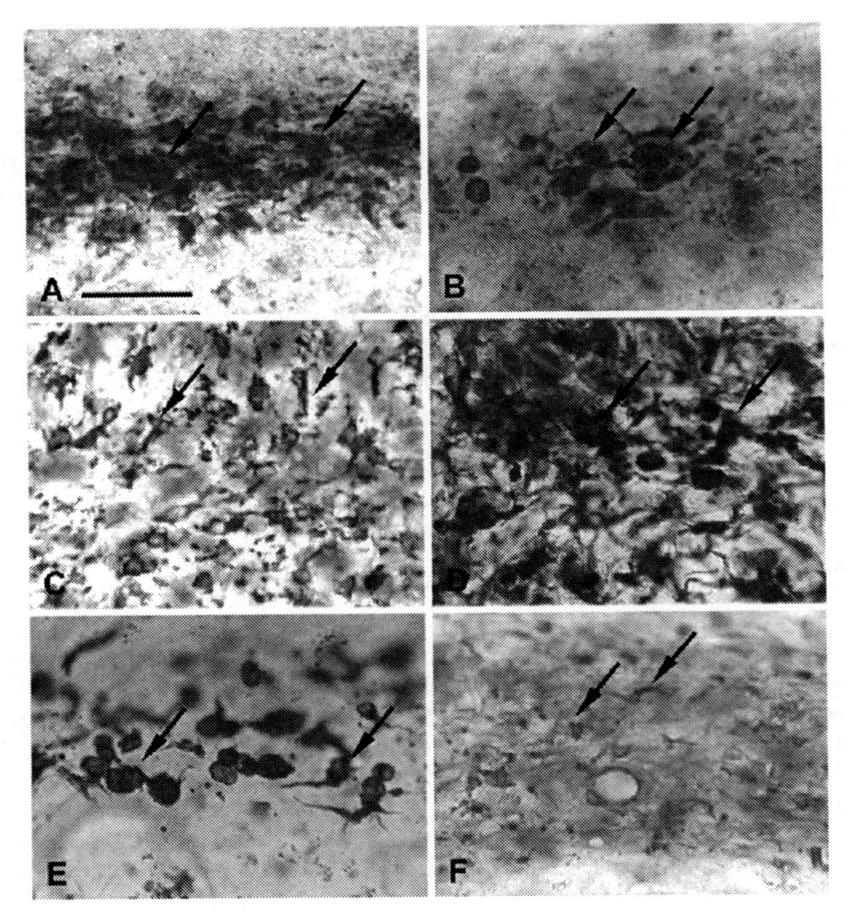

Fig. 9.2 A–C: Adjacent sections of field CA1, from a rat 4 weeks post-kainate injection. A is a section stained with Perl's stain, showing an increase in the number of ferric iron positive cells (arrows) in the degenerating CA field (see Fig. 9.1A). B is a section stained with Turnbull's blue stain, showing an increase in ferrous iron positive cells (arrows) in the degenerating CA field (see Fig. 9.1B). C is a ferritin immunostained section, showing an increased number of ferritin positive cells (arrows). D–F adjacent sections of field CA1, from a rat 8 weeks post-kainate injection. D is a section stained with Perl's stain, showing an increase in the number of ferric iron positive cells, compared to 4 weeks post-injection (see A). E is a section stained with Turnbull's blue, showing an increase in the number of ferrous iron-containing cells, compared to 4 weeks post-injection (see B). F is a ferritin-stained section, showing a decrease in the number of ferritin positive cells (arrows) compared to 4 weeks post-injection (see C). The spherical, heavily ferric or ferrous iron-laden cells shown in D and E are unlabeled for ferritin. Scale=50 μm. Reproduced with kind permission from Huang and Ong (2005) Experimental Brain Research 161:502–511. Springer.

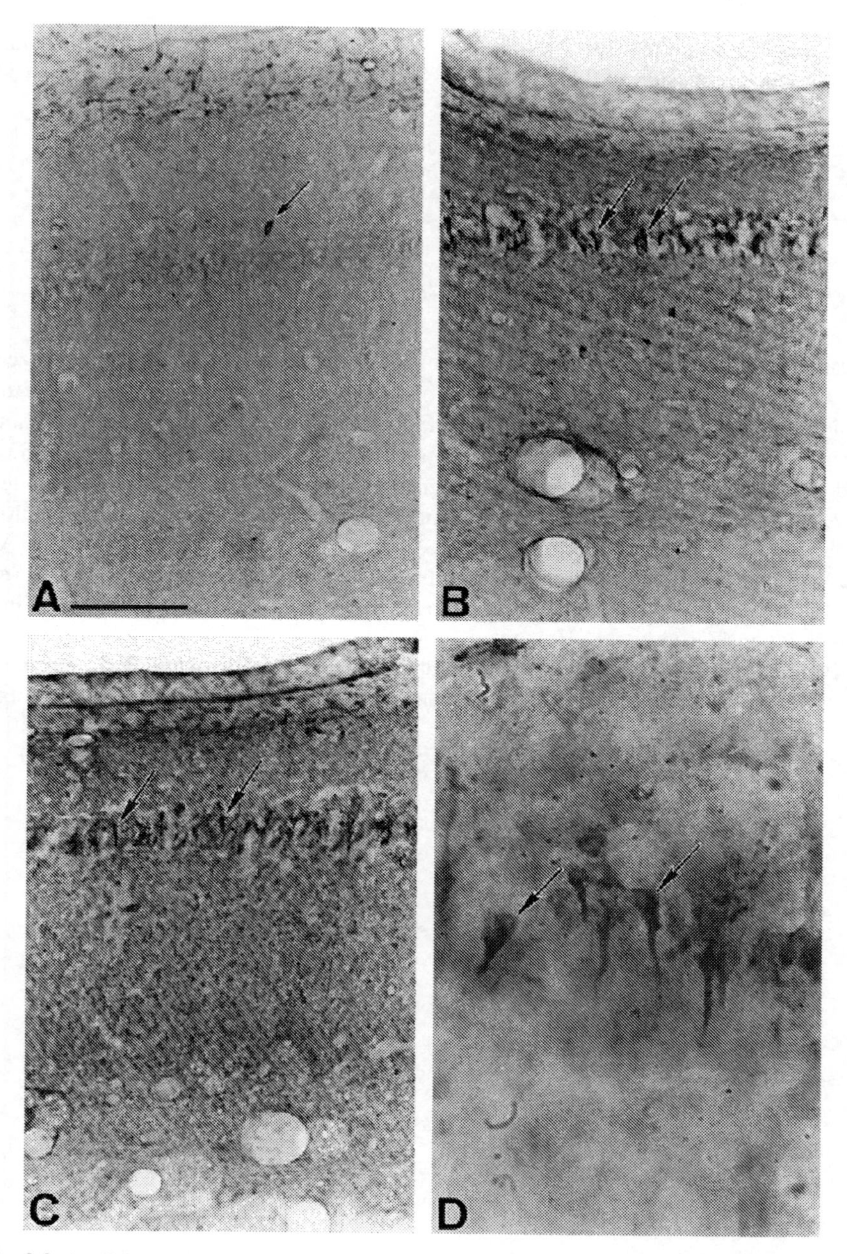

Fig. 9.3 Apolipoprotein D-immunolabeled sections of field CA1 of the hippocampus from a normal rat (A), and rats injected with kainate one (B) or 3 (C, D) days previously. Very few isolated neurons (arrow) are labeled for apolipoprotein D in the normal hippocampus (A). In contrast, large numbers of neurons are labeled for apolipoprotein D in the kainate-injected animals (B–D, arrows). The labeled neurons have features of pyramidal neurons (D, arrows). Scale bar=160 μm (A–C), 40 μm (D). Reproduced with kind permission from Ong et al., 1997, Neuroscience 79:359–367. Elsevier.

antioxidant defense protein. Further studies are necessary to elucidate factors that regulate the expression of this protein.

9.4 Low Molecular Weight Endogenous Antioxidants in Glutamate Neurotoxicity

9.4.1 Plasmalogens and Glutamate Neurotoxicity

Plasmalogens (PlsEtn) are a unique group of glycerophospholipids characterized by a long-chain enol ether (vinyl ether) bond at the *sn*-1 position of the glycerol moiety (Farooqui and Horrocks, 2004) (Fig. 9.4). The enol ether linked aliphatic moieties at the *sn*-1 position in PlsEtn consist almost exclusively of 16:0, 18:0, and 18:1 alkenyl groups. These alkenyl groups form the corresponding fatty aldehyde upon acid hydrolysis. The *sn*-2 position in the glycerol moiety of plasmalogens is enriched in arachidonic acid (AA) or docosahexaenoic acid (DHA). At the *sn*-3 position, ethanolamine and choline serve as the polar head group to form plasmenylethanolamine. ethanolamine plasmalogen, and plasmenylcholine, choline plasmalogen (Farooqui and Horrocks, 2004).

Plasmalogens protect biological structures against free radical attack (Maeba and Ueta, 2004; Engelmann, 2004) (Fig. 9.4). In neural membranes, transition metal

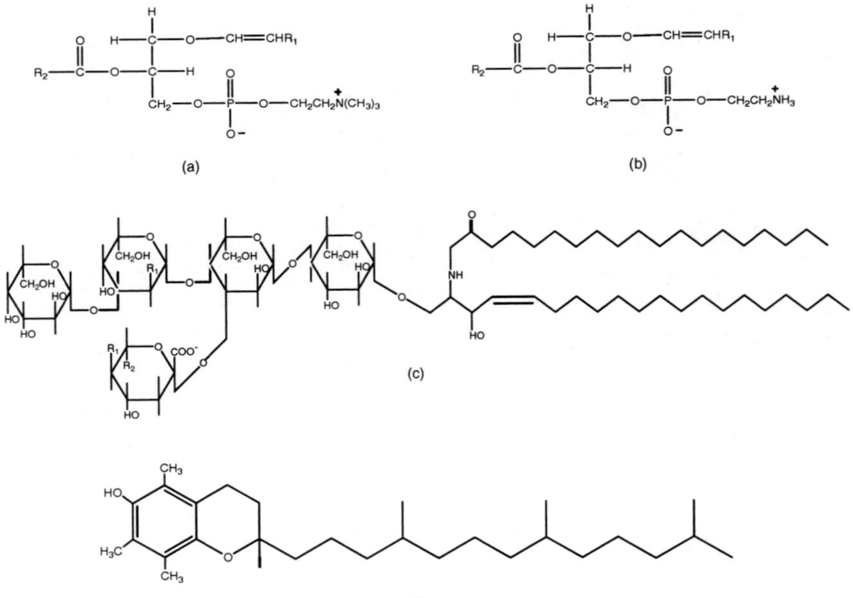

(a)

(b)

(c)

(d)

Fig. 9.4 Chemical structures of cellular lipids that act as endogenous antioxidants. Choline plasmalogens (a); ethanolamine plasmalogens (b); ganglioside (c) and vitamin E (α-tocopherol) (d)

ions (copper and iron) initiate lipid peroxidation by generating peroxyl and alkoxyl radicals from the decomposition of lipid hydroperoxides according to the following reactions.

$$LOOH + Cu^{2+}/Fe^{3+} ------- > LOO\bullet + H^{+} + Cu^{+}/Fe^{2+} \quad (9.1)$$
$$LOOH + Cu^{+}/Fe^{2+} --------- > LO\bullet + \bullet OH^{-} + Cu^{2+}/Fe^{3+} \quad (9.2)$$

Plasmalogen-containing liposomes have a strong ability to chelate transition metal ions and thereby prevent the formation of peroxyl and alkoxyl radicals (Zommara et al., 1995; Sindelar et al., 1999). Oxidation of the vinyl ether bond in plasmalogens markedly prevents the oxidation of highly polyunsaturated fatty acids. Products of plasmalogen degradation do not propagate lipid oxidation (Engelmann, 2004). This protection may also occur intramolecularly. Furthermore, ethanolamine plasmalogen protects cholesterol from oxidation by free radicals (Maeba and Ueta, 2004). Collectively these studies support the view that plasmalogens are major endogenous antioxidants that protect the polyunsaturated fatty acid and cholesterol pools at cellular and subcellular levels (Maeba and Ueta, 2004; Engelmann, 2004; Farooqui and Horrocks, 2007b). In neural membranes, the levels of plasmalogen are 25–100 times higher than that of vitamin E (Calzada et al., 1997; Hahnel et al., 1999). Both molecules are colocalized in lipoproteins and cellular membranes and are capable of scavenging peroxy radicals. However, vitamin E scavenges peroxy radicals with 20- to 25-fold higher efficiency than the plasmalogens (Hahnel et al., 1999), suggesting that vitamin E is the first line of cellular defense against oxidative stress. The other possibility is that neural cells contains a plasmalogen redox cycling system mediated by vitamin C (Yavin and Gatt, 1972) and this system plays an important role in defending neural cells from the oxidative stress that occurs during trauma and neurodegenerative diseases. Another possibility is that the protective effect of plasmalogen is due to docosahexaenoic acid (DHA) (Farooqui and Horrocks, 2004). This fatty acid is metabolized to the anti-inflammatory lipid mediators, resolvins and neuroprotectins (Serhan, 2005; Bazan, 2005a; Farooqui and Horrocks, 2007b). DHA also acts as an anti-excitotoxic agent in glutamate-mediated neurotoxicity and an antioxidant against β-amyloid-mediated oxidative stress (Högyes et al., 2003; Hashimoto et al., 2002; Hossain et al., 1999). DHA attenuates endothelial COX-2 induction through both NADPH oxidase and PKCε inhibition (Massaro et al., 2006). These studies indicate a mechanistic basis for not only for the anti-inflammatory effect of DHA but also for DHA involvement in plaque stabilization.

In brain tissue under normal conditions, neuroprotection activity (Maeba and Ueta, 2004; Engelmann, 2004) is performed by endogenous phospholipid (plasmalogens), DHA-derived lipid mediators (neuroprotectins and resolvins), and antioxidants (vitamin E and glutathione). Under pathological situations in which antioxidant systems are down regulated, plasmalogen levels are decreased, and oxidative stress overwhelms the above neuroprotective mechanisms, such as in KA-mediated neurotoxicity (Farooqui et al., 2003a), ischemia (Ray et al., 1994), and Alzheimer disease (Farooqui et al., 1997; Wells et al., 1995; Guan et al., 1999), the neurodegenerative process starts, brain damage accumulates, and the neural cell ultimately dies. Intense research efforts are underway to identify and characterize processes that induce the synthesis of plasmalogens, DHA, neuroprotectins, and resolvins (Bazan, 2005b; Farooqui et al., 2007b).

9.4.2 Gangliosides and Glutamate Neurotoxicity

Gangliosides are sialic acid-containing glycosphingolipids that are enriched in neural membranes. They contain sphingosine, fatty acid, and an oligosaccharide chain that varies in size from one to four or more monosaccharides (Fig. 9.4). They are located asymmetrically in the neural membranes with their hydrophilic oligosaccharide chains protruding from the extracellular surface and their ceramide portion inserted into the hydrophobic region of their membrane (Farooqui et al., 1988). Addition in vitro of exogenous GM1gangliosides to cell cultures and their injection *in vivo* results in their incorporation into neural membranes. This incorporation has a variety of biological effects including stabilization of neural membranes and induction of neuritogenesis. The molecular mechanism through which gangliosides exert their effect on neural membrane remains elusive. However, it is proposed that gangliosides not only regulate Ca^{2+} influx channels and Ca^{2+} exchange proteins (Ledeen and Wu, 2002), but also modulate activities of enzymes involved in signal transduction. These enzymes include adenylate cyclases, protein kinases, phospholipases A_2, PLC, and Na^+,K^+ATPases (Leon et al., 1981; Partington and Daly, 1979; Yang et al., 1994a,c; Yates et al., 1989). They inhibit IL-2-dependent T-cell proliferation (Molotkovskaya et al., 2002). Gangliosides are also involved in various receptor-mediated processes including induction of apoptosis (Bharti and Singh, 2000) and suppression of NK cytotoxicity (Grayson and Ladisch, 1992), thereby reducing anti-tumor immunity. Neuroprotective effects of GM1 ganglioside were demonstrated in glutamate-mediated neurotoxicity (Favaron et al., 1988; Phillis and O'Regan, 1995), acute neural trauma such as ischemia and spinal cord injury, and neurodegenerative diseases such as Alzheimer disease and Parkinson disease (Ala et al., 1990; Geisler et al., 1991; Flicker et al., 1994; Sanjurjo et al., 1997; Svennerholm, 1994; Levi et al., 1989; Uozumi et al., 1997) (Table 9.3). It up regulates catalase activity and scavenges free radicals generated during ischemic/reperfusion injury (Fighera et al., 2004). Collective evidence suggests that systemic administration of GM1 ganglioside reduces ischemia-evoked glutamate and aspartate release and oxidative stress. Furthermore, GM1 ganglioside rescues neuronal cells from apoptotic death through Trk nerve growth factor receptor as well as by trkB and tyrosine kinase receptors (Molaschi et al., 1996). Gangliosides inhibit lipid peroxidation in synaptosomes and phagocytic cells (Avrova et al., 2002). Thus it's tempting to speculate that GM1 ganglioside may act as a membrane stabilizer, an anti-excitotoxic, and antioxidant in brain tissue. In contrast, addition of GM3 ganglioside or overexpression of the GM3 synthase gene produces glutamate-mediated cell death in hippocampal cell line HT22 (Sohn et al., 2006). GM3 ganglioside strongly inhibits the plasmalogen-selective PLA_2 in brain. This prevents the release of DHA, the precursor of protective compounds (Yang et al., 1994b; Latorre et al.,

Table 9.3 Beneficial effects of DHA in neurological disorders

Neurological disorders	Oxidative stress	Changes in DHA levels	EPA/DHA treatment	Reference
Ischemia	Increased	Decreased	Beneficial	(Högyes et al., 2003)
Alzheimer disease	Increased	Decreased	Beneficial	(Hashimoto et al., 2005; Olivo and Hilakivi-Clarke, 2005; Cole et al., 2005; Puskás et al., 2003; Morris, 2006; Schaefer et al., 2006)
Parkinson disease	Increased	–	Beneficial	(Samadi et al., 2006)
Huntington disease	Increased	–	Beneficial	(Puri, 2005; Das and Vaddadi, 2004; Van Raamsdonk et al., 2005; Puri et al., 2005)
Epilepsy	Increased	Decreased	Beneficial	(Yuen et al., 2005)
Spinal cord injury	Increased	–	Beneficial	(King et al., 2006)
Multiple sclerosis	Increased	Decreased	Beneficial	(Nordvik et al., 2000)
AIDS dementia	Increased	–	Beneficial	(Das, 2005; Pocernich et al., 2005)
Peroxisomal disorders	Increased	Decreased	Beneficial	(Martínez et al., 2000)
Schizophrenia	Increased	Decreased	Beneficial	(Horrobin, 2003b; McNamara et al., 2006)
Cystic fibrosis	Increased	–	Beneficial	(Cawood et al., 2005; Guilbault et al., 2004)
Depression and mood disorders	–	Decreased	Beneficial	(Zanarini and Frankenburg, 2003; McNamara et al., 2006; Freeman, 2006; Leonard and Myint, 2006)

2003). RNA interference-mediated silencing of GM3 synthase protects HT22 cells from glutamate-mediated cell death. Collective evidence suggests that GM1 and GM3 gangliosides act very differently in brain tissue and more studies are required on the molecular mechanism of action of various gangliosides in brain tissue.

9.4.3 Vitamins C and E and Glutamate Neurotoxicity

Vitamin C (ascorbate) (Fig. 9.5) has the ability to act as a reducing agent, i.e. it will tend to reduce more reactive species. This ability to reduce Fe^{3+} to Fe^{2+}may be important in promoting iron uptake in the gut. Oxidation of ascorbate by reaction with reactive oxygen species or reactive nitrogen species seems to lead to its depletion. In vitro, vitamin C can also exert pro-oxidant properties. Fe^{3+} can react with ascorbate to form Fe^{2+} and the semi-dehydroascorbate or ascorbyl radical. The latter can react with hydrogen peroxide to form Fe^{3+}, the hydroxyl radical and a hydroxide anion. A key question with regard to the pro- or anti- oxidant effects of ascorbate may therefore be the availability of transition metal ions. Neurons main-

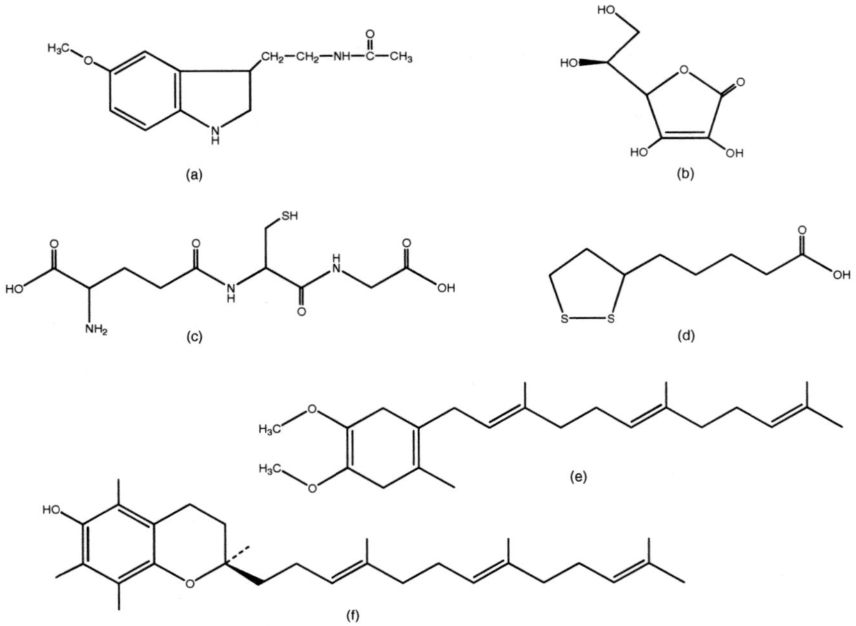

Fig. 9.5 Chemical structures of low molecular weight antioxidants Melatonin (a); ascorbic acid (Vitamin C) (b); glutathione (c); lipoic acid (d); α-tocopherol (Vitamin E) (e); and α-tocotrienol (f)

tain relatively high intracellular concentrations of ascorbate (Mori, 2006). Glutamate causes an acute concentration-dependent efflux of ascorbate from SH-SY5Y neuroblastoma cells. The anion channel blocker 4,4'-diisothiocyanatostilbene-2,2'-disulfonic acid can block this efflux (May et al., 2006). Intracellular ascorbate does not affect radiolabeled glutamate uptake, indicating absence of heteroexchange. Increased production of free radicals has been detected in the hippocampus after kainate injury (Masumizu et al., 2005). Administration of ascorbate reduces kainate-induced damage of the rat hippocampus, measured by means of the gliotic marker ligand [^{3}H]PK11195. Ascorbate is significantly effective at doses of 30 mg kg^{-1} and above, with total protection against kainate at 50 mg kg^{-1}. Histologically, ascorbate at 50 mg kg^{-1} prevented kainate-induced neuronal loss in the hippocampal CA1 and CA3a cell layers (MacGregor et al., 1996). The mechanism of how ascorbate treatment affects progression of kainate lesions at long time intervals after neural injury, when there is substantial accumulation of Fe^{3+} in the glial scar, is unknown.

Vitamin E includes naturally occurring substances, the α-tocopherols and α-tocotrienols (Fig. 9.5) (Khanna et al., 2003). The most biologically effective form in animals is RRR-α- tocopherol. α-Tocopherol and α-tocotrienols inhibit lipid peroxidation largely because they scavenge lipid peroxyl radicals at a faster rate than these radicals can react with fatty acid side chains or membrane proteins. In addition, α-tocopherols might protect membranes by quenching singlet oxygen. During its action as a chain breaking antioxidant, α-tocopherol is consumed and converted to the radical form. Mechanisms for reducing the radical back to α-tocopherol might involve synergy between vitamin E and vitamin C with ascorbate being converted to the semi-dehydroascorbate or ascorbyl radical and subsequently broken down. A study on use of α-tocopherol or γ-tocopherol in excitotoxic injury has shown that peripherally administered α-tocopherol and γ-tocopherol were equally effective at suppressing acute oxidative damage, F_2 and F_4 isoprostanes, from direct excitotoxicity caused by KA. In contrast, peripherally administered α-tocopherol, but not a γ-tocopherol, was effective at suppressing delayed neuronal oxidative damage from the innate immune response of activated glia (Milatovic et al., 2005). These data imply that AT may be more broadly protective of cerebrum from oxidative damage in different disease contexts. Vitamin E however, failed to show anticonvulsant effects after kainate injection (Levy et al., 1992).

α-Tocotrienol, the other form of vitamin E, possesses properties not shared by α-tocopherols (Packer et al., 2001; Roy et al., 2002). Nanomolar concentrations of α-tocotrienol are more effective antioxidants than α-tocopherol (Khanna et al., 2006). At concentrations of 0.1–10 μM, H_2O_2-tocotrienol significantly suppresses H_2O_2-induced cell death and provides protection against S-nitrocysteine-induced cell death and SIN-1 toxicity in a dose-dependent manner (Osakada et al., 2004). Primary neurons treated with α-tocotrienol maintained healthy growth and motility even in the presence of excess glutamate. Similarly, suppression of the early activation of c-Src kinase and 12-lipoxygenase by microinjection with nanomolar doses of α-tocotrienol into the cytosol, protects HT4 cells against glutamate-induced cell death (Khanna et al., 2005). In addition α-tocotrienol down-regulates 3-hydroxy-3-methylglutaryl-coenzyme A reductase, a key enzyme of the mevalonate pathway (Packer et al., 2001). In mice α-tocotrienol and γ-tocopherol also prevent cerebral infarction induced by middle cerebral artery occlusion (Horrobin, 2003a). Collective evidence suggests that α-tocotrienol is a naturally occurring antioxidant that protects neuronal cultures against glutamate-induced neurodegeneration (Roy et al., 2002; Khanna et al., 2003, 2006).

9.4.4 Melatonin and Glutamate Neurotoxicity

Melatonin (Fig. 9.5), in vitro, probably donates hydrogen from the NH- group, but at concentrations several orders of magnitude higher than those present *in vivo*. However, high doses of melatonin do seem to induce synthesis of antioxidant defense enzymes such as glutathione peroxidase in animals. Its antioxidant effects *in vivo* may be indirect rather than direct. Administration of melatonin resulted in reduction in the number of tunel-positive cells and loss of Nissl-stained cells after kainate treatment, indicating that it might be effective in reducing damage associated with excitotoxicity (Uz et al., 1996). Conversely, rats that were made melatonin deficient by pinealectomy showed greater neurodegeneration in two models of focal brain ischemia and glutamate receptor mediated epilepsy-like seizures (Manev et al., 1996). Melatonin was also effective in reducing seizures. Intracerebro-ventricular pre-treatment with melatonin attenuated the convulsant effects of ICV administered kainate, quinolinate, and glutamate (Lapin et al., 1998). Intraperitoneally injected melatonin at a single dose of 5 mg/kg before kainate administration also significantly reduced neurobehavioral changes such as tremors and seizures and almost completely attenuated the increased lipid peroxidation and morphological damage induced by kainate (Tan et al., 1998). Collective evidence suggests that melatonin may be protective against central nervous system pathologies involving excitotoxicity or where oxidative damage may contribute to mitochondrial dysfunction.

9.4.5 Glutathione and Glutamate Neurotoxicity

Glutathione (Fig. 9.5) is a co-factor for the glutathione peroxidase family. It is also involved in many metabolic processes including ascorbic acid metabolism, maintaining communication between cells, and generally preventing protein-SH from oxidizing and cross linking. In vitro, glutathione can react with free radicals to generate thiyl ($GS^{\cdot}$) radicals. Thiyl radicals can generate superoxide radicals. Hence SOD might co-operate with GSH in helping to remove free radicals *in vivo*, as might ascorbate. A decrease in glutathione immunoreactivity was observed in hippocampal pyramidal neurons at short time intervals after kainate injury. Conversely, increased glutathione was observed in astrocytes at longer post-lesion time intervals (Ong et al., 2000). Glutathione (GSH), injected by slow intravenous infusion, total dose, 1.5 g/kg, reduced the decrease in local cerebral glucose utilization and neuronal loss in the hippocampus at 48 h after kainate injection (Saija et al., 1994). Glutathione, however, has limited ability to cross the blood-brain barrier. Most recent studies concern the development of forms of glutathione that have improved permeability into the brain. Injection of glutathione ethyl ester into rats prior to middle cerebral artery occlusion resulted in substantially increased glutathione in mitochondria from the striatum in both the non-ischemic and the ischemic hemispheres. This treatment, however, did not alter the volume of tissue infarction assessed at 48 h after ischemia (Anderson et al., 2004). In contrast, intraperitoneal injections of glutathione monoethyl ester improve functional recovery, enhance neuronal survival, and stabilize spinal cord blood flow after spinal cord trauma in rats (Guízar-Sahagún et al., 2005), suggesting that more studies should be performed on the efficacy of reduced glutathione supplementation in spinal cord trauma and head injury.

9.4.6 Lipoic Acid and Glutamate Neurotoxicity

In its reduced form, α-lipoic acid (Fig. 9.5) is a powerful antioxidant. It can reduce GSSG to GSH, dehydroascorbate to ascorbate, and regenerate α-tocopherol from the α-tocopheryl radical either directly or via ascorbate. Supplementation with α-lipoic acid decreases oxidative stress and restores reduced levels of other antioxidants *in vivo*. However, α-lipoic acid and dihydrolipoic acid may exert pro-oxidant properties in vitro. α-Lipoic acid and dihydrolipoic acid promote the permeability transition in mitochondrial preparations (Moini et al., 2002). Dihydrolipoic acid also stimulated superoxide anion production in mitochondrial and submitochondrial preparations. There are marked differences in effects of α-lipoic acid *in vivo* and in vitro.

α-Lipoic acid is a naturally occurring precursor of an essential cofactor for mitochondrial enzymes, including pyruvate dehydrogenase and α-ketoglutarate dehydrogenase. It not only increases acetylcholine synthesis through the activation of choline acetyltransferase, but also up-regulates glucose uptake, thus supplying more acetyl-CoA for the synthesis of acetylcholine (Holmquist et al., 2007). α-Lipoic acid chelates redox-active transition metals, thus inhibiting the formation of hydroxyl radicals and also scavenges reactive oxygen species (ROS), thereby increasing the levels of reduced glutathione. The same mechanisms can also be used to down-regulate redox-sensitive inflammatory processes (Holmquist et al., 2007). Furthermore, α-lipoic acid scavenges lipid peroxidation products such as hydroxynonenal and acrolein. The antioxidant properties and ability of α-lipoic acid to regenerate glutathione have led to the use of α-lipoic acid as a therapeutic oxidant.

α-Lipoic acid has a low molecular weight and is absorbed from the diet and crosses the blood-brain barrier (Packer et al., 1997). α-Lipoic acid protects C6 cells from glutamate-mediated toxicity and cell death. α-Lipoic acid facilitates the extracellular supply of cysteine, the reduced form of cystine that is transported into the cell by a glutamate-insensitive transport mechanism (Han et al., 1997). It is proposed that α-lipoic acid-mediated protection of C6 cells may be due to restoration of cellular glutathione, which is usually depleted after glutamate treatment.

α-Lipoic acid also prevents and reverses age-related impairments in learning and memory (Wang and Chen, 2007). Although the molecular mechanism is not fully understood, lipoic acid may not only enhance Ca^{2+} entry through presynaptic N- and P/Q type calcium channels, but also produce glutamate release from rat cerebrocortical synaptosomes (Wang and Chen, 2007). Both Ca^{2+} and glutamate are associated with learning and memory through the activation of protein kinase A and protein kinase C.

9.4.7 Antioxidant Coenzyme Q_{10} and Glutamate Neurotoxicity

Coenzyme Q_{10} (CoQ_{10}) (ubiquinol) is a vital lipophilic molecule (Fig. 9.5) that transfers electrons from mitochondrial respiratory chain complexes I and II to complex III (Beal, 2004). It is a powerful antioxidant that buffers the potential adverse consequences of oxidative stress (Albano et al., 2002). It plays an essential role in the mitochondrial electron transport chain by undergoing oxidation and reduction via the free radical intermediate ubisemiquinone. In vitro, ubiquinol can scavenge peroxyl radicals and thus inhibit lipid peroxidation. Ubiquinol can also regenerate α-tocopherol from its radical. Thus CoQ_{10} acts as an antioxidant and protects mitochondrial inner-membrane proteins and DNA against oxidative damage during oxidative stress (Albano et al., 2002). After intraperitoneal injections of radioactive CoQ_{10} in rats, there is an efficient uptake into the circulation with high radioactivity found in liver, spleen, and RBC, lower radioactivity in adrenals, thymus, and heart, and lowest radioactivity in brain, kidney, and muscles (Bentinger et al., 2003). CoQ_{10} deficiency is associated with cerebellar ataxia either through the transprenylation pathway or in the metabolic steps after condensation of 4-hydroxybenzoate and the prenyl side chain of CoQ_{10} (Artuch et al., 2006). Statins, the HMG-CoA reductase inhibitors, markedly reduce the synthesis of CoQ_{10} and may thus cause a deficiency of this essential cofactor. Clinical improvement after CoQ_{10} supplementation has been remarkable, supporting the importance of an early diagnosis of this disorder. CoQ_{10} has been used for the treatment of oxidative damage in chronic inflammatory periodontal disease, age-related macular degeneration, and neurodegenerative diseases (Battino et al., 1999; Feher et al., 2003; Beal, 2004).

9.5 Antioxidants and Clinical Trials in Ischemic Injury

Despite numerous experimental studies and patents on the beneficial effects of antioxidants in acute neural trauma and neurodegenerative diseases (Contestabile, 2001), only four antioxidants have been used for clinical trials in ischemic injury. They include tirilazad mesylate (Freedox), an antioxidant (The RANTTAS Investigators, 1996; Haley, Jr., 1998), ebselen (Harmokisane), a selenium compound with glutathione peroxidase-like activity (Ogawa et al., 1999), edaravone (MCI-186, Radicut), a potent free radical scavenger (The Edaravone Acute Brain Infarction Study Group, 2003), and NXY-059, a nitrone-based free-radical–trapping agent (Cerovive) (Lees et al., 2006) (Fig. 9.6). Tirilazad mesylate had phase I, II, and III trials in patients with acute ischemic stroke. These trials were stopped because the drug did not improve overall functional outcome. The Phase III trials of ebselen, and edaravone are in progress in Japan, Australia, Europe, Canada, and United States. Phase II trials with NXY-059 failed so additional trials are needed (Doggrell, 2006; Lees et al., 2006). Furthermore, second generation nitrones (azulenyl nitrones) were synthesized and used in neuroprotection studies in animals (Doggrell, 2006). Based on the above trials, itseems unlikely that treatment with a single antioxidant can modulate neurochemical processes associated with ischemia and improve the outcome of ischemic injury.

Ischemia is a complex neurological disorder that involves multiple pathological factors including expression of genes, excitotoxicity, oxidative stress, and neuroinflammation. Most trials for the treatment of stroke aretargeted using one antioxidant for a specific mechanism of oxidative damage, but the pathogenesis of ischemic injury involves interplay among gene expression, excitotoxicity, oxidative stress, and neuroinflammation. The use of a cocktail of antioxidants and anti-inflammatory agents at the earliest stages of ischemic injury may be requiredto substantively and persistently alter gene expression and modulate the interplay among excitotoxicity, oxidative stress, and neuroinflammation (Morimoto et al., 2002; Farooqui and Horrocks, 2007b). Thus, a combination of antioxidants and anti-inflammatory agents may modulate the neurochemical events associated with ischemic injury and result in a successful outcome from the ischemicinsult.

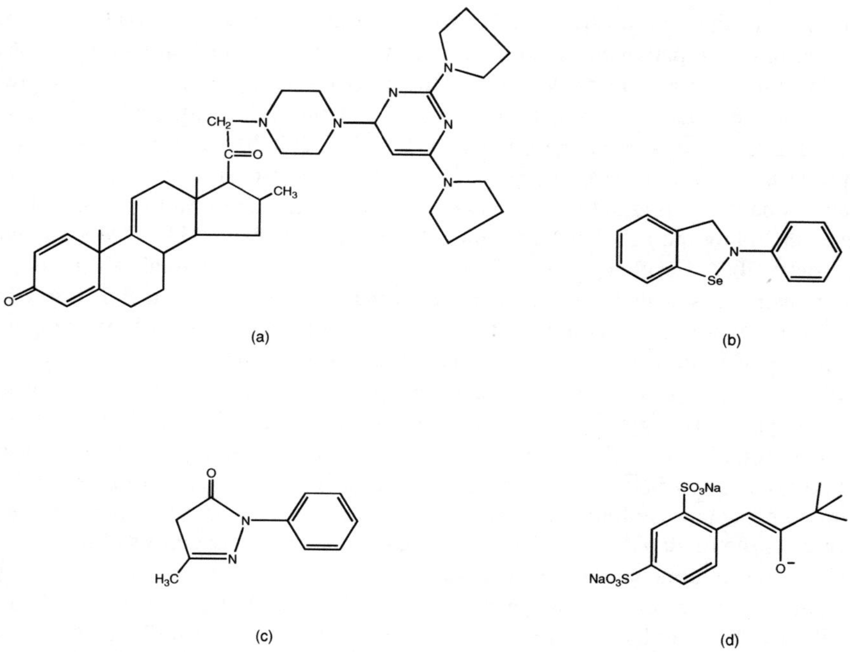

Fig. 9.6 Chemical structures of antioxidants in phase III clinical trials. Tirilazad mesylate (a); ebselen (b); edaravone (c); and NXY-095 (d)

9.6 Antioxidant Strategies and Therapeutic Aspects of Neurodegenerative Diseases

The pathogenesis of acute neural trauma and chronic neurodegenerative diseases involves a combination of genetic determinants, failure of neuroprotective mechanisms, and a wide range of environmental and metabolic factors that modulate the survival and regulate synaptic plasticity in brain tissue and regulate oxidative stress and neuroinflammation. The generation of lipid mediators through the release of arachidonic acid by PLA_2 and its oxidation by COX, LOX, EPOX pathways (Phillis et al., 2006; Farooqui et al., 2006) and the release of cytokines in neural and vascular cells play essential roles in vascular homeostasis and diseases through activating signal transduction pathways that control a variety of cellular functions, including vascular tone, gene expression, oxidative stress, and activation of leukocytes and platelets.

The mediators generated by the action of PLA_2, COX, LOX, and EPOX are predominantly arachidonate-derived and include lipid hydroxides, epoxides, hydroperoxides, and eicosanoids. These enzymes may also generate lipid-derived radicals in the vasculature following escape of substrate radicals from the active site. PLA_2, COX, LOX, and EPOX are often up-regulated in acute neural trauma and neurodegenerative diseases, with several lines of evidence suggesting that their interplay and relationships with upstream and downstream signal transduction pathways may be central to the pathogenesis of neurodegenerative process (Phillis et al., 2006). Based on these observations, we propose that interplay among excitotoxicity, oxidative stress, and neuroinflammation is closely associated with the pathogenesis of neurodegenerative diseases (Farooqui and Horrocks, 2007b). Neurons are most susceptible to direct oxidative injury by ROS. ROS can also indirectly contribute to brain damage by modulating the expression of inflammatory and stress-sensitive genes including the genes for cytokines (Palmer and Paulson, 1997; Farooqui and Horrocks, 2006). ROS also activate mechanisms that result in a glial cell-mediated inflammation associated with secondary neuronal damage (Farooqui and Horrocks, 2007b, Farooqui et al., 2007b). These activated glial cells are histopathological hallmarks of acute neural trauma and neurodegenerative diseases (Farooqui et al., 2007b). Although direct contact of activated glia with neurons may not necessarily be toxic per se, the immune mediators released by activated glial cells are endogenous neurotoxins, e.g. nitric oxide and reactive oxygen species, pro-inflammatory cytokines, and chemokines, that facilitate neurodegeneration. Therefore, the use of a mixture of anti-excitotoxic, antioxidant, and anti-inflammatory compounds for correcting the fundamental oxidant/antioxidant imbalance in patients suffering from acute neural trauma and neurodegenerative diseases opens important vistas. Their use has been reviewed (Gilgun-Sherki et al., 2001, 2002, 2006; Wang et al., 2006).

The mechanistic basis of the neuroprotective effects of anti-inflammatory and antioxidant agents may not only depend upon the general free radical trapping or antioxidant activity per se in neurons, but also on the down-regulation of NF-κB activity (Shen et al., 2003) and suppression of genes induced by pro-inflammatory cytokines and other mediators released by glial cells (Gilgun-Sherki et al., 2006; Wang et al., 2006). The efficacy of a cocktail of anti-inflammatory and antioxidant agents for protection in acute neural trauma and neurodegenerative diseases depends on their ability to cross the blood-brain barrier, their subcellular distribution in mitochondria, plasma membrane, and cytoplasm, their multifunctional capacity, as well as their synergistic actions (Gilgun-Sherki et al., 2001, 2002, 2006; Tan et al., 2003). Inclusion of agents that increase the production of ATP in degenerating neurons could improve the therapeutic outcome following acute neural trauma and neurodegenerative diseases. A clearer appreciation of the potential therapeutic ability of anti-inflammatory and antioxidant cocktails will emerge only when the importance *in vivo* of interactions among excitotoxicity, neuroinflammation, and oxidative stress is realized and fully understood at the molecular level (Farooqui and Horrocks, 2007b; Farooqui et al., 2007b).

References

Adams J. D., Jr., Klaidman L. K., Odunze I. N., Shen H. C., and Miller C. A. (1991). Alzheimer's and Parkinson's disease. Brain levels of glutathione, glutathione disulfide, and vitamin E. Mol. Chem. Neuropathol. 14:213–226.

Adams J., Collaco-Moraes Y., and De Belleroche J. (1996). Cyclooxygenase-2 induction in cerebral cortex: an intracellular response to synaptic excitation. J. Neurochem. 66:6–13.

Akbar M., Calderon F., Wen Z. M., and Kim H. Y. (2005). Docosahexaenoic acid: A positive modulator of Akt signaling in neuronal survival. Proc. Natl. Acad. Sci. USA 102: 10858–10863.

Ala T., Romero S., Knight F., Feldt K., and Frey II W. H. (1990). GM-1 treatment of Alzheimer's disease. A pilot study of safety and efficacy. Arch. Neurol. 47:1126–1130.

Albano C. B., Muralikrishnan D., and Ebadi M. (2002). Distribution of coenzyme Q homologues in brain. Neurochem. Res. 27:359–368.

Ambani L. M. and Van Woert M. H. (1973). Catalase and peroxidase in human brain. Trans. Am. Neurol. Assoc. 98:7–10.

Anderson M. F., Nilsson M., Eriksson P. S., and Sims N. R. (2004). Glutathione monoethyl ester provides neuroprotection in a rat model of stroke. Neurosci. Lett. 354:163–165.

Arab K., Rossary A., Flourie F., Tourneur Y., and Steghens J. P. (2006). Docosahexaenoic acid enhances the antioxidant response of human fibroblasts by upregulating γ-glutamyl-cysteinyl ligase and glutathione reductase. Br. J. Nutr. 95:18–26.

Artuch R., Brea-Calvo G., Briones P., Aracil A., Galván M., Espinós C., Corral J., Volpini V., Ribes A., Andreu A. L., Palau F., Sánchez-Alcázar J. A., Navas P., and Pineda M. (2006). Cerebellar ataxia with coenzyme Q_{10} deficiency: diagnosis and follow-up after coenzyme Q_{10} supplementation. J. Neurol. Sci. 246:153–158.

Avrova N. F., Zakharova I. O., Tyurin V. A., Tyurina Y. Y., Gamaley I. A., and Schepetkin I. A. (2002). Different metabolic effects of ganglioside GM1 in brain synaptosomes and phagocytic cells. Neurochem. Res. 27:751–759.

Ayasolla K., Singh A. K., and Singh I. (2002). Vitamin E restores amyloid peptide (25-35) induced changes in cellular redox in C6 glioma cells; Implications to Alzheimers disease. In: Pasquier C. (ed.), *XI Biennial Meeting of the Society for Free Radical Research International.* Monduzzi Editore, 40128 Bologna, pp. 503–508.

Balboa M. A., Perez R., and Balsinde J. (2003). Amplification mechanisms of inflammation: Paracrine stimulation of arachidonic acid mobilization by secreted phospholipase A_2 is regulated by cytosolic phospholipase A_2-derived hydroperoxyeicosatetraenoic acid. J. Immunol. 171:989–994.

Barbiroli B., Frassineti C., Martinelli P., Iotti S., Lodi R., Cortelli P., and Montagna P. (1997). Coenzyme Q10 improves mitochondrial respiration in patients with mitochondrial cytopathies. An *in vivo* study on brain and skeletal muscle by phosphorous magnetic resonance spectroscopy. Cell Mol. Biol. (Noisy. -le-grand) 43:741–749.

Barbiroli B., Medori R., Tritschler H. J., Klopstock T., Seibel P., Reichmann H., Iotti S., Lodi R., and Zaniol P. (1995). Lipoic (thioctic) acid increases brain energy availability and skeletal muscle performance as shown by *in vivo* ^{31}P-MRS in a patient with mitochondrial cytopathy. J. Neurol. 242:472–477.

Battino M., Bullon P., Wilson M., and Newman H. (1999). Oxidative injury and inflammatory periodontal diseases: the challenge of anti-oxidants to free radicals and reactive oxygen species. Crit Rev. Oral Biol. Med. 10:458–476.

Bazan N. G. (2005a). Lipid signaling in neural plasticity, brain repair, and neuroprotection. Mol. Neurobiol. 32:89–103.

Bazan N. G. (2005b). Synaptic signaling by lipids in the life and death of neurons. Mol. Neurobiol. 31:219–230.

Beal F. (2004). Therapeutic effects of coenzyme Q_{10} in neurodegenerative diseases. In: Sies H. and Packer L. (eds.), *Quinones and Quinone Enzymes, Pt B. Methods In Enzymology,* Academic Press Inc, San Diego, pp. 473–487.

Beckman J. S. (1994). Peroxynitrite versus hydroxyl radical: the role of nitric oxide in superoxide-dependent cerebral injury. Ann. N.Y Acad. Sci. 738:69–75.

Bentinger M., Dallner G., Chojnacki T., and Swiezewska E. (2003). Distribution and breakdown of labeled coenzyme Q_{10} in rat. Free Radical Biol. Med. 34:563–575.

Bharti A. C. and Singh S. M. (2000). Induction of apoptosis in bone marrow cells by gangliosides produced by a T cell lymphoma. Immunol. Lett. 72:39–48.

Boonplueang R., Akopian G., Stevenson F. F., Kuhlenkamp J. F., Lu S. C., Walsh J. P., and Andersen J. K. (2005). Increased susceptibility of glutathione peroxidase-1 transgenic mice to kainic acid-related seizure activity and hippocampal neuronal cell death. Exp. Neurol. 192: 203–214.

Brooksbank B. W. and Balazs R. (1983). Superoxide dismutase and lipoperoxidation in Down's syndrome fetal brain. Lancet 1:881–882.

Brooksbank B. W. L. and Balazs R. (1984). Superoxide dismutase, glutathione peroxidase and lipoperoxidation in Down's syndrome fetal brain. Dev. Brain Res. 16:37–44.

Bruce A. J. and Baudry M. (1995). Oxygen free radicals in rat limbic structures after kainate-induced seizures. Free Radic. Biol. Med. 18:993–1002.

Calon F., Lim G. P., Morihara T., Yang F. S., Ubeda O., Salem N. J., Frautschy S. A., and Cole G. M. (2005). Dietary n-3 polyunsaturated fatty acid depletion activates caspases and decreases NMDA receptors in the brain of a transgenic mouse model of Alzheimer's disease. Eur. J. Neurosci. 22:617–626.

Calzada C., Bruckdorfer K. R., and Rice-Evans C. A. (1997). The influence of antioxidant nutrients on platelet function in healthy volunteers. Atherosclerosis 128:97–105.

Cawood A. L., Carroll M. P., Wootton S. A., and Calder P. C. (2005). Is there a case for n-3 fatty acid supplementation in cystic fibrosis? Curr. Opin. Clin. Nutr. Metab. Care 8:153–159.

Choi D. W. (1988). Glutamate neurotoxicity and diseases of the nervous system. Neuron 1: 628–634.

Cole G. M., Lim G. P., Yang F., Teter B., Begum A., Ma Q., Harris-White M. E., and Frautschy S. A. (2005). Prevention of Alzheimer's disease: Omega-3 fatty acid and phenolic anti-oxidant interventions. Neurobiol. Aging 26(Suppl 1):133–136.

Contestabile A. (2001). Antioxidant strategies for neurodegenerative diseases. Expert Opin. Ther. Patents 11:573–585.

Dalle-Donne I., Rossi R., Giustarini D., Milzani A., and Colombo R. (2003). Protein carbonyl groups as biomarkers of oxidative stress. Clin. Chim. Acta 329:23–38.

Das U. N. (2005). Essential fatty acids and acquired immunodeficiency syndrome. Med. Sci. Monitor 11:RA206–RA211.

Das V. N. and Vaddadi K. S. (2004). Essential fatty acids in Huntington's disease. Nutrition 20:942–947.

Davis J. B. and Maher P. (1994). Protein kinase C activation inhibits glutamate-induced cytotoxicity in a neuronal cell line. Brain Res. 652:169–173.

Deeb R. S., Hao G., Gross S. S., Laine M., Qiu J. H., Resnick B., Barbar E. J., Hajjar D. P., and Upmacis R. K. (2006). Heme catalyzes tyrosine 385 nitration and inactivation of prostaglandin H_2 synthase-1 by peroxynitrite. J. Lipid Res. 47:898–911.

Doggrell S. A. (2006). Nitrone spin on cerebral ischemia. Curr. Opin. Invest. Drugs 7:20–24.

Dubovsky S. L., Christiano J., Daniell L. C., Franks R. D., Murphy J., Adler L., Baker N., and Harris R. A. (1989). Increased platelet intracellular calcium concentration in patients with bipolar affective disorders. Arch. Gen. Psychiatr. 46:632–638.

Duffield J. S., Hong S., Vaidya V. S., Lu Y., Fredman G., Serhan C. N., and Bonventre J. V. (2006). Resolvin D series and Protectin D1 mitigate acute kidney injury. J. Immunol. 177: 5902–5911.

Dumuis A., Sebben M., Haynes L., Pin J.-P., and Bockaert J. (1988). NMDA receptors activate the arachidonic acid cascade system in striatal neurons. Nature 336:68–70.

Ebadi M., Srinivasan S. K., and Baxi M. D. (1996). Oxidative stress and antioxidant therapy in Parkinson's disease. Prog. Neurobiol. 48:1–19.

Emerit J., Edeas M., and Bricaire F. (2004). Neurodegenerative diseases and oxidative stress. Biomed. Pharmacother. 58:39–46.

Engelmann B. (2004). Plasmalogens: targets for oxidants and major lipophilic antioxidants. Biochem. Soc. Trans. 32:147–150.

Engelmann B., Streich S., Schönthier U. M., Richter W. O., and Duhm J. (1992). Changes of membrane phospholipid composition of human erythrocytes in hyperlipidemias. I. Increased phosphatidylcholine and reduced sphingomyelin in patients with elevated levels of triacylglycerol-rich lipoproteins. Biochim. Biophys. Acta Lipids Lipid Metab. 1165:32–37.

Fam S. S. and Morrow J. D. (2003). The isoprostanes: unique products of arachidonic acid oxidation - A review. Curr. Medicinal Chem. 10:1723–1740.

Farooqui A. A. and Horrocks L. A. (1994). Excitotoxicity and neurological disorders: involvement of membrane phospholipids. Int. Rev. Neurobiol. 36:267–323.

Farooqui A. A. and Horrocks L. A. (2004). Plasmalogens, platelet activating factor, and other ether lipids. In: Nicolaou A. and Kokotos G. (eds.), *Bioactive Lipids*. Oily Press, Bridgwater, England, pp. 107–134.

Farooqui A. A. and Horrocks L. A. (2006). Phospholipase A_2-generated lipid mediators in the brain: the good, the bad, and the ugly. Neuroscientist 12:245–260.

Farooqui A. A. and Horrocks L. A. (2007a). Glutamate and cytokine-mediated alterations of phospholipids in head injury and spinal cord trauma. In: Banik N. (ed.), *Brain and Spinal Cord Trauma*. Handbook of Neurochemistry. Lajtha, A. (ed.). Springer, New York, in press.

Farooqui A. A. and Horrocks L. A. (2007b). *Glycerophospholipids in the Brain: Phospholipases A_2 in Neurological Disorders*, pp. 1–394. Springer, New York.

Farooqui A. A., Farooqui T., Yates A. J., and Horrocks L. A. (1988). Regulation of protein kinase C activity by various lipids. Neurochem. Res. 13:499–511.

Farooqui A. A., Rapoport S. I., and Horrocks L. A. (1997). Membrane phospholipid alterations in Alzheimer disease: deficiency of ethanolamine plasmalogens. Neurochem. Res. 22:523–527.

Farooqui A. A., Horrocks L. A., and Farooqui T. (2000). Glycerophospholipids in brain: their metabolism, incorporation into membranes, functions, and involvement in neurological disorders. Chem. Phys. Lipids 106:1–29.

Farooqui A. A., Ong W. Y., and Horrocks L. A. (2003a). Plasmalogens, docosahexaenoic acid, and neurological disorders. In: Roels F., Baes M., and de Bies S. (eds.), *Peroxisomal Disorders and Regulation of Genes*. Kluwer Academic/Plenum Publishers, London, pp. 335–354.

Farooqui A. A., Ong W. Y., and Horrocks L. A. (2003b). Stimulation of lipases and phospholipases in Alzheimer disease. In: Szuhaj B. and van Nieuwenhuyzen W. (eds.), *Nutrition and Biochemistry of Phospholipids*. AOCS Press, Champaign, pp. 14–29.

Farooqui A. A., Ong W. Y., Go M. L., and Horrocks L. A. (2005). Inhibition of brain phospholipase A2 by antimalarial drugs: Implications for neuroprotection in neurological disorders. Med. Chem. Rev. -Online 2:379–392.

Farooqui A. A., Ong W. Y., and Horrocks L. A. (2006). Inhibitors of brain phospholipase A_2 activity: their neuropharmacological effects and therapeutic importance for the treatment of neurologic disorders. Pharmacol. Rev. 58:591–620.

Farooqui A. A., Horrocks L. A., and Farooqui T. (2007a). Interactions between neural membrane glycerophospholipid and sphingolipid mediators: a recipe for neural cell survival or suicide. J. Neurosci. Res. 85:1834–1850.

Farooqui A. A., Horrocks L. A., and Farooqui T. (2007b). Modulation of inflammation in brain: a matter of fat. J. Neurochem. 101:577–599.

Favaron M., Manev H., Alho H., Bertolino M., Ferret B., Guidotti A., and Costa E. (1988). Gangliosides prevent glutamate and kainate neurotoxicity in primary neuronal cultures of neonatal rat cerebellum and cortex. Proc. Natl. Acad. Sci. USA 85:7351–7355.

Feher J., Papale A., Mannino G., Gualdi L., and Gabrieli C. B. (2003). Mitotropic compounds for the treatment of age-related macular degeneration - The metabolic approach and a pilot study. Ophthalmologica 217:351–357.

Ferrari G., Batistatou A., and Greene L. A. (1993). Gangliosides rescue neuronal cells from death after trophic factor deprivation. J. Neurosci. 13:1879–1887.

Fighera M. R., Bonini J. S., Frussa R., Dutra C. S., Hagen M. E. K., Rubin M. A., and Mello C. F. (2004). Monosialoganglioside increases catalase activity in cerebral cortex of rats. Free Radical Res. 38:495–500.

Flicker C., Ferris S. H., Kalkstein D., and Serby M. (1994). A double-blind, placebo-controlled crossover study of ganglioside GM_1 treatment for Alzheimer's disease. Am. J. Psychiat. 151:126–129.

Freeman M. P. (2006). Omega-3 fatty acids and perinatal depression: a review of the literature and recommendations for future research. Prostaglandins Leukot. Essent. Fatty Acids 75:291–297.

Geisler F. H., Dorsey F. C., and Coleman W. P. (1991). Recovery of motor function after spinal-cord injury - A randomized, placebo-controlled trial with GM-1 ganglioside. New Eng. J. Med. 324:1829–1887.

German O. L., Insua M., Gentili C., Rotstein N. P., and Politi L. E. (2006). Docosahexaenoic acid prevents apoptosis of retina photoreceptors by activating the ERK/MAPK pathway. J. Neurochem. 98:1507–1520.

Gilgun-Sherki Y., Melamed E., and Offen D. (2001). Oxidative stress induced-neurodegenerative diseases: the need for antioxidants that penetrate the blood brain barrier. Neuropharmacology 40:959–975.

Gilgun-Sherki Y., Rosenbaum Z., Melamed E., and Offen D. (2002). Antioxidant therapy in acute central nervous system injury: Current state. Pharmacol. Rev. 54:271–284.

Gilgun-Sherki Y., Barhum V., Atlas D., Melamed M., and Offen D. (2005). Analysis of gene expression in MOG-induced experimental autoimmune encephalomyelitis after treatment with a novel brain-penetrating antioxidant. J. Mol. Neurosci. 27:125–135.

Gilgun-Sherki Y., Melamed E., and Offen D. (2006). Anti-inflammatory drugs in the treatment of neurodegenerative diseases: current state. Curr. Pharmaceut. Design 12:3509–3519.

Grayson G. and Ladisch S. (1992). Immunosuppression by human gangliosides. II. Carbohydrate structure and inhibition of human NK activity. Cell Immunol. 139:18–29.

Guan Z., Wang Y., Cairns N. J., Lantos P. L., Dallner G., and Sindelar P. J. (1999). Decrease and structural modifications of phosphatidylethanolamine plasmalogen in the brain with Alzheimer disease. J. Neuropathol. Exp. Neurol. 58:740–747.

Guilbault C., Nelson A., Martin P., Stotland P., Boghdady M. L., Guiot M. C., DeSanctis J. B., and Radzioch D. (2004). Protective effect of DHA treatment in non-infected CFTR knockout mice with CF lung disease. In: Medimond (ed.), *Immunology 2004: Cytokine Network, Regulatory Cells, Signaling, and Apoptosis, Appendix*. Medimond Publishing Co, Bologna, pp. 79–88.

Guízar-Sahagún G., Ibarra A., Espitia A., Martínez A., Madrazo I., and Franco-Bourland R. E. (2005). Glutathione monoethyl ester improves functional recovery, enhances neuron survival, and stabilizes spinal cord blood flow after spinal cord injury in rats. Neuroscience 130: 639–649.

Haddad J. J. (2004). On the antioxidant mechanisms of Bcl-2: a retrospective of NF-κB signaling and oxidative stress. Biochem. Biophys. Res. Commun. 322:355–363.

Haddad J. J. and Harb H. L. (2005). L-γ-Glutamyl-L-cysteinyl-glycine (glutathione; GSH) and GSH-related enzymes in the regulation of pro- and anti-inflammatory cytokines: a signaling transcriptional scenario for redox(y) immunologic sensor(s)? Mol. Immunol. 42: 987–1014.

Hahnel D., Beyer K., and Engelmann B. (1999). Inhibition of peroxyl radical-mediated lipid oxidation by plasmalogen phospholipids and α-tocopherol. Free Radic. Biol. Med. 27: 1087–1094.

Haley E. C., Jr. (1998). High-dose tirilazad for acute stroke (RANTTAS II). RANTTAS II Investigators. Stroke 29:1256–1257.

Hall E. D., Yonkers P. A., Andrus P. K., Cox J. W., and Anderson D. K. (1992). Biochemistry and pharmacology of lipid antioxidants in acute brain and spinal cord injury. J. Neurotrauma 9(Suppl 2):S425–S442.

Han D., Sen C. K., Roy S., Kobayashi M. S., Tritschler H. J., and Packer L. (1997). Protection against glutamate-induced cytotoxicity in C6 glial cells by thiol antioxidants. Am. J. Physiol 273:R1771–R1778.

Hara E., Takahashi K., Tominaga T., Kumabe T., Kayama T., Suzuki H., Fujita H., Yoshimoto T., Shirato K., and Shibahara S. (1996). Expression of heme oxygenase and inducible nitric oxide synthase mRNA in human brain tumors. Biochem. Biophys. Res. Commun. 224: 153–158.

Hashimoto M., Hossain S., Shimada T., Sugioka K., Yamasaki H., Fujii Y., Ishibashi Y., Oka J. I., and Shido O. (2002). Docosahexaenoic acid provides protection from impairment of learning ability in Alzheimer's disease model rats. J. Neurochem. 81:1084–1091.

Hashimoto M., Hossain S., Agdul H., and Shido O. (2005). Docosahexaenoic acid-induced amelioration on impairment of memory learning in amyloid β-infused rats relates to the decreases of amyloid β and cholesterol levels in detergent-insoluble membrane fractions. Biochim. Biophys. Acta Mol. Cell Biol. Lipids 1738:91–98.

He X., Jennera M., Ong W. Y., Farooqui A. A., and Patel S. C. (2006). Lovastatin modulates increased cholesterol and oxysterol levels and has a neuroprotective effect on rat hippocampal neurons after kainate injury. J. Neurochem. 98:111.

Hernandez N. E., MacDonald J. S., Stier C. T., Belmonte A., Fernandez R., and Karpiak S. E. (1994). GM_1 ganglioside treatment of spontaneously hypertensive stroke prone rats. Exp. Neurol. 126:95–100.

Högyes E., Nyakas C., Kiliaan A., Farkas T., Penke B., and Luiten P. G. M. (2003). Neuroprotective effect of developmental docosahexaenoic acid supplement against excitotoxic brain damage in infant rats. Neuroscience 119:999–1012.

Holmquist L., Stuchbury G., Berbaum K., Muscat S., Young S., Hager K., Engel J., and Munch G. (2007). Lipoic acid as a novel treatment for Alzheimer's disease and related dementias. Pharmacol. Ther. 113:154–164.

Horrobin D. F. (2003a). Are large clinical trials in rapidly lethal diseases usually unethical? Lancet 361:695–697.

Horrobin D. F. (2003b). Omega-3 fatty acid for schizophrenia. Am. J. Psychiat. 160:188–189.

Hossain M. S., Hashimoto M., and Masumura S. (1998). Influence of docosahexaenoic acid on cerebral lipid peroxide level in aged rats with and without hypercholesterolemia. Neurosci. Lett. 244:157–160.

Hossain M. S., Hashimoto M., Gamoh S., and Masumura S. (1999). Antioxidative effects of docosahexaenoic acid in the cerebrum versus cerebellum and brainstem of aged hypercholesterolemic rats. J. Neurochem. 72:1133–1138.

Houdou S., Kuruta H., Hasegawa M., Konomi H., Takashima S., Suzuki Y., and Hashimoto T. (1991). Developmental immunohistochemistry of catalase in the human brain. Brain Res. 556:267–270.

Hsiao G., Fong T. H., Tzu N. H., Lin K. H., Chou D. S., and Sheu J. R. (2004). A potent antioxidant, lycopene, affords neuroprotection against microglia activation and focal cerebral ischemia in rats. *In Vivo* 18:351–356.

Huang E. and Ong W. Y. (2005). Distribution of ferritin in the rat hippocampus after kainate-induced neuronal injury. Exp. Brain Res. 161:502–511.

Huang E., Ong W. Y., Go M. L., and Garey L. J. (2005). Heme oxygenase-1 activity after excitotoxic injury: immunohistochemical localization of bilirubin in neurons and astrocytes and deleterious effects of heme oxygenase inhibition on neuronal survival after kainate treatment. J. Neurosci. Res. 80:268–278.

Ishikawa Y., Satoh T., Enokido Y., Nishio C., Ikeuchi T., and Hatanaka H. (1999). Generation of reactive oxygen species, release of L-glutamate and activation of caspases are required for oxygen-induced apoptosis of embryonic hippocampal neurons in culture. Brain Res. 824: 71–80.

Jiang D., Akopian G., Ho Y. S., Walsh J. P., and Andersen J. K. (2000). Chronic brain oxidation in a glutathione peroxidase knockout mouse model results in increased resistance to induced epileptic seizures. Exp. Neurol. 164:257–268.

Kaplan E. and Ansari K. (1984). Reduction of polyunsaturated fatty acid hydroperoxides by human brain glutathione peroxidase. Lipids 19:784–789.

Khanna S., Roy S., Ryu H., Bahadduri P., Swaan P. W., Ratan R. R., and Sen C. K. (2003). Molecular basis of vitamin E action - Tocotrienol modulates 12-lipoxygenase, a key mediator of glutamate-induced neurodegeneration. J. Biol. Chem. 278:43508–43515.

Khanna S., Roy S., Slivka A., Craft T. K. S., Chaki S., Rink C., Notestine M. A., DeVries A. C., Parinandi N. L., and Sen C. K. (2005). Neuroprotective properties of the natural vitamin E α-tocotrienol. Stroke 36:E144–E152.

Khanna S., Roy S., Parinandi N. L., Maurer M., and Sen C. K. (2006). Characterization of the potent neuroprotective properties of the natural vitamin E α-tocotrienol. J. Neurochem. 98:1474–1486.

Kim D. K., Rordorf G., Nemenoff R. A., Koroshetz W. J., and Bonventre J. V. (1995). Glutamate stably enhances the activity of two cytosolic forms of phospholipase A_2 in brain cortical cultures. Biochem. J. 310:83–90.

Kim H., Bing G., Jhoo W., Ko K. H., Kim W. K., Suh J. H., Kim S. J., Kato K., and Hong J. S. (2000a). Changes of hippocampal Cu/Zn-superoxide dismutase after kainate treatment in the rat. Brain Res. 853:215–226.

Kim H. C., Jhoo W. K., Kim W. K., Suh J. H., Shin E. J., Kato K., and Ho K. K. (2000b). An immunocytochemical study of mitochondrial manganese-superoxide dismutase in the rat hippocampus after kainate administration. Neurosci. Lett. 281:65–68.

King V. R., Huang W. L., Dyall S. C., Curran O. E., Priestley J. V., and Michael-Titus A. T. (2006). Omega-3 fatty acids improve recovery, whereas omega-6 fatty acids worsen outcome, after spinal cord injury in the adult rat. J. Neurosci. 26:4672–4680.

Kish S. J., Morito C., and Hornykiewicz O. (1985). Glutathione peroxidase activity in Parkinson's disease brain. Neurosci. Lett. 58:343–346.

Koistinaho J., Koponen S., and Chan P. H. (1999). Expression of cyclooxygenase-2 mRNA after global ischemia is regulated by AMPA receptors and glucocorticoids. Stroke 30: 1900–1905.

Kopp N., Claustrat B., and Tappaz M. (1980). Evidence for the presence of melatonin in the human brain. Neurosci. Lett. 19:237–242.

Lapin I. P., Mirzaev S. M., Ryzov I. V., and Oxenkrug G. F. (1998). Anticonvulsant activity of melatonin against seizures induced by quinolinate, kainate, glutamate, NMDA, and pentylenetetrazole in mice. J. Pineal Res. 24:215–218.

Latorre E., Collado M. P., Fernández I., Aragonés M. D., and Catalán R. E. (2003). Signaling events mediating activation of brain ethanolamine plasmalogen hydrolysis by ceramide. Eur. J. Biochem. 270:36–46.

Lazarewicz J. W., Wroblewski J. T., and Costa E. (1990). N-methyl-D-aspartate-sensitive glutamate receptors induce calcium-mediated arachidonic acid release in primary cultures of cerebellar granule cells. J. Neurochem. 55:1875–1881.

Ledeen R. W. and Wu G. (2002). Ganglioside function in calcium homeostasis and signaling. Neurochem. Res. 27:637–647.

Lees K. R., Zivin J. A., Ashwood T., Davalos A., Davis S. M., Diener H. C., Grotta J., Lyden P., Shuaib A., Hardemark H. G., and Wasiewski W. W. (2006). NXY-059 for acute ischemic stroke. N. Engl. J. Med. 354:588–600.

Leon A., Facci L., Toffano G., Sonnino S., and Tettamanti G. (1981). Activation of (Na^+, K^+)-ATPase by nanomolar concentrations of GM_1 ganglioside. J. Neurochem. 37:350–357.

Leonard B. E. and Myint A. (2006). Inflammation and depression: Is there a causal connection with dementia? Neurotox. Res. 10:149–160.

Levi A., Biocca S., Cataneo A., and Calissano P. (1989). The mode of action of nerve growth factor in PC12 cells. Mol. Neurobiol. 2:201–226.

Levy S. L., Burnham W. M., Bishai A., and Hwang P. A. (1992). The anticonvulsant effects of vitamin E: a further evaluation. Can. J. Neurol. Sci. 19:201–203.

Li Q., Wang M., Tan L., Wang C., Ma J., Li N., Li Y., Xu G., and Li J. S. (2005). Docosahexaenoic acid changes lipid composition and interleukin-2 receptor signaling in membrane rafts. J. Lipid Res. 46:1904–1913.

Loomis T. C., Yee G., and Stahl W. L. (1976). Regional and subcellular distribution of superoxide dismutase in brain. Experientia 32:1374–1376.

MacGregor D. G., Higgins M. J., Jones P. A., Maxwell W. L., Watson M. W., Graham D. I., and Stone T. W. (1996). Ascorbate attenuates the systemic kainate-induced neurotoxicity in the rat hippocampus. Brain Res. 727:133–144.

Maeba R. and Ueta N. (2004). A novel antioxidant action of ethanolamine plasmalogens in lowering the oxidizability of membranes. Biochem. Soc. Trans. 32:141–143.

Manev H., Uz T., Kharlamov A., and Joo J. Y. (1996). Increased brain damage after stroke or excitotoxic seizures in melatonin-deficient rats. FASEB J. 10:1546–1551.

Manev H., Uz T., and Qu T. (1998). Early upregulation of hippocampal 5-lipoxygenase following systemic administration of kainate to rats. Restor. Neurol. Neurosci. 12:81–85.

Margaill I., Plotkine M., and Lerouet D. (2005). Antioxidant strategies in the treatment of stroke. Free Radical Biol. Med. 39:429–443.

Martínez M., Vázquez E., García-Silva M. T., Manzanares J., Bertran J. M., Castelló F., and Mougan I. (2000). Therapeutic effects of docosahexaenoic acid ethyl ester in patients with generalized peroxisomal disorders. Am. J. Clin. Nutr. 71:376S-385S.

Massaro M., Habib A., Lubrano L., Del Turco S., Lazzerini G., Bourcier T., Weksler B. B., and De Caterina R. (2006). The omega-3 fatty acid docosahexaenoate attenuates endothelial cyclooxygenase-2 induction through both NADP(H) oxidase and PKCε inhibition. Proc. Natl. Acad. Sci. USA 103:15184–15189.

Masumizu T., Noda Y., Mori A., and Packer L. (2005). Electron spin resonance assay of ascorbyl radical generation in mouse hippocampal slices during and after kainate-induced seizures. Brain Res. Brain Res. Protoc. 16:65–69.

Matsuoka Y., Kitamura Y., Okazaki M., Kakimura J., Tooyama I., Kimura H., and Taniguchi T. (1998). Kainic acid induction of heme oxygenase *in vivo* and *in vitro*. Neuroscience 85: 1223–1233.

Matsuoka Y., Saito M., LaFrancois J., Saito M., Gaynor K., Olm V., Wang L. L., Casey E., Lu Y. F., Shiratori C., Lemere C., and Duff K. (2003). Novel therapeutic approach for the treatment of Alzheimer's disease by peripheral administration of agents with an affinity to beta-amyloid. J. Neurosci. 23:29–33.

Matthews R. T., Yang L., Browne S., Baik M., and Beal M. F. (1998). Coenzyme Q_{10} administration increases brain mitochondrial concentrations and exerts neuroprotective effects. Proc. Natl. Acad. Sci. U.S.A 95:8892–8897.

May J. M., Li L. Y., Hayslett K., and Qu Z. C. (2006). Ascorbate transport and recycling by SH-SY5Y neuroblastoma cells: Response to glutamate toxicity. Neurochem. Res. 31: 785–794.

McBean G. J. (2002). Cerebral cystine uptake: a tale of two transporters. Trends Pharmacol. Sci. 23:299–302.

McNamara R. K., Ostrander M., Abplanalp W., Richtand N. M., Benoit S. C., and Clegg D. J. (2006). Modulation of phosphoinositide-protein kinase C signal transduction by omega-3 fatty acids: Implications for the pathophysiology and treatment of recurrent neuropsychiatric illness. Prostaglandins Leukot. Essent. Fatty Acids 75:237–257.

Mefford I. N., Oke A. F., and Adams R. N. (1981). Regional distribution of ascorbate in human brain. Brain Res. 212:223–226.

Metcalfe T., Bowen D. M., and Muller D. P. (1989). Vitamin E concentrations in human brain of patients with Alzheimer's disease, fetuses with Down's syndrome, centenarians, and controls. Neurochem. Res. 14:1209–1212.

Milatovic D., VanRollins M., Li K., Montine K. S., and Montine T. J. (2005). Suppression of murine cerebral F_2-isoprostanes and F_4- neuroprostanes from excitotoxicity and innate immune response *in vivo* by α- or γ-tocopherol. J. Chromatogr. B-Anal. Technol. Biomed. Life Sci. 827:88–93.

Miyamoto M., Murphy T. H., Schnaar R. L., and Coyle J. T. (1989). Antioxidants protect against glutamate-induced cytotoxicity in a neuronal cell line. J. Pharmacol. Exp. Ther. 250: 1132–1140.

Moini H., Packer L., and Saris N. E. (2002). Antioxidant and prooxidant activities of ?-lipoic acid and dihydrolipoic acid. Toxicol. Appl. Pharmacol. 182:84–90.

Molaschi M., Ponzetto M., Bertagna B., Berrino E., and Ferrario E. (1996). Determination of selected trace elements in patients affected by dementia. Arch. Gerontol. Geriatr. (Suppl 5): 39–42.

Molotkovskaya I. M., Kholodenko R. V., and Molotkovsky J. G. (2002). Influence of gangliosides on the IL-2- and IL-4-dependent cell proliferation. Neurochem. Res. 27:761–770.

Montine T. J. and Morrow J. D. (2005). Fatty acid oxidation in the pathogenesis of Alzheimer's disease. Am. J. Pathol. 166:1283–1289.

Montine T. J., Montine K. S., Reich E. E., Terry E. S., Porter N. A., and Morrow J. D. (2003). Antioxidants significantly affect the formation of different classes of isoprostanes and neuroprostanes in rat cerebral synaptosomes. Biochem. Pharmacol. 65:611–617.

Mori T. A. (2006). Omega-3 fatty acids and hypertension in humans. Clin. Exp. Pharmacol. Physiol. 33:842–846.

Morimoto E., Murasugi T., and Oda T. (2002). Acute neuroinflammation exacerbates excitotoxicity in rat hippocampus *in vivo*. Exp. Neurol. 177:95–104.

Morris M. C. (2006). Docosahexaenoic acid and Alzheimer disease. Arch. Neurol. 63:1527–1528.

Morris M. C., Evans D. A., Bienias J. L., Tangney C. C., Bennett D. A., Aggarwal N., Wilson R. S., and Scherr P. A. (2002). Dietary intake of antioxidant nutrients and the risk of incident Alzheimer disease on a biracial community study. J. Am. Med. Assoc. 287:3230–3237.

Mukherjee P. K., Marcheselli V. L., Serhan C. N., and Bazan N. G. (2004). Neuroprotectin D1: A docosahexaenoic acid-derived docosatriene protects human retinal pigment epithelial cells from oxidative stress. Proc. Natl. Acad. Sci. USA 101:8491–8496.

Mullin B. R., Patrick D. H., Poore C. M., Rupp B. H., and Smith M. T. (1986). Prevention of experimental allergic encephalomyelitis by ganglioside G_{M4}. A follow-up study. J. Neurol. Sci. 73:55–60.

Musiek E. S., Gao L., Milne G. L., Han W., Everhart M. B., Wang D. Z., Backlund M. G., DuBois R. N., Zanoni G., Vidari G., Blackwell T. S., and Morrow J. D. (2005a). Cyclopentenone isoprostanes inhibit the inflammatory response in macrophages. J. Biol. Chem. 280:35562–35570.

Musiek E. S., Yin H. Y., Milne G. L., and Morrow J. D. (2005b). Recent advances in the biochemistry and clinical relevance of the isoprostane pathway. Lipids 40:987–994.

Nordvik I., Myhr K. M., Nyland H., and Bjerve K. S. (2000). Effect of dietary advice and n-3 supplementation in newly diagnosed MS patients. Acta Neurol. Scand. 102:143–149.

Ogawa A., Yoshimoto T., Kikuchi H., Sano K., Saito I., Yamaguchi T., and Yasuhara H. (1999). Ebselen in acute middle cerebral artery occlusion: a placebo-controlled, double-blind clinical trial. Cerebrovasc. Dis. 9:112–118.

Olivo S. E. and Hilakivi-Clarke L. (2005). Opposing effects of prepubertal low- and high-fat n-3 polyunsaturated fatty acid diets on rat mammary tumorigenesis. Carcinogenesis 26: 1563–1572.

Olney J. W., Fuller T., and de Gubareff T. (1979). Acute dendrotoxic changes in the hippocampus of kainate treated rats. Brain Res. 176:91–100.

Ong W. Y., He Y., Suresh S., and Patel S. C. (1997). Differential expression of apolipoprotein D and apolipoprotein E in the kainic acid-lesioned rat hippocampus. Neuroscience 79: 359–367.

Ong W. Y., Ren M. Q., Makjanic J., Lim T. M., and Watt F. (1999). A nuclear microscopic study of elemental changes in the rat hippocampus after kainate-induced neuronal injury. J. Neurochem. 72:1574–1579.

Ong W. Y., Hu C. Y., Hjelle O. P., Ottersen O. P., and Halliwell B. (2000). Changes in glutathione in the hippocampus of rats injected with kainate: depletion in neurons and upregulation in glia. Exp. Brain Res. 132:510–516.

Ong W. Y., Hu C. Y., and Patel S. C. (2002). Apolipoprotein D in the Niemann-Pick type C disease mouse brain: an ultrastructural immunocytochemical analysis. J. Neurocytol. 31: 121–129.

Ong W. Y., Goh E. W. S., Lu X. R., Farooqui A. A., Patel S. C., and Halliwell B. (2003). Increase in cholesterol and cholesterol oxidation products, and role of cholesterol oxidation products in kainate-induced neuronal injury. Brain Path. 13:250–262.

Ong, W. Y., He, X., and Patel, S. C. (2006). Neuroprotective effect of apoD on hippocampal neurons after kainic acid-induced injury. J. Neurochem. 96(Suppl 1):77

Ortiz G. G., Sánchez-Ruiz M. Y., Tan D. X., Reiter R. J., Benítez-King G., and Beas-Zárate C. (2001). Melatonin, vitamin E, and estrogen reduce damage induced by kainic acid in the hippocampus: potassium-stimulated GABA release. J. Pineal Res. 31:62–67.

Osakada F., Hashino A., Kume T., Katsuki H., Kaneko S., and Akaike A. (2004). α-Tocotrienol provides the most potent neuroprotection among vitamin E analogs on cultured striatal neurons. Neuropharmacology 47:904–915.

Packer L., Tritschler H. J., and Wessel K. (1997). Neuroprotection by the metabolic antioxidant α-lipoic acid. Free Radic. Biol. Med. 22:359–378.

Packer L., Weber S. U., and Rimbach G. (2001). Molecular aspects of alpha-tocotrienol antioxidant action and cell signalling. J. Nutr. 131:369S-373S.

Palmer H. J. and Paulson K. E. (1997). Reactive oxygen species and antioxidants in signal transduction and gene expression. Nutr. Rev. 55:353–361.

Park E., Velumian A. A., and Fehlings M. G. (2004). The role of excitotoxicity in secondary mechanisms of spinal cord injury: a review with an emphasis on the implications for white matter degeneration. J. Neurotrauma 21:754–774.

Partington C. R. and Daly J. W. (1979). Effect of gangliosides on adenylate cyclase activity in rat cerebral cortical membranes. Mol. Pharmacol. 15:484–491.

Pepicelli O., Fedele E., Bonanno G., Raiteri M., Ajmone-Cat M. A., Greco A., Levi G., and Minghetti L. (2002). *In vivo* activation of *N*-methyl-D-aspartate receptors in the rat hippocampus increases prostaglandin E_2 extracellular levels and triggers lipid peroxidation through cyclooxygenase-mediated mechanisms. J. Neurochem. 81:1028–1034.

Pepicelli O., Fedele E., Berardi M., Raiteri M., Levi G., Greco A., Ajmone-Cat M. A., and Minghetti L. (2005). Cyclo-oxygenase-1 and -2 differently contribute to prostaglandin E_2 synthesis and lipid peroxidation after *in vivo* activation of N-methyl-D-aspartate receptors in rat hippocampus. J. Neurochem. 93:1561–1567.

Pereira C. M. F. and Resende de Oliveira C. (1997). Glutamate toxicity on a PC12 cell line involves glutathione (GSH) depletion and oxidative stress. Free Radic. Biol. Med. 23:637–647.

Pereira C. F. and Resende de Oliveira C. (2000). Oxidative glutamate toxicity involves mitochondrial dysfunction and perturbation of intracellular Ca^{2+} homeostasis. Neurosci. Res. 37: 227–236.

Perry T. L., Hansen S., and Jones K. (1988). Brain amino acids and glutathione in progressive supranuclear palsy. Neurology 38:943–946.

Perry G., Taddeo M. A., Nunomura A., Zhu X. W., Zenteno-Savin T., Drew K. L., Shimohama S., Avila J., Castellani R. J., and Smith M. A. (2002). Comparative biology and pathology of oxidative stress in Alzheimer and other neurodegenerative diseases: beyond damage and response. Comp. Biochem. Physiol. C-Toxicol. Pharmacol. 133:507–513.

Phillis J. W. and O'Regan M. H. (1995). GM1 ganglioside inhibits ischemic release of amino acid neurotransmitters from rat cortex. NeuroReport 6:2010–2012.

Phillis J. W., Horrocks L. A., and Farooqui A. A. (2006). Cyclooxygenases, lipoxygenases, and epoxygenases in CNS: their role and involvement in neurological disorders. Brain Res. Rev. 52:201–243.

Pocernich C. B., Sultana R., Mohmmad-Abdul H., Nath A., and Butterfield D. A. (2005). HIV-dementia, Tat-induced oxidative stress, and antioxidant therapeutic considerations. Brain Res Brain Res Rev. 50:14–26.

Purdon A. D., Rosenberger T. A., Shetty H. U., and Rapoport S. I. (2002). Energy consumption by phospholipid metabolism in mammalian brain. Neurochem. Res. 27:1641–1647.

Puri B. K. (2005). Treatment of Huntington's disease with eicosapentaenoic acid. In: Yehuda S. and Mostofsky D. I. (eds.), *Nutrients, Stress and Medical Disorders.* Nutrition and Health (Series) Humana Press Inc, Totowa, pp. 279–286.

Puri B. K., Leavitt B. R., Hayden M. R., Ross C. A., Rosenblatt A., Greenamyre J. T., Hersch S., Vaddadi K. S., Sword A., Horrobin D. F., Manku M., and Murck H. (2005). Ethyl-EPA in Huntington disease - A double-blind, randomized, placebo-controlled trial. Neurology 65:286–292.

Puskás L. G., Kitajka K., Nyakas C., Barcelo-Coblijn G., and Farkas T. (2003). Short-term administration of omega 3 fatty acids from fish oil results in increased transthyretin transcription in old rat hippocampus. Proc. Natl. Acad. Sci. USA 100:1580–1585.

Rahman I. and MacNee W. (2000). Regulation of redox glutathione levels and gene transcription in lung inflammation: therapeutic approaches. Free Radic. Biol. Med. 28:1405–1420.

Ray P., Ray R., Broomfield C. A., and Berman J. D. (1994). Inhibition of bioenergetics alters intracellular calcium, membrane composition, and fluidity in a neuronal cell line. Neurochem. Res. 19:57–63.

Regan R. F. and Guo Y. P. (1999). Potentiation of excitotoxic injury by high concentrations of extracellular reduced glutathione. Neuroscience 91:463–470.

Rickhag M., Wieloch T., Gido G., Elmer E., Krogh M., Murray J., Lohr S., Bitter H., Chin D. J., von Schack D., Shamloo M., and Nikolich K. (2006). Comprehensive regional and temporal

gene expression profiling of the rat brain during the first 24 h after experimental stroke identifies dynamic ischemia-induced gene expression patterns, and reveals a biphasic activation of genes in surviving tissue. J. Neurochem. 96:14–29.
Roberts L. J., II, Montine T. J., Markesbery W. R., Tapper A. R., Hardy P., Chemtob S., Dettbarn W. D., and Morrow J. D. (1998). Formation of isoprostane-like compounds (neuroprostanes) *in vivo* from docosahexaenoic acid. J. Biol. Chem. 273:13605–13612.
Roy S., Lado B. H., Khanna S., and Sen C. K. (2002). Vitamin E sensitive genes in the developing rat fetal brain: a high-density oligonucleotide microarray analysis. FEBS Lett. 530:17–23.
Saija A., Princi P., Pisani A., Lanza M., Scalese M., Aramnejad E., Ceserani R., and Costa G. (1994). Protective effect of glutathione on kainic acid-induced neuropathological changes in the rat brain. Gen. Pharmacol. 25:97–102.
Salinska E., Danysz W., and Lazarewicz J. W. (2005). The role of excitotoxicity in neurodegeneration. Folia Neuropathol. 43:322–339.
Samadi P., Gregoire L., Rouillard C., Bedard P. J., Di Paolo T., and Levesque D. (2006). Docosahexaenoic acid reduces levodopa-induced dyskinesias in 1-methyl-4-phenyl-1,2,3,6-tetrahydropyridine monkeys. Ann. Neurol. 59:282–288.
Sanfeliu C., Hunt A., and Patel A. J. (1990). Exposure to N-methyl-D-aspartate increases release of arachidonic acid in primary cultures of rat hippocampal neurons and not in astrocytes. Brain Res. 526:241–248.
Sanjurjo P., Ruiz J. I., and Montejo M. (1997). Inborn errors of metabolism with a protein-restricted diet: effect on polyunsaturated fatty acids. J. Inherited Metab. Dis. 20:783–789.
Sano M., Ernesto C., Thomas R. G., Klauber M. R., Schafer K., Grundman M., Woodbury P., Growdon J., Cotman D. W., Pfeiffer E., Schneider L. S., and Thal L. J. (1997). A controlled trial of selegiline, alpha-tocopherol, or both as treatment for Alzheimer's disease. New Eng. J. Med. 336:1216–1222.
Schaefer E. J., Bongard V., Beiser A. S., Lamon-Fava S., Robins S. J., Au R., Tucker K. L., Kyle D. J., Wilson P. W. F., and Wolf P. A. (2006). Plasma phosphatidylcholine docosahexaenoic acid content and risk of dementia and Alzheimer disease - The Framingham heart study. Arch. Neurol. 63:1545–1550.
Schiller H. J., Reilly P. M., and Bulkley G. B. (1993). Antioxidant therapy. Crit. Care Med. 21: S92-S102.
Schipper H. M., Cisse S., and Stopa E. G. (1995). Expression of heme oxygenase-1 in the senescent and Alzheimer-diseased brain. Ann. Neurol. 37:758–768.
Schneider J. S. (1992). MPTP-induced Parkinsonism: Acceleration of biochemical and behavioral recovery by GM_1 ganglioside treatment. J. Neurosci. Res. 31:112–119.
Schwedhelm E., Bartling A., Lenzen H., Tsikas D., Maas R., Brummer J., Gutzki F. M., Berger J., Frolich J. C., and Boger R. H. (2004). Urinary 8-iso-prostaglandin $F_2\alpha$ as a risk marker in patients with coronary heart disease - A matched case-control study. Circulation 109:843–848.
Serhan C. N. (2005). Novel ω-3-derived local mediators in anti-inflammation and resolution. Pharmacol. Ther. 105:7–21.
Shen W. H., Zhang C. Y., and Zhang G. Y. (2003). Antioxidants attenuate reperfusion injury after global brain ischemia through inhibiting nuclear factor-kappa B activity in rats. Acta Pharmacol. Sinica 24:1125–1130.
Sindelar P. J., Guan Z. Z., Dallner G., and Ernster L. (1999). The protective role of plasmalogens in iron-induced lipid peroxidation. Free Radic. Biol. Med. 26:318–324.
Slivka A., Spina M. B., and Cohen G. (1987). Reduced and oxidized glutathione in human and monkey brain. Neurosci. Lett. 74:112–118.
Sofic E., Riederer P., Killian W., and Rett A. (1987). Reduced concentrations of ascorbic acid and glutathione in a single case of Rett syndrome: a postmortem brain study. Brain Dev. 9: 529–531.
Sohn H., Kim Y. S., Kim H. T., Kim C. H., Cho E. W., Kang H. Y., Kim N. S., Kim C. H., Ryu S. E., Lee J. H., and Ko J. H. (2006). Ganglioside GM3 is involved in neuronal cell death. FASEB J. 20:1248–1250.
Soule J., Messaoudi E., and Bramham C. R. (2006). Brain-derived neurotrophic factor and control of synaptic consolidation in the adult brain. Biochem. Soc. Trans. 34:600–604.

Sun G. Y., Horrocks L. A., and Farooqui A. A. (2007). The role of NADPH oxidase and phospholipases A_2 in mediating oxidative and inflammatory responses in neurodegenerative diseases. J. Neurochem. in press. doi: 10.1111/j.1471-4159.2007.04670.x
Surai P. F. and Sparks N. H. C. (2001). Designer eggs: from improvement of egg composition to functional food. Trends Food Sci. Technol. 12:7–16.
Svennerholm L. (1994). Gangliosides–A new therapeutic agent against stroke and Alzheimer's disease. Life Sci. 55:2125–2134.
Takahashi K., Hara E., Suzuki H., Sasano H., and Shibahara S. (1996). Expression of heme oxygenase isozyme mRNAs in the human brain and induction of heme oxygenase-1 by nitric oxide donors. J. Neurochem. 67:482–489.
Tan D. X., Manchester L. C., Reiter R. J., Qi W., Kim S. J., and El Sokkary G. H. (1998). Melatonin protects hippocampal neurons *in vivo* against kainic acid-induced damage in mice. J. Neurosci. Res. 54:382–389.
Tan D. X., Manchester L. C., Sainz R., Mayo J. C., Alvares F. L., and Reiter R. J. (2003). Antioxidant strategies in protection against neurodegenerative disorders. Expert Opin. Ther. Patents 13:1513–1543.
Tanaka A., Kobayashi S., and Fujiki Y. (2006). Peroxisome division is impaired in a CHO cell mutant with an inactivating point-mutation in dynamin-like protein 1 gene. Exp. Cell Res. 312:1671–1684.
Terpstra M. and Gruetter R. (2004). ^{1}H NMR detection of vitamin C in human brain *in vivo*. Magn Reson. Med. 51:225–229.
The Edaravone Acute Brain Infarction Study Group (2003). Effect of a novel free radical scavenger, edaravone (MCI-186), on acute brain infarction. Randomized, placebo-controlled, double-blind study at multicenters. Cerebrovasc. Dis. 15:222–229.
The RANTTAS Investigators (1996). A randomized trial of tirilazad mesylate in patients with acute stroke (RANTTAS). Stroke 27:1453–1458.
Thomas E. A., Danielson P. E., Nelson P. A., Pribyl T. M., Hilbush B. S., Hasel K. W., and Sutcliffe J. G. (2001). Clozapine increases apolipoprotein D expression in rodent brain: towards a mechanism for neuroleptic pharmacotherapy. J. Neurochem. 76:789–796.
Thwin M. M., Ong W. Y., Fong C. W., Sato K., Kodama K., Farooqui A. A., and Gopalakrishnakone P. (2003). Secretory phospholipase A_2 activity in the normal and kainate injected rat brain, and inhibition by a peptide derived from python serum. Exp. Brain Res. 150:427–433.
Uozumi N., Kume K., Nagase T., Nakatani N., Ishii S., Tashiro F., Komagata Y., Maki K., Ikuta K., Ouchi Y., Miyazaki J., and Shimizu T. (1997). Role of cytosolic phospholipase A_2 in allergic response and parturition. Nature 390:618–622.
Uz T., Giusti P., Franceschini D., Kharlamov A., and Manev H. (1996). Protective effect of melatonin against hippocampal DNA damage induced by intraperitoneal administration of kainate to rats. Neuroscience 73:631–636.
Van der Veen R. C. and Roberts L. J. (1999). Contrasting roles for nitric oxide and peroxynitrite in the peroxidation of myelin lipids. J. Neuroimmunol. 95:1–7.
Van Raamsdonk J. M., Pearson J., Rogers D. A., Lu G., Barakauskas V. E., Barr A. M., Honer W. G., Hayden M. R., and Leavitt B. R. (2005). Ethyl-EPA treatment improves motor dysfunction, but not neurodegeneration in the YAC128 mouse model of Huntington disease. Exp. Neurol. 196:266–272.
Walker D. W., Muffat J., Rundel C., and Benzer S. (2006). Overexpression of a Drosophila homolog of apolipoprotein D leads to increased stress resistance and extended lifespan. Curr. Biol. 16:674–679.
Wang X. S., Ong W. Y., and Connor J. R. (2002a). A light and electron microscopic study of divalent metal transporter-1 distribution in the rat hippocampus, after kainate-induced neuronal injury. Exp. Neurol. 177:193–201.
Wang X. S., Ong W. Y., and Connor J. R. (2002b). Increase in ferric and ferrous iron in the rat hippocampus with time after kainate-induced excitotoxic injury. Exp. Brain Res. 143:137–148.
Wang H., Cheng E., Brooke S., Chang P., and Sapolsky R. (2003). Over-expression of antioxidant enzymes protects cultured hippocampal and cortical neurons from necrotic insults. J. Neurochem. 87:1527–1534.

Wang J. Y., Wen L. L., Huang Y. N., Chen Y. T., and Ku M. C. (2006). Dual effects of antioxidants in neurodegeneration: Direct neuroprotection against oxidative stress and indirect protection via suppression of glia-mediated inflammation. Curr. Pharmaceut. Design 12:3521–3533.

Wang S. J. and Chen H. H. (2007). Presynaptic mechanisms underlying the α-lipoic acid facilitation of glutamate exocytosis in rat cerebral cortex nerve terminals. Neurochem. Int. 50:51–60.

Warner D. S., Sheng H. X., and Batinic-Haberle I. (2004). Oxidants, antioxidants and the ischemic brain. J. Exp. Biol. 207:3221–3231.

Wells K., Farooqui A. A., Liss L., and Horrocks L. A. (1995). Neural membrane phospholipids in Alzheimer disease. Neurochem. Res. 20:1329–1333.

Willcox J. K., Ash S. L., and Catignani G. L. (2004). Antioxidants and prevention of chronic disease. Crit. Rev. Food Sci. Nutr. 44:275–295.

Wullner U., Seyfried J., Groscurth P., Beinroth S., Winter S., Gleichmann M., Heneka M., Loschmann P., Schulz J. B., Weller M., and Klockgether T. (1999). Glutathione depletion and neuronal cell death: the role of reactive oxygen intermediates and mitochondrial function. Brain Res. 826:53–62.

Yang H.-C., Farooqui A. A., and Horrocks L. A. (1994a). Effects of glycosaminoglycans and glycosphingolipids on cytosolic phospholipases A_2 from bovine brain. Biochem. J. 299:91–95.

Yang H. C., Farooqui A. A., and Horrocks L. A. (1994b). Effects of sialic acid and sialoglycoconjugates on cytosolic phospholipases A_2 from bovine brain. Biochem. Biophys. Res. Commun. 199:1158–1166.

Yang H.-C., Farooqui A. A., and Horrocks L. A. (1994c). Effects of sialic acid and sialoglycoconjugates on cytosolic phospholipases A_2 from bovine brain. Biochem. Biophys. Res. Commun. 199:1158–1166.

Yates A. J., Walters J. D., Wood C. L., and Johnson J. D. (1989). Ganglioside modulation of cyclic AMP-dependent protein kinase and cyclic nucleotide phosphodiesterase in vitro. J. Neurochem. 53:162–167.

Yavin E. and Gatt S. (1972). Oxygen-dependent cleavage of the vinyl ether linkage of plasmalogens. 2. Identification of the low molecular weight active component and the reaction mechanism. Eur. J. Biochem. 25:437–446.

Yoneda Y., Kuramoto N., Kitayama T., and Hinoi E. (2001). Consolidation of transient ionotropic glutamate signals through nuclear transcription factors in the brain. Prog. Neurobiol. 63: 697–719.

Yuen A. W. C., Sander J. W., Fluegel D., Patsalos P. N., Bell G. S., Johnson T., and Koepp M. J. (2005). Omega-3 fatty acid supplementation in patients with chronic epilepsy: A randomized trial. Epilepsy & Behavior 7:253–258.

Zanarini M. C. and Frankenburg F. R. (2003). Omega-3 fatty acid treatment of women with borderline personality disorder: A double-blind, placebo-controlled pilot study. Am. J. Psychiat. 160:167–169.

Zemlyak I., Nimon V., Brooke S., Moore T., McLaughlin J., and Sapolsky R. (2006). Gene therapy in the nervous system with superoxide dismutase. Brain Res. 1088:12–18.

Zhang B., Tanaka J., Yang L., Yang L., Sakanaka M., Hata R., Maeda N., and Mitsuda N. (2004). Protective effect of vitamin E against focal brain ischemia and neuronal death through induction of target genes of hypoxia-inducible factor-1. Neuroscience 126:433–440.

Zhang J., Dawson V. L., Dawson T. M., and Snyder S. H. (1994). Nitric oxide activation of poly(ADP-ribose) synthetase in neurotoxicity. Science 263:687–689.

Zhao B. L. (2005). Natural antioxidants for neurodegenerative diseases. Mol. Neurobiol. 31: 283–293.

Zoeller R. A., Lake A. C., Nagan N., Gaposchkin D. P., Legner M. A., and Lieberthal W. (1999). Plasmalogens as endogenous antioxidants: somatic cell mutants reveal the importance of the vinyl ether. Biochem. J. 338:769–776.

Zommara M., Tachibana N., Mitsui K., Nakatani N., Sakono M., Ikeda I., and Imaizumi K. (1995). Inhibitory effect of ethanolamine plasmalogen on iron- and copper-dependent lipid peroxidation. Free Radical Biol. Med. 18:599–602.

Chapter 10
Glutamate Receptor Antagonists and the Treatment of Neurological Disorders

Glutamate receptor-mediated toxicity is a major cause of neurodegeneration in acute neural trauma and neurodegenerative diseases. Acute neural trauma includes ischemia, spinal cord trauma, and head injury. Neurodegenerative diseases include Alzheimer disease (AD), Parkinson disease (PD), Huntington disease (HD), amyotrophic lateral sclerosis (ALS), HIV-associated dementia, and multiple sclerosis (MS). Although the primary causes of these neurological disorders may be quite different, these diseases share excitotoxicity as a final and common mechanism of neurodegeneration. Thus the sustained activation of glutamate receptors, and especially of the NMDA receptor, is a causal phenomenon that leads to neuronal death (Farooqui and Horrocks, 1994; Lipton and Rosenberg, 1994).

A key factor in recognizing the importance of NMDA receptors in ischemia has been the consistent finding that NMDA receptor antagonists reduce the volume of cerebral infarction resulting from experimental stroke. Consistent neuroprotection by NMDA antagonists holds great promise for the development of viable therapy for stroke in humans, but there are concerns over their potential neurotoxicity and behavioral side effects, particularly when longer-term therapy in chronic CNS diseases is contemplated (see below). Theoretically the pharmacological intervention of glutamate receptors is a valuable therapeutic strategy for the treatment of ischemia, epilepsy, AD, PD, and HD (Choi, 1990; Choi et al., 2000; Ratan et al., 1994).

10.1 NMDA Antagonists for the Treatment of Neurological Disorders

The complex structure of the NMDA receptor provides multiple sites for therapeutic inhibition. NMDA receptors contain several antagonist-binding sites. They include (a) a NMDA or L-glutamate binding site; (b) a cation binding site; (c) a PCP binding site; and (d) a glycine binding site (Fagg and Baud, 1988). Competitive NMDA antagonists bind directly to the glutamate binding site of the NMDA receptor and block the action of glutamate. Noncompetitive antagonists block the NMDA-associated ion channel in a use-dependent manner and prevent the Ca^{2+}

influx. Other sites on the NMDA receptor susceptible to antagonism include the glycine site and the polyamine site. NMDA receptor antagonists have been synthesized for competitive (glutamate), non-competitive (PCP), polyamine, and glycine binding sites. Many NMDA receptor antagonists have been synthesized (Fig. 10.1). The challenge facing investigators trying to design NMDA antagonists to treat excitotoxic insult is that the same processes which, in excess, cause neurotoxicity are, at lower levels, absolutely necessary for normal neural cell function (Sonkusare et al., 2005; Ratan et al., 1994). The sensitivity of NMDA receptor to dizocilpine (MK-801), phencyclidine (PCP) and ketamine is quite similar.

Mutation of asparagine residues at N sites in segment M2 of the $\varepsilon 2$ and $\zeta 1$ subunit of the NMDA receptor channel decreases the affinity to MK-801, suggesting that binding site of this antagonist overlaps the Mg^{2+} block site (Mori et al., 1992). Detailed molecular pharmacological studies indicate that the molecular determinants for the PCP block site are confined to the M2 segment, whilst structural interactions between the M2 and M3 segments are necessary for MK-801 binding (Ferrer-Montiel et al., 1995). PCP and MK-801 have similar behavioral effects. Instead of protecting neurons, these drugs produce their neurotoxic effect on neurons in the posterior cingulate cortex (Ellison, 1994, 1995). Both of these drugs,

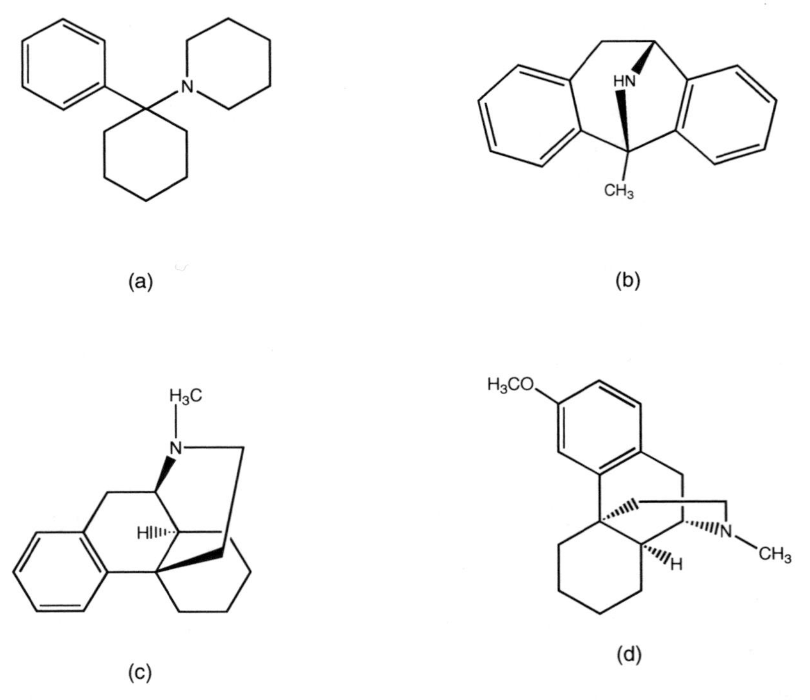

Fig. 10.1 Chemical structures of NMDA and AMPA antagonists used for the treatment of ischemic injury in animal models. PCP (a); MK-801 (b); Dextrorphan (c); and Dextromethorphan (d)

when given continuously for several days, induce further neuronal degeneration in other limbic structures. These include the brain regions of rats related to olfaction, associated limbic structures such as piriform cortex, and posterior regions of entorhinal cortex and its projections through the perforant pathway to the dentate gyrus and other cells in the ventral hippocampus. These degenerative consequences may be excitatory neurotoxic effects. PCP and MK-801 also elevate glucose metabolism to the maximal level in just those structures where degeneration is observed and the degeneration involves entire cells, with all of their processes. The molecular mechanism underlying PCP and MK-801 neurotoxicity in parahippocampus, hippocampus, olfactory regions, and the posterior cingulate cortex, and the upregulation in glucose metabolism is not understood. However, the molecular mechanism may involve nitric oxide synthesis.

PCP inhibits brain nitric oxide synthase irreversibly (Osawa and Davila, 1993; Jewett et al., 1996; Klamer et al., 2005). Depending upon its levels, nitric oxide acts as a neuroprotective or neurodestructive molecule (Lipton, 1993; Lipton et al., 1998). NMDA receptor antagonists that have treated ischemic injury of the brain in animal models with some benefit are presented in Tables 10.1 and 10.2. All studies on their use in humans have been unsuccessful because they not only block normal neuronal function, but also produce serious side effects such as headache, anxiety, agitation, nausea, vomiting, hallucinations, dizziness, and coma (Schehr, 1996; Koroshetz and Moskowitz, 1996; Ratan et al., 1994). Clinical trials of NMDA antagonists for stroke and traumatic brain injury have been abandoned (Kemp and McKernan, 2002; Lees et al., 2000; Sacco et al., 2001).

Table 10.1 Use of NMDA receptor antagonists for the treatment of ischemia in animal models

NMDA antagonist	Nature	Effect on ischemia	Reference
PCP	Non-competitive antagonist	Beneficial with harmful side effects	(Ogren and Goldstein, 1994)
Ketamine	Non-competitive antagonist	Beneficial with harmful side effects	(Reeker et al., 2000)
MK-801	Non-competitive antagonist	Beneficial with harmful side effects	(Sarraf-Yazdi et al., 1998)
GPI-3000	Competitive antagonist	No effect	(Helfaer et al., 1998)
HU-211	Non competitive antagonist	Beneficial with harmful side effects	(Shohami et al., 1993)
CPP-ene	Competitive antagonist	Beneficial with harmful side effects	(Davis et al., 1995)
Gelsolin		Beneficial	(Endres et al., 1999)

CPP-ene is 3-(2-carboxy-piperazine-4-yl)-1-propenyl-1-phosphonic acid.

Table 10.2 Treatment of stroke by NMDA antagonists during phase trials

NMDA antagonist	Mechanism of action	Manufacturer	Developing stage	Results	Reference
Eliprodil	Polyamine site competitive blocker	SyntheLabo Lorex	Phase III	Abandoned	(Lees, 1997)
Selfotel (CGS 19755)	Competitive antagonist	Ciba-Geigy	Phase III	Abandoned	(Morris et al., 1999)
Remacemide*	Noncompetitive antagonist	Astra Zeneca	Phase I	Abandoned	(Dyker and Lees, 1999)
Aptiganel	Competitive antagonist	Boehringer-Ingelheim	Phase III	Abandoned	(Dyker et al., 1999)
Dextrorphan	Noncompetitive antagonist	Hoffman-LarRoche	Phase I	Abandoned	(Albers et al., 1995)
Amantadine	Uncompetitive antagonist	DuPont	Launched	Unknown	(Teramoto et al., 1999)
Methoxy idazoxan	Competitive	Pfizer, Bristol-Myers Squibb	Launched in 1966	Abandoned	(Gustafson et al., 1990)
**PhencyclidineHCl	Uncompetitive	Pfizer	Unspecified	Unknown	(Rajdev et al., 1998)

*Used for the treatment of epilepsy, PD, HD, and trauma. **Street drug “angel dust”.

10.2 NMDA Antagonists for the Treatment of Ischemic Injury

Non-competitive NMDA antagonists interact with a specific site on the NMDA receptor complex to block or retard the Ca^{2+} influx. They cross the blood-brain barrier easily. Noncompetitive NMDA antagonists include drugs like MK 801, selfotel, dextrorphan, dextrometorphan, and aptiganel (Cerestat). These drugs have been used for the treatment of stroke in animal models and also in human patients. In humans these drugs produce similar adverse clinical and behavioral effects regardless of the molecular mechanism of their pharmacological action. Low doses produce altered sensory perception, dysphoria, nystagmus and hypotension, whereas higher doses may cause psychological adverse events such as excitement, paranoia, hallucinations, and severe motor retardation, leading ultimately to catatonia, which may occur with the highest doses (Labiche and Grotta, 2004).

10.2.1 Selfotel (CGS 19755)

Selfotel (CGS19755) is a competitive NMDA receptor antagonist that limits neuronal damage in animal stroke models (Grotta et al., 1990). Studies on the efficacy of selfotel have been performed in a randomized, double-blind, placebo-controlled, ascending dose phase 2a trial to determine its safety and tolerability, and to obtain pharmacokinetic data in stroke patients (Grotta et al., 1995). During this study, patients were injected with selfotel within 12 h of ischemic hemispheric stroke. Stroke patients showed common neuro-psychiatric adverse affects related to selfotel that lasted an average of 24 h. Neuropsychiatric symptoms include hallucinations, agitation, confusion, dysarthria, ataxia, delirium, paranoia, and somnolence. Low doses of selfotel, 1.5 mg/kg, caused mild adverse experiences whereas a higher dose of selfotel, 2 mg/kg, given once or twice produced severe adverse experiences in all patients. Based on this information, phase 3 parallel trials of a single dose of 1.5 mg/kg of selfotel given within 6 h of the onset of acute hemispheric stroke were performed in the United States and Europe, but these trials were suspended after 31% of the patients had a severe unfavorable reaction to the drug (Davis et al., 2000). Moreover, in stroke patients the progression of adverse neurologic affects was higher in the selfotel treated group than in control and placebo groups. There were no differences between selfotel and placebo groups in the primary endpoint of functional independence. In animal models, a plasma level of 40 μg/ml selfotel may be cytoprotective. However, the highest tolerated level in human stroke patients is only half of this target cytoprotective concentration, 21 μg/ml, and even these "subtherapeutic" levels produce marked neurological and psychiatric effects (Grant et al., 1994; Labiche and Grotta, 2004). Collectively these trials indicate that selfotel is not an efficacious drug for ischemic injury (Labiche and Grotta, 2004). It may in fact exert a neurotoxic effect in patients with severe stroke.

10.2.2 Dextrorphan

The efficacy of the noncompetitive NMDA antagonist, dextrorphan, was evaluated in a pilot study (Albers et al., 1995). Within 48 h of the onset of hemispheric cerebral infarction, patients were treated with dextrorphan. Dextrorphan treatment caused severe neurologic affects in human subjects. The adverse effects of dextrorphan occur in a dose-dependent manner and include agitation, confusion, hallucinations, nystagmus, somnolence, nausea, and vomiting. Hypotension also occurs at the highest loading doses (>200 mg/h), but is not associated with neurologic deterioration. No apparent differences in the outcome between patients treated with placebo or low-, medium-, or high-dose dextrorphan patients were observed (Labiche and Grotta, 2004). Unlike selfotel, plasma concentrations of dextrorphan can be achieved. These concentrations were comparable to the cytoprotective level determined in cell culture and animal models of ischemic injury. At the present time, no further clinical trials of dextrorphan are planned.

10.2.3 Aptiganel (Cerestat)

Aptiganel (Cerestat, CNS1102) is a noncompetitive NMDA antagonist. In an animal model of stroke, this drug produced beneficial clinical and behavioral effects (Schabitz et al., 2000). Phase IIa trials were performed to judge its tolerability and efficacy in stroke patients within 18 h of stroke onset. Many toxic side effects were reported in patients following aptiganel treatment. These effects include hypertension, agitation, confusion, headache, sedation, nausea, vomiting, disorientation, and paresthesias. The severity of side effects in patients treated with aptiganel was less than in human subjects treated with selfotel and dextrorphan (Minematsu et al., 1993). This trial was followed by a multi-center placebo-controlled, double-blind randomized trial with higher doses of aptiganel (Dyker et al., 1999). Forty-six patients with ischemic stroke in the territory of the carotid artery were enrolled in this study within 24 h of onset of stroke. In group A, patients were randomized 3:1 to a single bolus of aptiganel or placebo. A dose of 6.0 mg was well tolerated, but at 7.5 mg aptiganel produced sedation, hallucination, and confusion. In group B, a dose of aptiganel, 6.0 mg, was followed by a 6- to 12-h continuous infusion of 1 mg/hr. The dosing regimen in group B was stopped due to hypertension and severe sedation. Lower doses of aptiganel were adopted, 4.5 mg, followed by 0.75 mg/hr infusion. Using this dose, a successful target cytoprotective plasma concentration of >10 ng/ml was achieved. This dose produced moderately increased systolic blood pressure, ~30 mm Hg, which was controlled by anti-hypertensive medication. Patients tolerated neurologic adverse experiences, such as mild sedation and confusion. However, no signs of recovery from ischemic injury were observed in groups A or B.

Another nested phase 2/phase 3 study was also performed evaluating low-dose and high-dose aptiganel regimens compared to placebo (Albers et al., 2001).

Patients with a clinical diagnosis of ischemic stroke were randomized to 1 of 3 treatment groups within 6 h of symptom onset. Phase 3 enrollment was abandoned early, based on analysis of the phase 2 results that indicated an increase in mortality within the aptiganel cohort. Analysis of available phase 3 results from 628 patients showed no difference in 90-day outcome measured with a modified Rankin Scale among the 3 groups (Albers et al., 2001). There was no significant difference in mortality after 90-days; however, there was a marginal increase in mortality at 120 days in the high-dose group. Other secondary analyses showed a significant difference favoring the placebo group over high-dose aptiganel in multiple outcome measures. Based on these studies it was concluded that aptiganel is not efficacious within 6 h of stroke onset and may be harmful at higher doses (Labiche and Grotta, 2004).

10.2.4 Gavestinel

Gavestinel (GV150526) is a novel glycine site antagonist at the NMDA receptor complex. It has neuroprotective effects in experimental models of ischemia at a very low plasma level (10–30 μg/ml) at 6 h after the ischemic injury (Bordi et al., 1997). A phase II randomized, placebo-controlled, two-part ascending dose trial was used to evaluate the safety, tolerability, and pharmacokinetics. The loading dose of this drug, 800 mg, was followed by the continuous infusion of this drug within 12 h of stroke onset (Dyker et al., 1999). To evaluate the tolerability of gavestinel, the first part of this trial required the administration of the loading doses, 800 mg. In the second part, the maximum tolerated loading dose from part one, 800 mg, was followed by five maintenance infusions of 200 mg every 12 h. No drug-related side effects were observed following drug administration compared to placebo. However, gavestinel administration resulted in significantly higher serum glucose levels. Thus, the 800-mg loading dose followed by five 200-mg infusions twice daily was identified as the optimal dosage to be used for phase 3 efficacy studies. At this dosage, two large phase 3 randomized placebo-controlled, double-blind trials failed to demonstrate efficacy of gavestinel despite adequate powering to detect even small differences (Sacco et al., 2001). The GAIN International trial recruited 1804 patients within 6 h of stroke onset and used the same dosing regimen of gavestinel. It was concluded that this drug had no effect on primary or secondary outcome measures in stroke patients compared with placebo. These observations resulted in cancellation of further gavestinel trials in humans (Sacco et al., 2001; Lees et al., 2000).

10.2.5 Ifenprodil and Eliprodil

Initially, ifenprodil was reported to bind to the polyamine site of NMDA receptors, but recent studies indicate that it acts on a site distinct from the polyamine site (Gallagher et al., 1996; Yamakura and Shimoji, 1999). Arginine-337 in the NR2B subunit is absolutely required for high affinity ifenprodil inhibition, but not for polyamine stimulation (Gallagher et al., 1996). Furthermore, glutamate-201 on the NR2B subunit is critical for polyamine stimulation, but not for ifenprodil inhibition. Collectively these studies suggest that ifenprodil and polyamines interact at discrete sites on the NR2B subunit of NMDA receptor (Fig. 10.2).

In contrast, eliprodil (SL82.0715), Ro 256981, Ro 8-4304 and CP 101 606 are highly selective NR1/NR2B inhibitors. Although these antagonists share subunit selectivity and overlapping in the extent of inhibition with ifenprodil, mutation studies indicate that they do not have a common structural determinant (Yamakura and Shimoji, 1999). In animal models of stroke, these drugs have neuroprotective efficacy. Eliprodil also has significant voltage-sensitive Ca^{2+} antagonist properties and is a high-affinity ligand for the sigma receptor, whose function remains unknown. In normal volunteers, a 60 mg oral dose of eliprodil causes vertigo, and drowsiness (The STIPAS Investigators, 1994). Stroke patients who received IV doses of 1.5 and 3 mg twice daily for 3 days or oral doses of 5–10 mg twice daily for 11 days show no psychological effects, cardiovascular changes, or cardiac repolarization abnormalities. Eliprodil was tried in phase 2 and 3 trials in stroke patients (Lees, 1997) but investigations were stopped due to an unsatisfactory risk/benefit ratio (Lees, 1997).

10.2.6 YM872

YM872 is an AMPA antagonist that reduced infarct volume in animal models (Shimizu-Sasamata et al., 1996; Kawasaki-Yatsugi et al., 2000). Two clinical trials of YM872 were terminated prematurely. The first trial, ARTIST (AMPA Receptor

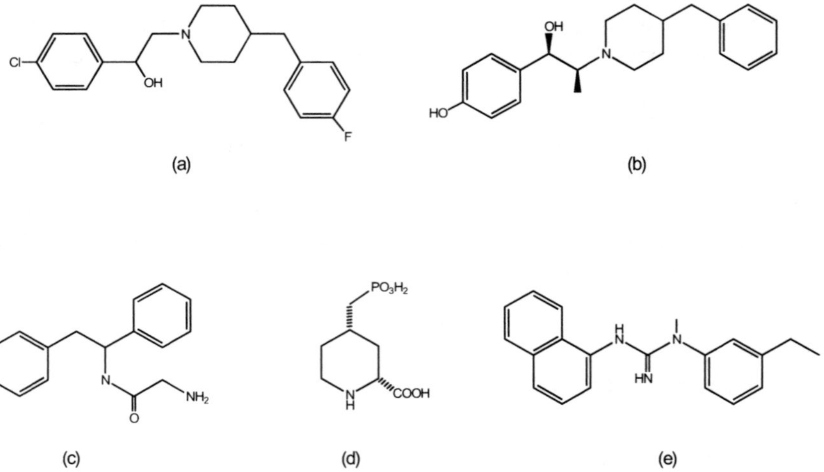

(a) (b)

(c) (d) (e)

Fig. 10.2 Chemical structures of NMDA and AMPA antagonists used for the treatment of ischemic injury, epilepsy and head injury in human subjects. Eliprodil (a); ifenprodil (b); remacemide (c); selfotel (d); and aptiganel (e)

Antagonist Treatment in Ischemic Stroke), was based on an interim futility analysis. In this trial, the efficacy of YM872 plus TPA was compared with placebo plus TPA. Preclinical data indicated that the co-administration of TPA and YM872 within 2 h of stroke imparts a greater degree of cytoprotection than either agent alone (Suzuki et al., 2003). The second trial was ARTIST MRI. This trial was performed to evaluate the safety and potential efficacy of YM872 in stroke patients within 6 h of onset of stroke using MRI as a surrogate marker of outcome. In both trials patients had multiple severe side effects, such as hallucination, agitation, and catatonia, resulting in abandonment of trials.

10.3 NMDA Antagonists for the Treatment of Alzheimer Disease

Alzheimer disease (AD) is a complex and multifactorial disease. Its pathophysiology involves defective beta-amyloid (Aβ) protein metabolism, abnormalities of cholinergic and glutamatergic neurotransmission, and interactions among excitotoxicity, oxidative stress, and inflammation (Farooqui and Horrocks, 2007b; Palmer, 2001; Grossberg et al., 2006). The cognitive symptoms of AD may be caused not only by the loss of cholinergic and glutamatergic neurons, but also by irregular functioning of surviving neurons in the cholinergic and glutamatergic neurotransmitter systems (Wenk, 2006). Thus aberrant cholinergic functioning in AD has been linked to deficits of the neurotransmitter acetylcholine, whilst AD-related abnormalities in glutamatergic signaling are attributed to abnormalities in learning and memory processes as well as to excitotoxicity caused by persistent, low-level stimulation of glutamatergic neurons via the chronic influx of Ca^{2+} ions through the NMDA receptor Ca^{2+} channel.

Physiological NMDA receptor activity is also essential for normal neuronal function in brain tissue. This means that potential non-competitive NMDA antagonists that block virtually all NMDA receptor activity, phencyclidine and MK-801, cannot be used for the treatment of neurodegenerative diseases. In addition, the high affinity noncompetitive NMDA antagonists not only produce differential effects on neurons from various regions of brain tissue, but also interact with neurotransmitter receptors other than NMDA receptors (Kornhuber and Weller, 1997). Their low doses are associated with altered sensory perception, dysphoria, nystagmus, and hypotension, whilst higher doses may cause psychological adverse events such as excitement, paranoia, and hallucinations; in addition to severe motor retardation, leading ultimately to catatonia, which may occur with the highest doses.

For these reasons many previous NMDA receptor antagonists have disappointingly failed advanced clinical trials for a number of neurodegenerative disorders (Ratan et al., 1994; Sonkusare et al., 2005). Competitive NMDA antagonists, midafotel and selfotel, interact at the same site at which glutamate binds to the NMDA receptor complex. They weakly cross the blood-brain barrier, but have high specificity and potency. These antagonists produce dose-dependent psychotomimetic side effects and therefore cannot be used for the treatment of neurodegenerative diseases.

The aminoadmantanes represent a novel class of drugs that are largely free of side effects. Aminoadmantanes have been used clinically as anti-viral and anti-parkinsonian agents for the past 10 years in Germany and other European countries (Koch et al., 2005). Memantine, 1-amino-3,5-dimethyladmantane, is a moderately low affinity and uncompetitive NMDA receptor antagonist (Fig. 10.3). It enters the NMDA receptor channel preferentially when the channel is open excessively. Importantly, the off-rate of memantine is relatively fast so that it does not accumulate substantially in the channel to interfere with normal synaptic transmission (Doraiswamy, 2002; Sonkusare et al., 2005; Ratan et al., 1994). Thus low concentrations of memantine prevent excessive glutamatergic stimulation while still maintaining normal glutamate-mediated neurotransmission. Memantine has good tolerability, low side-effect profiles, and a positive therapeutic impact in moderate-to-severe AD, both as monotherapy and in conjunction with donepezil (Schmitt et al., 2007). In addition, low levels of this drug promote neuroplasticity and memory formation (Rogawski and Wenk, 2003). Memantine treatment, 20 mg/day, for a period of 28 weeks reduced clinical deterioration in moderate to severe AD (Reisberg et al., 2003; Rogawski and Wenk, 2003).

Aβ deposition promotes oxidative stress and neuroinflammation, which in turn impair glutamatergic receptor function in such a way that the ability of these receptors to prevent the influx of Ca^{2+} in the absence of an appropriate presynaptic signal may be compromised. This process can promote neurodegeneration in AD (Farooqui and Horrocks, 2007b). It should be noted that glutamate-mediated abnormalities in neural membrane phospholipid metabolism appear long before the symptomatic onset of AD (Pettegrew et al., 1995). Memantine, with its ability to alleviate glutamatergic receptor over stimulation, may allow normal signaling between neu-

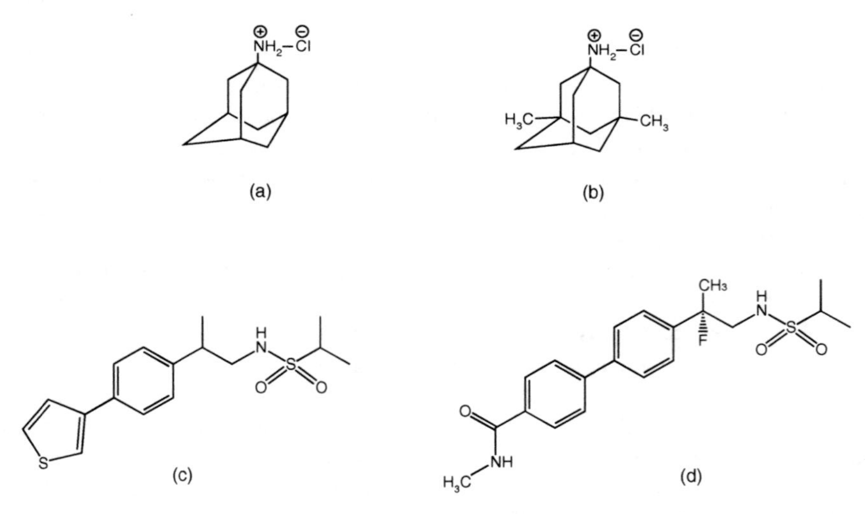

Fig. 10.3 Chemical structures of low affinity NMDA antagonists used for the treatment of AD, AD, AIDS dementia, and migrane. Memantine (a); Amantadine (b); (R,S)-N-2-(4-(3-thienyl)phenyl)-2-propanesulfonamide (LY392098) (c); and (R)-4'-[1-fluoro-1-methyl-2-(propane-2-sulphonylamino)-ethyl]-biphenyl-4-carboxylic acid methylamide (LY503430) (d)

rons. These processes may be associated with the therapeutic benefits of memantine in the earliest stages of the disease (Rogawski and Wenk, 2003; Wenk, 2006; Planells-Cases et al., 2006). In addition, memantine also blocks β-amyloid (Aβ) production and thus abrogates Aβ toxicity (Sonkusare et al., 2005). Collective evidence suggests that memantine is a neuroprotective agent in various animal models, based on both neurodegenerative and vascular processes, as it ameliorates cognitive and memory deficits (Sonkusare et al., 2005). Memantine may not stop or reverse AD, but through its moderating effect on glutamate metabolism, it may facilitate and restore normal neural cell signaling. This process may promote neuroprotection in AD patients. The compound is completely absorbed after oral intake and undergoes little metabolism. Having a low probability for drug-drug interactions, memantine, in principle, is suited for elderly patients exposed to multiple therapeutic therapies. Based on the above conclusion, several clinical trials for treatment of AD and vascular dementia were performed (Smith et al., 2006; Tariot et al., 2004; Wenk, 2006). Results are very encouraging. Several studies demonstrate that treatment with memantine has cognitive and functional benefits through all disease stages, and it is safe and well tolerated.

Second generation memantines (nitromemantines) were also synthesized (Ratan et al., 1994; Chen and Lipton, 2006). According to Lipton, these drugs use memantine as a homing signal to target NO to hyperactivated NMDAR in order to avoid systemic side-effects of NO such as hypotension (low blood pressure). The nitromemantines have enhanced neuroprotective efficacy in animal models of neurological disorders in vitro and *in vivo*. Nitromemantines lack the blood pressure lowering effect of nitroglycerin (Chen and Lipton, 2006). They interact with sulfhydryl groups associated with the cysteine residue of NMDA receptor channel through S-nitrosylation and down-regulate, but not completely shut off, NMDA receptor activity (Ratan et al., 1994; Chen and Lipton, 2006). Detailed structural and pharmacological investigations indicate that the NMDA receptor contains 5 cysteine residues. Cysteine residue 399 on the NR2 subunit of the NMDA receptor mediates 90% of the effects of nitric oxide (Choi et al., 2000). The binding of nitric oxide at cysteine 399 may produce a conformational change in the NMDA receptor protein that facilitates tight binding of glutamate and zinc to the NMDA receptor channel. The tight binding of glutamate and zinc results in desensitization of the NMDA receptor causing its channel to close (Lipton et al., 2002).

In addition to direct effects on the NMDA receptor, memantine up-regulates the levels of brain-derived neurotropic factor (BDNF) in the limbic cortex in a dose-dependent manner and induces isoforms of the BDNF receptor trkB (Lu, 2003; Marvanova et al., 2001). Both processes may potentiate the neuroprotective and memory enhancing effects of memantine. Other neuropharmacological effects of memantine include non-competitive and voltage-independent inhibition of $5HT_3$ receptor current (Rammes et al., 2001) and inhibition of human neuronal nicotinic cholinergic receptors (Buisson and Bertrand, 1998). This suggests that detailed neurochemical and neuropharmacological investigations are required on the molecular mechanism of action of memantine in mammalian brain.

Besides AD, memantine has been used for the treatment of stroke, Parkinson disease, amyotrophic lateral sclerosis, epilepsy, HIV associated dementia,

Huntington disease, depression, severe neuropathic pain, and glaucoma (Palmer, 2001; Kilpatrick and Tilbrook, 2002; Kohl and Dannhardt, 2001; Paul and Bolton, 2002; Lee et al., 2006; Wu et al., 2006). Collectively, these studies suggest that the approval of memantine, a safe, well-tolerated uncompetitive NMDA antagonist for the treatment of moderate to severe Alzheimer dementia, validates the NMDA type of glutamate receptors as key therapeutic targets of neurodegenerative diseases in humans. Consequently, at the present time attempts are made to identify and develop safe and potent NMDA antagonists for the treatment of AD and for other neurological disorders. The side effects of long-term oral administration of high doses (40 mg/day) of memantine include vertigo, agitation, fatigue, and rarely dizziness.

10.4 NMDA Receptor Antagonists for the Treatment of Huntington Disease

Huntington disease (HD) is an autosomal dominant, inherited neurodegenerative disorder for which there is no effective treatment or cure. Striatal medium spiny neurons are the most sensitive in HD. Dysregulation of the glutamate/Ca^{2+} signaling pathway is a possible cause of striatal MSN neurodegeneration in HD. NMDA receptor antagonists, memantine and riluzole, were used for the treatment of HD in animal models (Wu et al., 2006; Schiefermeier and Yavin, 2002). Memantine may act by exerting its effect at the micro-calpain level thus reducing the levels of huntingtin proteolytic fragments (Wu et al., 2006; Lee et al., 2006). In addition, memantine increased Bcl-xl and decreased Bax levels in 3NP rats. Similarly, riluzole prolongs survival time and alters nuclear inclusion formation in the R6/2 transgenic mouse model of HD (Schiefermeier and Yavin, 2002), suggesting profound changes in ubiquitination and proteolytic activity in HD.

10.5 NMDA Antagonists for the Treatment of Epilepsy

Epilepsy is a common neurological disorder in which recurrent seizures are caused by abnormal electrical discharges from the brain. Because glutamate is involved in the initiation of seizures and their propagation, attempts have been made to use NMDA receptor antagonists for the treatment of epilepsy in animal models (Yen et al., 2004; Palmer, 2001; Kohl and Dannhardt, 2001). Remacemide hydrochloride is a low-affinity, non-competitive NMDA receptor channel blocker that has been used for the treatment of epilepsy. A double-blind, placebo-controlled, and multicenter study was used to assess the safety and efficacy of remacemide hydrochloride in 252 adult patients with refractory epilepsy who were already on three antiepileptic drugs (Chadwick et al., 2002). Patients were randomized to one of three doses of remacemide hydrochloride, 300, 600 or 1200 mg/day, or placebo Q.I.D., for up to 15 weeks. An increasing percentage of responders were observed with increasing remacemide hydrochloride dose. At 1200 mg/day, 23% of patients were responders compared with 7% on placebo. This difference was significant, as was the overall difference between treatments. Adverse affects, including dizziness, abnormal gait, gastrointestinal problems, somnolence, diplopia, and fatigue, were mild or moderate in severity. A dose-dependent significant increase in responders was observed following adjunctive remacemide hydrochloride compared with placebo. Remacemide hydrochloride was well tolerated. Phase III clinical trials are ongoing with remacemide for epilepsy.

10.6 NMDA Antagonists for the Treatment of Head Injury

Excessive release of glutamate following head injury contributes to progressive injury in animal models and human studies (Farooqui and Horrocks, 2007a). Several pharmacological agents that act as antagonists to the glutamate receptor have promise in limiting this progression in animal models of head injury (Wang et al., 2006a; Parton et al., 2005). The efficacy of the NMDA receptor antagonist selfotel (CGS 19755) was evaluated in two parallel studies of severely head injured patients (Morris et al., 1999). Both trials were aborted prematurely because of increased deaths and other adverse events. Data analysis on these patients revealed no statistically significant difference in mortality rates in all cases between the two treatment groups in the head injury trials. A major issue was the lack of information on blood-brain barrier permeability and the availability of selfotel at the injury site. Furthermore, events associated with secondary injury, worsening of neurological symptoms, were a powerful predictor of poor outcome during these trials (Farin and Marshall, 2004).

This information was utilized for a subsequent clinical trial of CP101-606, a novel NR2B subunit antagonist of the NMDA receptor complex, in consultation with the Pfizer Company, the US Brain Injury Consortium, and the San Diego Clinical Trial Group (Farin and Marshall, 2004). Based on this trial, several recommendations were made. They include (a) the development of therapeutic regimens targeted towards the mechanisms of brain injury, (b) availability of adequate preclinical data, (c) directing treatment towards an appropriate patient population, (d) central gathering and interpretation of the neuroradiological findings, (e) monitoring of trial center performance, (f) stratification and pre-trial prognostic analysis for identification of subgroups (Farin and Marshall, 2004). Phase III clinical trials are ongoing for head trauma with dexanabinol (HU-211), 7-hydroxy-tetrahydrocannabinol 1, 1-dimethylheptyl (Palmer, 2001).

10.7 NMDA Receptor Antagonists for the Treatment of Chronic Pain

Chronic pain due to nerve injury results in sensitization of the central nervous system, maintained by a state of sensitization that is mediated in part by glutamate binding to the NMDA receptor (Dickenson, 1990; Coderre, 1993). A key step in the development of sensitization is the removal of the Mg^{2+} blockade from the NMDA receptor channel and the expression of Na^{+} channels in primary sensory neurons (Max and Hagen, 2000). These processes result in depolarization that is associated with enhanced excitability of the pain pathways (Hewitt, 2000). NMDA antagonists are apparently more potent in experimental models of neuropathic pain, whereas AMPA antagonists are more effective in abolition of the hyperalgesia seen during experimental inflammation (Weber, 1998; Hewitt, 2000; Gilron and Max, 2005). Among known NMDA antagonists, amantadine, dextromethorphan, and ketamine have been tested in animal models of pain as well as in human subjects. Ketamine was quite active in certain cases of neuropathic pain. It reduces opiate demand when used for postoperative analgesia (Hewitt, 2000). Ketamine, however, produces severe cognitive and affective side effects, making it unattractive for chronic treatment in human subjects (Klepstad and Borchgrevink, 1997). Clinical trials with dextromethorphan, 400 mg/day, in patients with painful diabetic neuropathy and postherpetic neuralgia reduced the intensity of pain by 25% relative to placebo in 32 patients with diabetic neuropathy, but had little effect on post-herpetic neuralgia (Sang et al., 2002; Nelson et al., 1997). Collective evidence suggests that improving the efficacy of opioids by blocking NMDA receptor-mediated activity constitutes another clinically relevant concept for pain management. Thus NMDA-receptor antagonists represent a new class of analgesics and may have potential as co-analgesics when used in combination with opioids (Weber, 1998; Sang et al., 2002; Nelson et al., 1997).

The involvement of AMPA/KA receptors in pain and hyperalgesia was tested in animal models and human volunteers. The administration of LY293558 reduced allodynia and hyperalgesia in a dose-dependent manner with no side effects (Gilron et al., 2000; Lee et al., 2006). These promising results led Eli Lilly to perform more studies on a new class of glutamate antagonists. The anti-hyperalgesic effect for these compounds may be due to the blockade of GluR5 kainate-type receptors, whereas sedative effects result from GluR1-4 AMPA receptors. If studies on LY 293558 are confirmed in patients with chronic pain, then kainate antagonists may offer a powerful and nontoxic treatment for this condition (Gilron et al., 2000; Lee et al., 2006; Gilron and Max, 2005).

10.8 Conclusion

Excessive release of glutamate and the over stimulation of glutamate receptors play a pivotal role in the pathogenesis of stroke, head injury, spinal cord trauma, epilepsy, and neurodegenerative diseases such as AD, PD, ALS, HD, MS, and AIDS dementia (Farooqui and Horrocks, 1994; Lipton and Rosenberg, 1994). Antagonists of glutamate receptor function have neuroprotective effects in animal models and preclinical models of stroke, head injury, spinal cord trauma, epilepsy, and many neurodegenerative diseases (Sonkusare et al., 2005; Chen and Lipton, 2006; Ratan et al., 1994). Both noncompetitive and competitive antagonists have undergone tolerability studies in acute stroke and traumatic brain injury. These drugs not only antagonize excitotoxicity, but also block normal neuronal function. In addition they produce neuropsychological symptoms and have important cardiovascular side effects. NMDA receptor antagonists against glycine, polyamine, and magnesiumion sites were also used in clinical trials. These trials were abandoned owing to concerns about drug toxicity, particularly in stroke and head injury. A systematic review on the use of these drugs for stroke and head injury indicates that definitive conclusions cannot be drawn for most drugs because of the early termination of clinical trials.

In dementia, memantine has shown some benefit in moderate-to-severe Alzheimer's disease. It is currently the only drug that has been approved for use in more advanced stages of AD (Sonkusare et al., 2005; Chen and Lipton, 2006; Tanovic and Alfaro, 2006). Significant reductions in functional and cognitive decline were observed with memantine relative to placebo in randomized double-blind trials in an AD patient population. Clinical trial and post-marketing surveillance data indicate that memantine is well tolerated. To date, no clear benefits have been reported for milder stages of AD or for vascular dementia. It should be noted that most of the memantine trials have been small and not long enough to detect clinically important benefits. Attempts to synthesize second-generation derivatives of memantine are proceeding well. Several laboratory tests compare their ability to affect the function of the NMDA receptor and to dilate blood vessels. These tests provide an initial indication of which compounds may have enhanced activity relative to memantine. The results also provide guidance for the synthesis of additional compounds. Some second-generation memantine compounds may have greater neuroprotective properties than memantine (Ratan et al., 1994; Chen and Lipton, 2006). These second-generation of memantine-like drugs take advantage of the fact that the NMDA receptor has other modulatory sites in addition to its ion channel that potentially may be used for safe but effective clinical intervention. Thus a new generation of glutamate receptor antagonists that block excessive activation of NMDA receptors, but allow intact physiological NMDA receptor function, are needed for the treatment of acute neural trauma and neurodegenerative diseases (Muir, 2006; Wang et al., 2006b; Ratan et al., 1994).

References

Albers G. W., Atkinson R. P., Kelley R. E., and Rosenbaum D. M. (1995). Safety, tolerability, and pharmacokinetics of the *N*-methyl-D-aspartate antagonist dextrorphan in patients with acute stroke. Dextrorphan Study Group. Stroke 26:254–258.

Albers G. W., Goldstein L. B., Hall D., and Lesko L. M. (2001). Aptiganel hydrochloride in acute ischemic stroke: a randomized controlled trial. J. Am. Med. Assoc. 286:2673–2682.

Bordi F., Pietra C., Ziviani L., and Reggiani A. (1997). The glycine antagonist GV150526 protects somatosensory evoked potentials and reduces the infarct area in the MCAo model of focal ischemia in the rat. Exp. Neurol. 145:425–433.

Buisson B. and Bertrand D. (1998). Open-channel blockers at the human $\alpha 4\beta 2$ neuronal nicotinic acetylcholine receptor. Mol. Pharmacol. 53:555–563.

Chadwick D. W., Betts T. A., Boddie H. G., Crawford P. M., Lindstrom P., Newman P. K., Soryal I., Wroe S., and Holdich T. A. (2002). Remacemide hydrochloride as an add-on therapy in epilepsy: a randomized, placebo-controlled trial of three dose levels (300, 600 and 1200 mg/day) in a Q.I.D. regimen. Seizure. 11:114–123.

Chen H. S. and Lipton S. A. (2006). The chemical biology of clinically tolerated NMDA receptor antagonists. J. Neurochem. 97:1611–1626.

Choi D. W. (1990). Cerebral hypoxia: some new approaches and unanswered questions. J. Neurosci. 10:2493–2501.

Choi Y. B., Tenneti L., Le D. A., Ortiz J., Bai G., Chen H. S., and Lipton S. A. (2000). Molecular basis of NMDA receptor-coupled ion channel modulation by *S*-nitrosylation. Nat. Neurosci. 3:15–21.

Coderre T. J. (1993). The role of excitatory amino acid receptors and intracellular messengers in persistent nociception after tissue injury in rats. Mol. Neurobiol 7:229–246.

Davis M., Mendelow A. D., Perry R. H., Chambers I. R., and James O. F. (1995). Experimental stroke and neuroprotection in the aging rat brain. Stroke 26:1072–1078.

Davis S. M., Lees K. R., Albers G. W., Diener H. C., Markabi S., Karlsson G., and Norris J. (2000). Selfotel in acute ischemic stroke: possible neurotoxic effects of an NMDA antagonist. Stroke 31:347–354.

Dickenson A. H. (1990). A cure for wind up: NMDA receptor antagonists as potential analgesics. Trends Pharmacol. Sci. 11:307–309.

Doraiswamy P. M. (2002). Non-cholinergic strategies for treating and preventing Alzheimer's disease. CNS Drugs 16:811–824.

Dyker A. G., Edwards K. R., Fayad P. B., Hormes J. T., and Lees K. R. (1999). Safety and tolerability study of aptiganel hydrochloride in patients with an acute ischemic stroke. Stroke 30:2038–2042.

Dyker A. G. and Lees K. R. (1999). Remacemide hydrochloride: a double-blind, placebo-controlled, safety and tolerability study in patients with acute ischemic stroke. Stroke 30: 1796–1801.

Ellison G. (1994). Competitive and non-competitive NMDA antagonists induce similar limbic degeneration. NeuroReport 5:2688–2692.

Ellison G. (1995). The N-methyl-D-aspartate antagonists phencyclidine, ketamine and dizocilpine as both behavioral and anatomical models of the dementias. Brain Res. Rev. 20:250–267.

Endres M., Fink K., Zhu J., Stagliano N. E., Bondada V., Geddes J. W., Azuma T., Mattson M. P., Kwiatkowski D. J., and Moskowitz M. A. (1999). Neuroprotective effects of gelsolin during murine stroke. J. Clin. Invest 103:347–354.

Fagg G. E. and Baud J. (1988). Characterization of NMDA receptor-ionophore complexes in the brain. In: Lodge D. (ed.), *Excitatory Amino Acids in Health and Disease*. John Wiley & Sons, New York, pp. 63–90.

Farin A. and Marshall L. F. (2004). Lessons from epidemiologic studies in clinical trials of traumatic brain injury. Acta Neurochir. Suppl 89:101–107.

Farooqui A. A. and Horrocks L. A. (1994). Excitotoxicity and neurological disorders: involvement of membrane phospholipids. Int. Rev. Neurobiol. 36:267–323.

Farooqui A. A. and Horrocks L. A. (2007a). Glutamate and cytokine-mediated alterations of phospholipids in head injury and spinal cord trauma. In: Banik N. (ed.), *Brain and Spinal Cord Trauma*. Handbook of Neurochemistry. Lajtha, A. (ed.). Springer, New York.

Farooqui A. A. and Horrocks L. A. (2007b). *Glycerophospholipids in the Brain: Phospholipases A_2 in Neurological Disorders*, pp. 1–394. Springer, New York.

Ferrer-Montiel A. V., Sun W., and Montal M. (1995). Molecular design of the N-methyl-D-aspartate receptor binding site for phencyclidine and dizolcipine. Proc. Natl. Acad. Sci. U.S.A 92:8021–8025.

Gallagher M. J., Huang H., Pritchett D. B., and Lynch D. R. (1996). Interactions between ifenprodil and the NR2B subunit of the *N*-methyl-D-aspartate receptor. J. Biol. Chem. 271:9603–9611.

Gilron I. and Max M. B. (2005). Combination pharmacotherapy for neuropathic pain: current evidence and future directions. Expert Rev. Neurother. 5:823–830.

Gilron I., Max M. B., Lee G., Booher S. L., Sang C. N., Chappell A. S., and Dionne R. A. (2000). Effects of the 2-amino-3-hydroxy-5-methyl-4-isoxazole-proprionic acid/kainate antagonist LY293558 on spontaneous and evoked postoperative pain. Clin. Pharmacol. Ther. 68:320–327.

Grant G., Cadossi R., and Steinberg G. (1994). Protection against focal cerebral ischemia following exposure to a pulsed electromagnetic field. Bioelectromagnetics 15:205–216.

Grossberg G. T., Edwards K. R., and Zhao Q. (2006). Rationale for combination therapy with galantamine and memantine in Alzheimer's disease. J. Clin. Pharmacol. 46:17S–26S.

Grotta J., Clark W., Coull B., Pettigrew L. C., Mackay B., Goldstein L. B., Meissner I., Murphy D., and LaRue L. (1995). Safety and tolerability of the glutamate antagonist CGS 19755 (Selfotel) in patients with acute ischemic stroke: results of a phase IIa randomized trial. Stroke 26:602–605.

Grotta J. C., Picone C. M., Ostrow P. T., Strong R. A., Earls R. M., Yao L. P., Rhoades H. M., and Dedman J. R. (1990). CGS-19755, a competitive NMDA receptor antagonist, reduces calcium-calmodulin binding and improves outcome after global cerebral ischemia. Ann. Neurol. 27:612–619.

Gustafson I., Westerberg E., and Wieloch T. (1990). Protection against ischemia-induced neuronal damage by the alpha 2-adrenoceptor antagonist idazoxan: influence of time of administration and possible mechanisms of action. J. Cereb. Blood Flow Metab 10:885–894.

Helfaer M. A., Ichord R. N., Martin L. J., Hurn P. D., Castro A., and Traystman R. J. (1998). Treatment with the competitive NMDA antagonist GPI 3000 does not improve outcome after cardiac arrest in dogs. Stroke 29:824–829.

Hewitt D. J. (2000). The use of NMDA-receptor antagonists in the treatment of chronic pain. Clin. J. Pain 16:S73–S79.

Jewett D. C., Butelman E. R., and Woods J. H. (1996). Nitric oxide synthase inhibitors produce phencyclidine-like behavioral effects in pigeons. Brain Res. 715:25–31.

Kawasaki-Yatsugi S., Ichiki C., Yatsugi S., Takahashi M., Shimizu-Sasamata M., Yamaguchi T., and Minematsu K. (2000). Neuroprotective effects of an AMPA receptor antagonist YM872 in a rat transient middle cerebral artery occlusion model. Neuropharmacology 39:211–217.

Kemp J. A. and McKernan R. M. (2002). NMDA receptor pathways as drug targets. Nat. Neurosci. 5:1039–1042.

Kilpatrick G. J. and Tilbrook G. S. (2002). Memantine. Merz. Curr. Opin. Investig. Drugs 3: 798–806.

Klamer D., Zhang J., Engel J. A., and Svensson L. (2005). Selective interaction of nitric oxide synthase inhibition with phencyclidine: behavioural and NMDA receptor binding studies in the rat. Behav. Brain Res. 159:95–103.

Klepstad P. and Borchgrevink P. C. (1997). Four years' treatment with ketamine and a trial of dextromethorphan in a patient with severe post-herpetic neuralgia. Acta Anaesthesiol. Scand. 41:422–426.

Koch H. J., Uyanik G., and Fischer-Barnicol D. (2005). Memantine: a therapeutic approach in treating Alzheimer's and vascular dementia. Curr. Drug Targets CNS Neurol. Disord. 4: 499–506.

Kohl B. K. and Dannhardt G. (2001). The NMDA receptor complex: a promising target for novel antiepileptic strategies. Curr. Med. Chem. 8:1275–1289.

Kornhuber J. and Weller M. (1997). Psychotogenicity and N-methyl-D-aspartate receptor antagonism: implications for neuroprotective pharmacotherapy. Biol. Psychiatry 41:135–144.

Koroshetz W. J. and Moskowitz M. A. (1996). Emerging treatments for stroke in humans. Trends Pharmacol. Sci. 17:227–233.

Labiche L. A. and Grotta J. C. (2004). Clinical trials for cytoprotection in stroke. NeuroRx. 1: 46–70.

Lee S. T., Chu K., Park J. E., Kang L., Ko S. Y., Jung K. H., and Kim M. (2006). Memantine reduces striatal cell death with decreasing calpain level in 3-nitropropionic model of Huntington's disease. Brain Res. 1118:199–207.

Lees K. R. (1997). Cerestat and other NMDA antagonists in ischemic stroke. Neurology 49: S66–S69.

Lees K. R., Asplund K., Carolei A., Davis S. M., Diener H. C., Kaste M., Orgogozo J. M., and Whitehead J. (2000). Glycine antagonist (gavestinel) in neuroprotection (GAIN International) in patients with acute stroke: a randomised controlled trial. GAIN International Investigators. Lancet 355:1949–1954.

Lipton S. A. (1993). Prospects for clinically tolerated NMDA antagonists: open-channel blockers and alternative redox states of nitric oxide. Trends Neurosci. 16:527–532.

Lipton S. A., Choi Y. B., Sucher N. J., and Chen H. S. (1998). Neuroprotective versus neurodestructive effects of NO-related species. BioFactors 8:33–40.

Lipton S. A., Choi Y. B., Takahashi H., Zhang D., Li W., Godzik A., and Bankston L. A. (2002). Cysteine regulation of protein function–as exemplified by NMDA-receptor modulation. Trends Neurosci. 25:474–480.

Lipton S. A. and Rosenberg P. A. (1994). Excitatory amino acids as a final common pathway for neurologic disorders. N. Engl. J. Med. 330:613–622.

Lu B. (2003). BDNF and activity-dependent synaptic modulation. Learn. Mem. 10:86–98.

Marvanova M., Lakso M., Pirhonen J., Nawa H., Wong G., and Castren E. (2001). The neuroprotective agent memantine induces brain-derived neurotrophic factor and trkB receptor expression in rat brain. Mol. Cell Neurosci. 18:247–258.

Max M. B. and Hagen N. A. (2000). Do changes in brain sodium channels cause central pain? Neurology 54:544–545.

Minematsu K., Fisher M., Li L., Davis M. A., Knapp A. G., Cotter R. E., McBurney R. N., and Sotak C. H. (1993). Effects of a novel NMDA antagonist on experimental stroke rapidly and quantitatively assessed by diffusion-weighted MRI. Neurology 43:397–403.

Mori H., Masaki H., Yamakura T., and Mishina M. (1992). Identification by mutagenesis of a Mg^{2+}-block site of the NMDA receptor channel. Nature 358:673–675.

Morris G. F., Bullock R., Marshall S. B., Marmarou A., Maas A., and Marshall L. F. (1999). Failure of the competitive *N*-methyl-D-aspartate antagonist selfotel (CGS 19755) in the treatment of severe head injury: results of two phase III clinical trials. The Selfotel Investigators. J. Neurosurg. 91:737–743.

Muir K. W. (2006). Glutamate-based therapeutic approaches: clinical trials with NMDA antagonists. Curr. Opin. Pharmacol. 6:53–60.

Nelson K. A., Park K. M., Robinovitz E., Tsigos C., and Max M. B. (1997). High-dose oral dextromethorphan versus placebo in painful diabetic neuropathy and postherpetic neuralgia. Neurology 48:1212–1218.

Ogren S. O. and Goldstein M. (1994). Phencyclidine- and dizocilpine-induced hyperlocomotion are differentially mediated. Neuropsychopharmacology 11:167–177.

Osawa Y. and Davila J. C. (1993). Phencyclidine, a psychotomimetic agent and drug of abuse, is a suicide inhibitor of brain nitric oxide synthase. Biochem. Biophys. Res. Commun. 194: 1435–1439.

Palmer G. C. (2001). Neuroprotection by NMDA receptor antagonists in a variety of neuropathologies. Curr. Drug Targets 2:241–271.

Parton A., Coulthard E., and Husain M. (2005). Neuropharmacological modulation of cognitive deficits after brain damage. Curr. Opin. Neurol. 18:675–680.

Paul C. and Bolton C. (2002). Modulation of blood-brain barrier dysfunction and neurological deficits during acute experimental allergic encephalomyelitis by the *N*-methyl-D-aspartate receptor antagonist memantine. J. Pharmacol. Exp. Ther. 302:50–57.

Pettegrew J. W., Klunk W. E., Kanal E., Panchalingam K., and McClure R. J. (1995). Changes in brain membrane phospholipid and high-energy phosphate metabolism precede dementia. Neurobiol. Aging 16:973–975.

Planells-Cases R., Lerma J., and Ferrer-Montiel A. (2006). Pharmacological intervention at ionotropic glutamate receptor complexes. Curr. Pharm. Des 12:3583–3596.

Rajdev S., Fix A. S., and Sharp F. R. (1998). Acute phencyclidine neurotoxicity in rat forebrain: induction of haem oxygenase-1 and attenuation by the antioxidant dimethylthiourea. Eur. J. Neurosci. 10:3840–3852.

Rammes G., Rupprecht R., Ferrari U., Zieglgänsberger W., and Parsons C. G. (2001). The *N*-methyl-D-aspartate receptor channel blockers memantine, MRZ 2/579 and other amino-alkyl-cyclohexanes antagonise 5-HT_3 receptor currents in cultured HEK-293 and N1E-115 cell systems in a non-competitive manner. Neurosci. Lett. 306:81–84.

Ratan R. R., Murphy T. H., and Baraban J. M. (1994). Oxidative stress induces apoptosis in embryonic cortical neurons. J. Neurochem. 62:376–379.

Reeker W., Werner C., Mollenberg O., Mielke L., and Kochs E. (2000). High-dose S(+)-ketamine improves neurological outcome following incomplete cerebral ischemia in rats. Can. J. Anaesth. 47:572–578.

Reisberg B., Doody R., Stoffler A., Schmitt F., Ferris S., and Mobius H. J. (2003). Memantine in moderate-to-severe Alzheimer's disease. N. Engl. J. Med. 348:1333–1341.

Rogawski M. A. and Wenk G. L. (2003). The neuropharmacological basis for the use of memantine in the treatment of Alzheimer's disease. CNS Drug Rev. 9:275–308.

Sacco R. L., DeRosa J. T., Haley E. C., Jr., Levin B., Ordronneau P., Phillips S. J., Rundek T., Snipes R. G., and Thompson J. L. (2001). Glycine antagonist in neuroprotection for patients with acute stroke: GAIN Americas: a randomized controlled trial. J. Am. Med. Assoc. 285:1719–1728.

Sang C. N., Booher S., Gilron I., Parada S., and Max M. B. (2002). Dextromethorphan and memantine in painful diabetic neuropathy and postherpetic neuralgia: efficacy and dose-response trials. Anesthesiology 96:1053–1061.

Sarraf-Yazdi S., Sheng H., Miura Y., McFarlane C., Dexter F., Pearlstein R., and Warner D. S. (1998). Relative neuroprotective effects of dizocilpine and isoflurane during focal cerebral ischemia in the rat. Anesth. Analg. 87:72–78.

Schabitz W. R., Li F., and Fisher M. (2000). The N-methyl-D-aspartate antagonist CNS 1102 protects cerebral gray and white matter from ischemic injury following temporary focal ischemia in rats. Stroke 31:1709–1714.

Schehr R. S. (1996). New treatments for acute stroke. Nat. Biotechnol. 14:1549–1554.

Schiefermeier M. and Yavin E. (2002). n-3 Deficient and docosahexaenoic acid-enriched diets during critical periods of the developing prenatal rat brain. J. Lipid Res. 43:124–131.

Schmitt F., Ryan M., and Cooper G. (2007). A brief review of the pharmacologic and therapeutic aspects of memantine in Alzheimer's disease. Expert Opin. Drug Metab Toxicol. 3: 135–141.

Shimizu-Sasamata M., Kawasaki-Yatsugi S., Okada M., Sakamoto S., Yatsugi S., Togami J., Hatanaka K., Ohmori J., Koshiya K., Usuda S., and Murase K. (1996). YM90K: pharmacological characterization as a selective and potent alpha-amino-3-hydroxy-5-methylisoxazole-4-propionate/kainate receptor antagonist. J. Pharmacol. Exp. Ther. 276:84–92.

Shohami E., Novikov M., and Mechoulam R. (1993). A nonpsychotropic cannabinoid, HU-211, has cerebroprotective effects after closed head injury in the rat. J. Neurotrauma 10:109–119.

Smith M., Wells J., and Borrie M. (2006). Treatment effect size of memantine therapy in Alzheimer disease and vascular dementia. Alzheimer Dis. Assoc. Disord. 20:133–137.

Sonkusare S. K., Kaul C. L., and Ramarao P. (2005). Dementia of Alzheimer's disease and other neurodegenerative disorders–memantine, a new hope. Pharmacol. Res 51:1–17.

Suzuki M., Sasamata M., and Miyata K. (2003). Neuroprotective effects of YM872 coadministered with t-PA in a rat embolic stroke model. Brain Res. 959:169–172.

Tanovic A. and Alfaro V. (2006). Neuroprotecci6n con memantina (antagonista no competitivo del receptor NMDA-glutamato) frente a la excitotoxicidad asociada al glutamato en la enfermedad de Alzheimer y en la demencia vascular [Glutamate-related excitotoxicity neuroprotection with memantine, an uncompetitive antagonist of NMDA-glutamate receptor, in Alzheimer's disease and vascular dementia]. Rev. Neurol. 42:607–616.

Tariot P. N., Farlow M. R., Grossberg G. T., Graham S. M., McDonald S., and Gergel I. (2004). Memantine treatment in patients with moderate to severe Alzheimer disease already receiving donepezil: a randomized controlled trial. J. Am. Med. Assoc. 291:317–324.

Teramoto S., Matsuse T., and Ouchi T. (1999). Amantadine and pneumonia in elderly stroke patients. Lancet 353:2156–2157.

The STIPAS Investigators (1994). Safety study of tirilazad mesylate in patients with acute ischemic stroke (STIPAS). Stroke 25:418–423.

Wang K. K., Larner S. F., Robinson G., and Hayes R. L. (2006a). Neuroprotection targets after traumatic brain injury. Curr. Opin. Neurol. 19:514–519.

Wang Y., Eu J., Washburn M., Gong T., Chen H. S., James W. L., Lipton S. A., Stamler J. S., Went G. T., and Porter S. (2006b). The pharmacology of aminoadamantane nitrates. Curr. Alzheimer Res. 3:201–204.

Weber C. (1998). [NMDA-receptor antagonist in pain therapy]. Anasthesiol. Intensivmed. Notfallmed. Schmerzther. 33:475–483.

Wenk G. L. (2006). Neuropathologic changes in Alzheimer's disease: potential targets for treatment. J. Clin. Psychiatry 67(Suppl 3):3–7.

Wu J., Tang T., and Bezprozvanny I. (2006). Evaluation of clinically relevant glutamate pathway inhibitors in *in vitro* model of Huntington's disease. Neurosci. Lett. 407:219–223.

Yamakura T. and Shimoji K. (1999). Subunit- and site-specific pharmacology of the NMDA receptor channel. Prog. Neurobiol. 59:279–298.

Yen W., Williamson J., Bertram E. H., and Kapur J. (2004). A comparison of three NMDA receptor antagonists in the treatment of prolonged status epilepticus. Epilepsy Res. 59:43–50.

Chapter 11
Future Perspectives: New Strategies for Antagonism of Excitotoxicity, Oxidative Stress and Neuroinflammation in Neurodegenerative Diseases

Excitotoxicity, oxidative stress, and inflammation are major components central to the pathogenesis of acute neural trauma (ischemia, spinal cord trauma, and head injury) and neurodegenerative diseases such as Alzheimer disease (AD), Parkinson disease (PD), Huntington disease (HD), and amyotrophic lateral sclerosis (ALS) (Farooqui and Horrocks, 2007b; Farooqui et al., 2007). Neurons undergoing severe ischemia, spinal cord trauma, and head injury die rapidly (minutes to hours) by necrotic cell death at the core of injury, whereas in penumbral region or surrounding area neurons display delayed vulnerability and die through apoptotic cell death (McIntosh et al., 1998; Farooqui et al., 2004; Farooqui and Horrocks, 2007a). Furthermore, cerebral ischemia, head trauma, and spinal cord injuries trigger similar auto-protective mechanisms including the induction of heat shock proteins, anti-inflammatory cytokines, and production of endogenous antioxidants (Leker and Shohami, 2002). In contrast, neurodegenerative diseases are a complex group of neurological disorders that involve site-specific premature and slow death of certain neuronal populations in specific area of brain tissue (Farooqui and Horrocks, 1994; Graeber and Moran, 2002). Interplay among the second messengers generated following excitotoxicity, oxidative stress, neuroinflammation, and related processes may be closely associated with neurodegeneration in acute neural trauma and neurodegenerative diseases (Farooqui and Horrocks, 2006a). Our emphasis on the interplay among excitotoxicity, oxidative stress, and neuroinflammation does not rule out the participation of other mechanisms involved in neurodegeneration. However, it is timely and appropriate to apply the concept of interplay among excitotoxicity, oxidative stress, and neuroinflammation to neural cell injury in acute neural trauma and neurodegenerative diseases. Lack of coordination among the above parameters may control the time taken by neural cells to die. For example, in ischemic brain, traumatized brain, and injured spinal cord, neuronal cell death occurs within hours to days, whereas in neurodegenerative diseases, neuronal damage take years to develop. For ischemia and traumatic brain and spinal cord injuries, there is a therapeutic window of 4–6 h (Leker and Shohami, 2002; Farooqui and Horrocks, 2007a) for partial restoration of many functions. In contrast, there is no therapeutic window for neurodegenerative diseases and drug intervention should start as soon as first detected.

11.1 Sources and Mechanism of Glutamate Release

Glutamate is the major excitatory amino acid transmitter within the CNS. It is present in 30–40% of synapses throughout the brain and 80–90% of synapses in those brain regions that are involved in cognitive processing, cerebral cortex and hippocampus. On one hand, the glutamatergic neurotransmitter system is connected to the acetylcholinergic neurotransmitter system through the glutamatergic/aspartatergic-acetylcholinergic circuit (Riederer and Hoyer, 2006) and on the other hand it interacts with dopaminergic neurotransmission in the nigrostriatal tract (Page et al., 2001; Ohishi et al., 1993; Testa et al., 1994). Several postsynaptic ionotropic types, NMDA, KA, and AMPA, and metabotropic types of glutamate receptors mediate glutamatergic signaling. The activation of glutamate receptors leads to depolarization and neuronal excitation.

In normal synaptic functioning, activation of glutamate receptors is transitory and is coupled with interplay among neurotransmitters and second messengers associated with glutamatergic, acetylcholinergic, and dopaminergic neurotransmission. Lipid mediators that play important roles in learning and memory and cognition processes associated with normal brain function modulate this interplay or cross talk. Under pathological situations, large amounts of glutamate are released in the synaptic cleft. This results in an excessive and prolonged stimulation of glutamate receptors. This process is excitotoxicity (Olney et al., 1979; Choi, 1988).

Excitotoxicity is mediated by Ca^{2+} influx that initiates a cascade of events involving free radical generation, mitochondrial dysfunction, and activation of many enzymes including those involved in the generation and metabolism of arachidonic acid (Farooqui and Horrocks, 1991, 1994; Farooqui et al., 2001; Wang et al., 2005). These enzymes include isoforms of PLA_2, cyclooxygenase-2 (COX-2), lipoxygenases (LOX), epoxygenases (EPOX), and nitric oxide synthase (Phillis et al., 2006). The COX-2, LOX, and EPOX-generated products, such as prostaglandins, thromboxanes, and leukotrienes, and nonenzymic oxidation products of arachidonic acid, such as isoprostanes and 4-hydroxynonenal, accumulate. This, together with changes in cellular redox potential, and lack of energy generation, is associated with neural cell injury and cell death in acute neural trauma, ischemia, epilepsy, head injury, and spinal cord trauma, and neurodegenerative diseases, such as AD, PD, multiple sclerosis (MS), Creutzfeldt-Jakob disease (CJD) and AIDS dementia complex (Farooqui and Horrocks, 2006; Phillis et al., 2006; Farooqui et al., 2006).

The sources and mechanisms through which glutamate is released in the synaptic cleft still remain controversial (Danbolt, 2001). Some investigators believe that released glutamate comes from intracellular stores while others suggest the involvement of phosphate-activated mitochondrial glutaminase in glutamate production (Newcomb et al., 1997; Zhao et al., 2004). Another source of glutamate production may be N-acetyl-aspartyl-glutamate (NAAG) (Slusher et al., 1999). This metabolite is hydrolyzed by N-acetylated-alpha-linked-acidic dipeptidase into *N*-acetylaspartate (NAA) and glutamate. If all these sources are involved in glutamate release, then information on their relative contribution to glutamate release is needed in excitotoxic cell death. Like glutamate sources, molecular mechanisms of its release in early phases of excitotoxicity remain controversial.

The molecular mechanisms include (a) the release of glutamate from synaptic vesicles through exocytosis (Tuz et al., 2004; Katchman and Hershkowitz, 1993), dependent on both Ca^{2+} and ATP; (b) release of glutamate from cytosol through swelling-activated anion channels(Kimelberg and Mongin, 1998); (c) release of glutamate from vesicles in astrocytes, implicating vesicle proteins such as (SNARE proteins) in mediating the release of glutamate from astrocytes (Araque et al., 2000; Zhang et al., 2004; Liu et al., 2006); and (d) release of glutamate from cytosol via reversed operation of the glutamate transporters (Rossi et al., 2000; Roettger and Lipton, 1996; Jabaudon et al., 2000). Cellular sources of glutamate (neurons vs glial cells) also remain controversial. At present, the evidence is in favor of neurons because most glutamate in brain is stored in neurons rather than glial cells. The highest concentrations of glutamate are associated with glutamatergic nerve terminals. Thus investigations are required on the mechanisms associated with glutamate release in normal and pathological conditions.

11.2 Interplay Among Excitotoxicity, Oxidative Stress, and Neuroinflammation as a Possible Cause of Neurodegeneration

Neurodegenerative diseases are a heterogeneous group of slowly progressive disorders whose causes and pathogenesis remain unknown. Most chronic neurodegenerative diseases occur in both sporadic and familial forms. Additional mutations are being found in families with familial forms of these diseases. The events leading to the pathogenesis of sporadic neurodegenerative diseases remain unclear, although excitotoxicity, oxidative stress, and inflammatory processes following the activation of glial cells may be among the major processes closely associated with the pathogenesis of AD, PD, HD, ALS, and AIDS dementia complex (Farooqui and Horrocks, 2007b).

Excitotoxicity, inflammation, and oxidative stress are interrelated processes that may bring about neural cell demise independently or synergistically. An upregulation of interplay among excitotoxicity, oxidative stress, and neuroinflammation may involve increased vulnerability of neurons in acute neural trauma and neurodegenerative diseases (Facheris et al., 2004; Farooqui and Horrocks, 2007b). In acute neural trauma, because of the faster rate of upregulation of interplay among excitotoxicity, oxidative stress and neuroinflammation, neurons die rapidly in a matter of hours to days, following the sudden lack of oxygen, decreased ATP level, and sudden collapse of ion gradients. In contrast, in neurodegenerative diseases, oxygen, nutrients, and ATP continue to be available to the nerve cells and ionic homeostasis is maintained to a limited extent. The interplay among excitotoxicity, oxidative stress, and neuroinflammation occurs at a slow rate, resulting in a neurodegenerative process that takes several years to develop (Farooqui and Horrocks, 2007b).

In neurodegenerative diseases pain perception is below the threshold of detection. The immune system continues to attack brain tissue at the cellular and subcellular levels. Thus chronic inflammation lingers for years causing continued insult to the brain tissue, ultimately reaching the threshold of detection many years after the onset of the neurodegenerative diseases (Wood, 1998). Morphologically in neurodegenerative diseases, major hallmarks of neuroinflammation are phenotypic changes of glial cells, mainly activation and transformation of microglial cells into phagocytic cells, and to a lesser extent, reactive astrocytosis. The molecular mechanisms and internal and external factors (cytokines and growth factors) that modulate the dynamic aspects of chronic inflammation in neurodegenerative diseases remain unclear. Furthermore, it remains unclear to what extent neuroinflammation is beneficial for the injured brain tissue (Correale and Villa, 2004) and how it contributes to secondary brain injury and progressive neuronal loss in neurodegenerative diseases (Farooqui et al., 2007).

Although each neurodegenerative disease has a separate etiology with distinct morphological and pathophysiological characteristics, they may also share the same terminal neurochemical mechanism of cell damage and death (Farooqui and Horrocks, 1994, 2007b; Facheris et al., 2004; Polidori, 2004). Diet (Kidd, 2005), genetic, and environmental factors may also play a prominent role in modulating the interplay among excitotoxicity, oxidative stress, and neuroinflammation. The most important risk factors for neurodegenerative diseases are old age and a positive family history. The onset of neurodegenerative diseases is often subtle and usually occurs in mid to late life. Progression depends not only on genetic, but also on environmental factors (Graeber and Moran, 2002), leading to progressive cognitive and motor disabilities with devastating consequences. Aging can be looked upon as a disease process that includes aberrations associated with the development of other pathologies such as AD, PD, HD, and ALS (Berlett and Stadtman, 1997). Perhaps in these neurodegenerative diseases, the intensity of interplay and cross talk among lipid mediators generated during excitotoxicity, oxidative stress, and neuroinflammation may lead to unique manifestations that are characteristic features of AD, PD, HD, and ALS. Another factor that modulates neurodegenerative processes in acute trauma and neurodegenerative diseases is the energy status of degenerating neurons (Farooqui et al., 2004; Leist and Nicotera, 1998). It remains controversial whether excitotoxicity, oxidative stress, and neuroinflammation are the cause or consequence of neurodegeneration (Andersen, 2004; Juranek and Bezek, 2005). Similarly, very little information is available on the rate of neurodegeneration and clinical expression of neurodegenerative diseases with age. Studies should be planed to characterize the specific and sensitive markers to monitor excitotoxicity, oxidative stress, and neuroinflammation *in vivo* in living normal subjects and patients with neurodegenerative diseases.

If interplay among excitotoxicity, oxidative stress, and neuroinflammation is a key mediator involved in the onset and development and progression of neurodegenerative diseases, then what is responsible for the specific premature and slow death of certain neuronal populations in brain tissue in neurological disorders (Graeber and Moran, 2002)? For example in Alzheimer disease (AD), neuronal degeneration occurs in the nucleus basalis, whereas in Parkinson disease (PD), neurons in the substantia nigra die. The most severely affected neurons in Huntington disease (HD) are striatal medium spiny neurons. ALS is characterized by selective death of motor neurons in corticospinal tracts. Although factors and neurochemical events modulating the regional specificity of neuronal loss in neurodegenerative diseases remain unknown, two scenarios are possible.

In the first scenario, interplay among excitotoxicity, oxidative stress, and neuroinflammation may be necessary, but insufficient, for the development of neurodegenerative diseases and additional factor(s) are required for the onset of pathogenesis (Zhu et al., 2004; Smith et al., 2005). In the second scenario, interplay among excitotoxicity, oxidative stress, and neuroinflammation may be omnipotent but the disease process turns on specific genes that affect only a specific neuronal population in those regions where neuronal degeneration occurs (Dwyer et al., 2005; Migliore et al., 2005). This proposal is supported by the hypothesis that the nature of neuron-neuron connections as well as interactions between neurons and glial cells are essential for determining the selective neuronal vulnerability of neurons in neurodegenerative diseases (Wilde et al., 1997). Selective degeneration of neurons in stroke models also depends upon the nature of ROS. For example the selective vulnerability of the CA1 pyramidal neurons following hypoxic/ischemic injury depends upon superoxide radicals, rather than hydroxyl radicals (Wilde et al., 1997; Bernaudin et al., 1998). Discovering factors and neurochemical events that modulate the specificity of regional neurodegeneration in neurodegenerative diseases remains a most challenging area of neuroscience research (Graeber and Moran, 2002; Farooqui and Horrocks, 2007b). More studies are urgently required on this important topic.

Many investigations have been performed to determine the molecular mechanism associated with the pathogenesis of neurodegenerative diseases (Jellinger, 2001), but the heterogeneity of the etiologic factors makes it difficult to define the most important clinical factor determining the onset and progression of the disease. However, increasing evidence indicates that interplay among excitotoxicity, oxidative stress, and neuroinflammation may disturb neuronal lipid, protein, nucleic acid, and carbohydrate metabolism that leads to irreversible neuronal damage (Esposito et al., 2002). Earlier attempts to search for soluble peripheral biomarkers of oxidative stress and neuroinflammation in biological fluids, mainly cerebrospinal fluids (CSF) using classical biochemical procedures have not been successful in AD patients (Kim et al., 2004; Migliore et al., 2005; Floyd and Hensley, 2002). Lipidomics and proteomics have emerged as important technologies (German et al., 2007; Watson, 2006; Soule et al., 2006; Bowers-Gentry et al., 2006) for the identification and full characterization of *in vivo* markers for oxidative stress and neuroinflammation such as F_2-isoprostanes, prostaglandins, leukotrienes, lipoxins, hydroxyeicosatetraenoic acids, nitrotyrosine, carbonyls in proteins, oxidized DNA bases, and 4-HNE in CSF (Serhan, 2005a, 2006; Adibhatla et al., 2006; Milne et al., 2006; Hunt and Postle, 2006; Morrow, 2006; Lu et al., 2006; Perluigi et al., 2005).

Establishment of automatic systems including databases and accurate analyses of various lipid mediators derived from enzymic and non-enzymic metabolism of neuronal membrane polyunsaturated fatty acids (eicosanoids, resolvins, neuroprotectins, isoprostanes, neuroprostanes, and isofurans) would facilitate identification of key biomarkers associated with neurodegenerative diseases (Lu et al., 2006). A comprehensive lipidomics analysis using liquid chromatography-tandem mass spectrometry in rat hippocampal tissue indicates that systemic KA administration produces large amounts of $PGF_2\alpha$ and PGD_2 and smaller amounts of other prostaglandins and hydroxyeicosatetraenoic acids (Yoshikawa et al., 2006). This increase can be blocked by intracerebroventricular administration of KA receptor antagonists. Lipidomics can be applied to small tissue and biological fluid samples from patients with neurodegenerative diseases (Butterfield et al., 2006). Microarray analysis of tissue samples from brain regions associated with neurodegenerative diseases can provide information on candidate genes that influence excitotoxicity, oxidative stress, and neuroinflammatory responses. This would not only help in understanding molecular mechanisms associated with the development of neurodegenerative diseases, but would also facilitate molecular diagnostics and targets for drug therapy based on gene expression in body fluids such as CSF and blood (Facheris et al., 2004).

Another technique that can provide useful information on rates of generation of lipid mediators is quantitative autoradiography or positron emission tomography (PET) (Rapoport, 1999, 2005; Hampel et al., 2002). This procedure utilizes incorporation of fatty acids labeled with a radioactive or heavy isotope followed by determination of its distribution in brain regions and their lipid compartments as a function of time. From measurements, fluxes, turnover rates, half-lives, and ATP consumption rates, fatty acid metabolism *in vivo* can be determined and its incorporation can be imaged. Based on the *in vivo* metabolism of these fatty acids, PET is used to image brain signaling and neuroplasticity in humans in health and diseases. Initial experiments on animal models of Alzheimer and Parkinson disease with chronic unilateral lesions of nucleus basalis or substantia nigra indicate that PET and $[^{11}C]$arachidonic acid can be used with drug activation to image signal transduction. Detailed investigations are required on the use of MRI along with PET imaging to judge the severity and progression of dementia during the course of various neurodegenerative diseases in patients and normal human subjects (Rapoport, 2001; Hampel et al., 2002; Wendum et al., 2003; Masters et al., 2006).

Interplay among excitotoxicity, oxidative stress, and neuroinflammation facilitates neuronal death via apoptosis as well as necrosis (Nicotera and Lipton, 1999; Sastry and Rao, 2000; Farooqui et al., 2004). In neurons, intracellular energy levels and mitochondrial function are rapidly compromised in necrosis, but not in apoptosis, suggesting that the ATP level is a prominent factor in determining the neurochemical events associated with apoptotic and necrotic neural cell death (Nicotera and Leist, 1997; Liu et al., 2001). Apoptosis and necrosis represent the extreme ends of the spectrum of mechanisms for neural cell death (Williamson and Schlegel, 2002; Leist and Nicotera, 1998; Nicotera and Lipton, 1999).

The expression of genes associated with excitotoxicity, oxidative stress, and inflammation-mediated apoptosis is controlled through cytokines and chemokines at transcriptional and post-transcriptional levels (Allan and Rothwell, 2003; Lucas et al., 2006; Rothwell, 1999). The released cytokines act through their receptors causing activation of cascades of protein kinases and the pathway leading to activation of the transcription factor nuclear factor kappa B, NF-κB (Yamamoto and Gaynor, 2004). Similarly ROS also interacts and stimulates NF-κB . Upon stimulation IκB is rapidly phosphorylated, ubiquinated, and then degraded by proteasomes resulting in the release and subsequent nuclear translocation of active NF-κB. In the nucleus NFκB mediates the transcription of many genes implicated in inflammatory and immune responses. These genes include COX-2, intracellular adhesion molecule-1 (ICAM-1), vascular adhesion molecule-1 (VCAM-1), E-selectin, TNF-α, IL-1β, IL-6, sPLA$_2$, inducible nitric oxide synthase (iNOS), and matrix metalloproteinases (MMPs). ROS-mediated NF-κB activation involves NADPH oxidase, which is an important component of the innate immune response against toxic agents, and is involved in shaping the cellular response to a variety of physiological and pathological signals in neurodegenerative diseases (Anrather et al., 2006; Zhang et al., 2004; Frey et al., 2006; Rubin et al., 2005; Miller et al., 2006; Sun et al., 2007).

11.3 NMDA Receptor Antagonists, Antioxidants, and Anti-Inflammatory Drugs for Treatment of Neural Trauma and Neurodegenerative Diseases

Although neurodegenerative diseases have different primary causes, dysfunctions, and losses of selective groups of neurons in specific brain regions, they may utilize interplay among excitotoxicity, oxidative stress, and neuroinflammation as the terminal mechanism of cell death (Farooqui and Horrocks, 2007b).This concept may allow the use of NMDA antagonists, antioxidants and anti-inflammatory nutraceuticals as therapeutic drugs to delay the onset and to treat neurodegenerative disease. Efforts to treat neurodegenerative diseases with glutamate antagonists, antioxidants and anti-inflammatory nutraceuticals have not been successful (Prasad et al., 1999; Berman and Brodaty, 2004; Willcox et al., 2004; Esposito et al., 2002; Gilgun-Sherki et al., 2002, 2006; Gasparini et al., 2004; Yamazaki et al., 2002; Craft et al., 2005; Imbimbo, 2004; Luchsinger et al., 2003). The key for successful treatment of neurodegenerative diseases is early detection and intervention to prevent excitotoxicity, oxidative stress, and neuroinflammation. A cocktail of specific antagonists that inhibit the overstimulation of glutamate receptors but maintain their normal function (the uncompetitive NMDA antagonist, memantine), antioxidants that cross the blood-brain barrier and reach into mitochondria, the seat of ROS production (mitoQ, MitoVitE, MCI-186 etc), and provide optimal gene regulation (Gilgun-Sherki et al., 2001, 2002; Esposito et al., 2002; Reddy, 2006), and anti-inflammatory nutraceuticals (curcumin, resveratrol, lycopene etc) (Craft et al., 2005; Gilgun-Sherki et al., 2006; Wang et al., 2006; Huang et al., 2004; Luers et al., 2006) without side effects, is needed to dissect the molecular mechanism of glutamate release during excitotoxicity and for treating neurodegenerative diseases. Sustained use of this cocktail of drugs may not only slow down the progression of neurodegenerative diseases and provide relief from symptoms, but also retard changes that occur during the course of neurodegenerative diseases.

The approach for designing NMDA antagonists, antioxidants, and anti-inflammatory agents should be based on our rapidly emerging concept of signal transduction pathways in neurological disorders. Better drug delivery systems that target brain need to be developed to protect NMDA receptor antagonists, antioxidants, and anti-inflammatory agents from *in vivo* degradation or detoxification. Drugs delivered through these drug delivery systems must reach the site where inflammatory and neurodegenerative processes are taking place (Yoshikawa et al., 1999; Andresen and Jorgensen, 2005). This would enhance the *in vivo* efficacy of these drugs. The effects of these drugs on genes can be monitored by microarray procedures (Colangelo et al., 2002; Bosetti et al., 2005) and changes in lipid mediator levels can be determined by lipidomics (German et al., 2007; Watson, 2006; Serhan et al., 2006; Butterfield et al., 2006). These studies can lead to better understanding of the molecular mechanisms associated with the pathogenesis and the mechanism of action of therapeutic agents for the treatment of neurological disorders involving interplay among excitotoxicity, oxidative stress, and neuroinflammation.

11.4 n-3 Fatty Acids as Anti-Excitotoxic, Antioxidant, and Anti-Inflammatory Agents

Brain tissue is enriched in the n-6 arachidonic acid (AA) and the n-3 docosahexaenoic acid (DHA). Despite their abundance in the nervous system, AA and DHA cannot be synthesized de novo by mammals; they, or their precursors, must be ingested from dietary sources and transported to the brain (Horrocks and Farooqui, 2004; Marszalek and Lodish, 2005). The present day Western diet has a ratio of n-6 to n-3 fatty acids of about 15:1. The Paleolithic diet on which human beings evolved, and lived for most of their existence, had a ratio of 1:1 and was high in antioxidants (Simopoulos, 2002b, 2004, 2006; Cordain et al., 2005). Changes in eating habits, natural versus processed food, and agriculture development within the past 100–200 years caused these changes in the n-6 to n-3 ratio. The present western diet also has decreased levels of antioxidants and micronutrients.

The decreased consumption of DHA-enriched foods and increased consumption of n-6 enriched vegetable oils is responsible for the 15:1 n-6:n-3 ratio (Weylandt and Kang, 2005; Simopoulos, 2006). The present western diet promotes the pathogenesis of many chronic diseases such as cardiovascular disease, inflammatory and autoimmune diseases, and neurodegenerative diseases, whereas a diet enriched in n-3 fatty acids exerts cardioprotective, immunosuppressive, and neuroprotective effects (Simopoulos, 2006). A lower AA: DHA ratio suppresses the human disorders mentioned above. The richest source of DHA is fish oil.

The consumption of DHA has numerous beneficial effects on the health of the human brain (Horrocks and Yeo, 1999; Horrocks and Farooqui, 2004). The beneficial effects of DHA may be due not only to its effect on the physicochemical properties of neural membranes, but also to modulation of neurotransmission (Chalon et al., 1998; Zimmer et al., 2000; Itokazu et al., 2000; Hőgyes et al., 2003; Chalon, 2006; Joardar et al., 2006), gene expression (Farkas et al., 2000; Kitajka et al., 2002; Barceló-Coblijn et al., 2003; Puskás et al., 2003; De Caterina and Massaro, 2005; Deckelbaum et al., 2006), activities of enzymes, ion channels, receptors, and immunity (Yehuda et al., 2002, 2005; Tsutsumi et al., 1995; Isbilen et al., 2006).

A substantial body of epidemiological, biochemical and clinical data supports the use of n-3 fatty acids as anti-excitotoxic, antioxidant, and anti-inflammatory agents (see Chapter 9) (Hőgyes et al., 2003; Mori and Beilin, 2004; Calder, 2004, 2006; Roland et al., 2004; Farooqui and Horrocks, 2004). The most important dietary supplement that can reduce oxidative stress and neuroinflammation is fish oil because it is rich in eicosapentaenoic acid (EPA) and docosahexaenoic acid (DHA). Fish oil, which is now approved by the U.S. FDA under the commercial name, Omacor^{TM}, reduces neuroinflammation in several ways. Firstly; it decreases the formation of AA by blocking the activity of Δ5-desaturase; secondly, it inhibits the synthesis of eicosanoids (Calder, 2005); it induces the synthesis of resolvins and neuroprotectins; and lastly it decreases the production of cytokines, ROS, and the expression of adhesion molecules (Serhan, 2005b, 2006; Bazan, 2005a,c; Bazan, 2006; Calder, 2006; Arita et al., 2006). DHA modulates T-cell activation via protein kinase C-α and the NF-κB signaling pathway (Denys et al., 2005). DHA also promotes the differentiation of neural stem cells into neurons by promoting the cell cycle exit and suppressing cell death (Kawakita et al., 2006). DHA enrichment in membrane glycerophospholipids increases gap junction coupling capacity in cultured astrocytes (Champeil-Potokar et al., 2006). The collective evidence suggests that the ratio of n-6 to n-3 fatty acid is an important dietary factor in reducing oxidative stress and inflammation in brain tissue.

DHA also acts as an anti-excitotoxic agent (Hőgyes et al., 2003) and antioxidant (Hossain et al., 1998). It induces antioxidant defenses by enhancing cerebral activities of catalase, glutathione peroxidase, and levels of glutathione (Hossain et al., 1999). Thus, DHA exerts neuroprotective effects by modulating the secretion of cytokines and inhibiting neuroinflammation and oxidative stress. It reduces plasma levels of F_2-isoprostanes without affecting COX-mediated prostaglandin formation in healthy human subjects (Nalsen et al., 2006). Furthermore, in brain tissue, DHA-derived metabolites promote resolution and protect neural cells from neurodegeneration (Fig. 11.1) (Serhan, 2005b; Lukiw et al., 2005; Bazan, 2005c). Collectively these studies suggest that the generation of resolvins and docosatrienes may be an internal neuroprotective mechanism for preventing brain damage (Lukiw et al., 2005; Bazan, 2005b, 2006; Serhan, 2005b; Bazan et al., 2005).

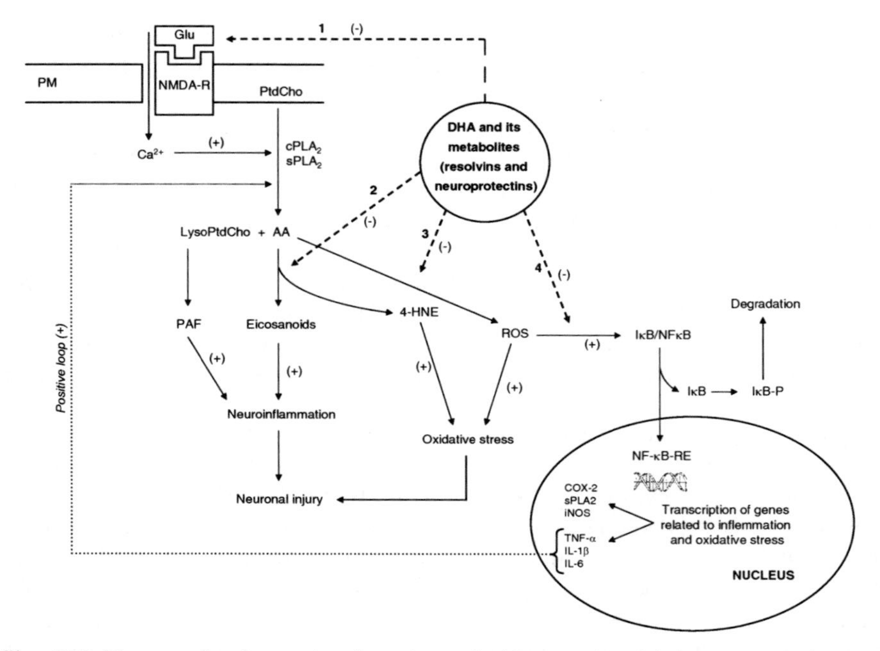

Fig. 11.1 Diagram showing anti-excitotoxic, antioxidant, and anti-inflammatory effects of DHA on lipid mediators in brain. PM, Plasma membrane; NMDA-R, *N*-methyl-D-aspartate receptor; Glu, glutamate; PtdCho, phosphatidylcholine; lyso-PtdCho, lysophosphatidylcholine; $cPLA_2$, cytosolic phospholipase A_2; $sPLA_2$, secretory phospholipase A_2; COX, cyclooxygenase; LOX, lipoxygenase; EPOX, epoxygenase; AA, arachidonic acid; DHA, docosahexaenoic acid; PAF, platelet-activating factor; 4-HNE, 4-hydroxynonenal; ROS, reactive oxygen species; NF-κB, nuclear factor kappaB; NF-κB-RE, nuclear factor kappaB response element; IκB, inhibitory subunit of NFκB; TNF-α, tumor necrosis factor-α; IL-1β, interleukin-1β; IL-6, interleukin-6; COX-2, cyclooxygenase-2; iNOS, inducible nitric oxide synthase; MMPs, matrix metalloproteinases; VCAM-1, vascular adhesion molecule-1; 1, anti-excitotoxic effect; 2 and 4, anti-inflammatory effect; 3, antioxidant effect; positive sign (+) indicates upregulation and negative sign (–) indicates downregulation

Thus DHA supplementation may restore signal transduction processes by protecting neurons from harmful effects of oxidative stress and neuroinflammation. Collectively these studies suggest that DHA plays an important role in normal neurological and cognitive functions (Horrocks and Farooqui, 2004). Levels of DHA are markedly decreased in neural membranes obtained from brains of aged healthy elderly people and also from patients with neurological disorders (Bechoua et al., 2003; Horrocks and Farooqui, 2004). Numerous epidemiological studies indicate that increased fatty fish consumption and high DHA intake are associated with a reduced risk of AD (Kalmijn et al., 2004; Freund-Levi et al., 2006).

Dietary DHA incorporates into neuronal membranes and restores neuronal functions (Rapoport, 1999). Chronic pre-administration of DHA prevents β-amyloid-induced impairment of an avoidance ability-related memory function in a rat model of AD (Hashimoto et al., 2005; Park et al., 2004) and protects mice from synaptic loss and dendritic pathology in another model of AD (Calon et al., 2004). DHA and its metabolite neuroprotectin D1 prevent apoptosis and β-amyloid secretion from aging brain cells (Lukiw et al., 2005; Park et al., 2004; Lukiw and Bazan, 2006; Bazan et al., 2005). Collectively these studies suggest that DHA is beneficial in preventing learning deficiencies in animal AD models (Cole and Frautschy, 2006; Lukiw and Bazan, 2006; Freund-Levi et al., 2006). Similarly DHA protects against other types of dementia (Lim et al., 2006).

DHA administration reduces L-DOPA induced dyskinesias in 1-methyl-4-phenyl-1,2,3,6-tetrahydropyridine (MPTP)-treated monkeys (Samadi et al., 2006). EPA and DHA have also been used for the treatment of HD (Puri, 2005). In spinal cord injury, DHA improves locomotor performance, decreases lesion size and apoptosis, and increases neuronal and oligodendrocyte survival (King et al., 2006). DHA protects brain from ischemic injury (Högyes et al., 2003; Mori, 2006). DHA also slows the progression of retinitis pigmentosa and macular degeneration (Johnson and Schaefer, 2006; Hodge et al., 2006). Intravenous use of fish oil has positive effects on immunological function and clinical parameters in surgical and septic patients (Mayer et al., 2006; Nakamura et al., 2005). Numerous clinical trials found benefits of DHA-enriched diets in several inflammatory and autoimmune diseases in humans, including rheumatoid arthritis, Crohn's disease, ulcerative colitis, psoriasis, lupus erythematosus, multiple sclerosis, and migraine headaches. Many of these placebo-controlled trials of fish oil in chronic inflammatory diseases reveal significant benefits, including decreased disease activity and a lowered use of anti-inflammatory drugs (Simopoulos, 2002a). Collective evidence suggests that well-planned, better-designed, and larger trials of purified n-3 fatty acid preparations like Omacor ™are required for judging the therapeutic potentials of DHA and EPA for chronic neurodegenerative diseases.

We hope that in the coming 50 years, studies on the effect of diets, particularly with lower dietary n-6/n-3 fatty acid ratios, on gene-expression patterns, organization of the chromatin, protein expression and lipid mediator profiles, using genomics, proteomics, and lipidomics of tissue and biological fluid samples from patients with neurodegenerative diseases, will show major progress for the treatment of neurodegenerative diseases. New classes of therapeutic agents based on diet, antioxidants and lower n-6/n-3 fatty acid ratio, must be developed. Thus future progress on nutrigenomics, the study of interactions between nutritional factors, genetic factors, and health outcomes (Ordovas and Corella, 2004; Afman and Muller, 2006; Miggiano and De Sanctis, 2006), will increase our understanding of the influence of nutrition on genetic expression, metabolic pathways, and homeostatic control in normal human subjects and how this regulation is disturbed in the early phases of neurodegenerative diseases (Valenzuela and Nieto, 2001; Grant, 1997). Dysregulation of metabolism in early human development may result in enhanced susceptibility to neurodegenerative diseases caused by a sustained deficiency of certain dietary factors, such as antioxidants and n-3 fatty acids, in the daily Western lifestyle, with a high calorie diet, from childhood to old age in human subjects.

References

Adibhatla R. M., Hatcher J. F., and Dempsey R. J. (2006). Lipids and lipidomics in brain injury and diseases. AAPS J. 8:E314–E321.
Afman L. and Muller M. (2006). Nutrigenomics: from molecular nutrition to prevention of disease. J. Am. Diet. Assoc. 106:569–576.
Allan S. M. and Rothwell N. J. (2003). Inflammation in central nervous system injury. Philos. Trans. R. Soc. Lond B Biol. Sci. 358:1669–1677.
Andersen J. K. (2004). Oxidative stress in neurodegeneration: cause or consequence? Nature Med. 10:S18–S25.
Andresen T. L. and Jorgensen K. (2005). Synthesis and membrane behavior of a new class of unnatural phospholipid analogs useful as phospholipase A2 degradable liposomal drug carriers. Biochim. Biophys. Acta Biomembr. 1669:1–7.
Anrather J., Racchumi G., and Iadecola C. (2006). NF-κB regulates phagocytic NADPH oxidase by inducing the expression of gp91^{phox}. J. Biol. Chem. 281:5657–5667.
Araque A., Li N., Doyle R. T., and Haydon P. G. (2000). SNARE protein-dependent glutamate release from astrocytes. J. Neurosci. 20:666–673.
Arita M., Oh S. F., Chonan T., Hong S., Elangovan S., Sun Y. P., Uddin J., Petasis N. A., and Serhan C. N. (2006). Metabolic inactivation of resolvin E1 and stabilization of its anti-inflammatory actions. J. Biol. Chem. 281:22847–22854.
Barceló-Coblijn G., Kitajka K., Puskás L. G., Högyes E., Zvara A., Hackler L., Jr., and Farkas T. (2003). Gene expression and molecular composition of phospholipids in rat brain in relation to dietary *n*-6 to *n*-3 fatty acid ratio. Biochim. Biophys. Acta 1632:72–79.
Bazan N. G. (2005a). Lipid signaling in neural plasticity, brain repair, and neuroprotection. Mol. Neurobiol. 32:89–103.
Bazan N. G. (2005b). Neuroprotectin D1 (NPD1): a DHA-derived mediator that protects brain and retina against cell injury-induced oxidative stress. Brain Path. 15:159–166.
Bazan N. G. (2005c). Synaptic signaling by lipids in the life and death of neurons. Mol. Neurobiol. 31:219–230.
Bazan N. G. (2006). The onset of brain injury and neurodegeneration triggers the synthesis of docosanoid neuroprotective signaling. Cell. Molec. Neurobiol. 26:901–913.
Bazan N. G., Marcheselli V. L., and Cole-Edwards K. (2005). Brain response to injury and neurodegeneration-Endogenous neuroprotective signaling. In: Slikker W., Andrews R. J., and Trembly B. (eds.), *Neuroprotective Agents*. Annals of the New York Academy of Sciences, New York, pp. 137–147.
Bechoua S., Dubois M., Vericel E., Chapuy P., Lagarde M., and Prigent A. F. (2003). Influence of very low dietary intake of marine oil on some functional aspects of immune cells in healthy elderly people. Br. J. Nutr. 89:523–531.
Berlett B. S. and Stadtman E. R. (1997). Protein oxidation in aging, disease, and oxidative stress. J. Biol. Chem. 272:20313–20316.
Berman K. and Brodaty H. (2004). Tocopherol (vitamin E) in Alzheimer's disease and other neurodegenerative disorders. CNS Drugs 18:807–825.
Bernaudin M., Nouvelot A., MacKenzie E. T., and Petit E. (1998). Selective neuronal vulnerability and specific glial reactions in hippocampal and neocortical organotypic cultures submitted to ischemia. Exp. Neurol. 150:30–39.
Bosetti F., Bell J. M., and Manickam P. (2005). Microarray analysis of rat brain gene expression after chronic administration of sodium valproate. Brain Res. Bull. 65:331–338.
Bowers-Gentry R. C., Deems R. A., Harkewicz R., and Dennis E. A. (2006). Eicosanoid lipidomics. In: Feng L. and Prestwich G. D. (eds.), *Functional Lipidomics*. CRC Press-Taylor & Francis Group, Boca Raton, pp. 79–100.
Butterfield D. A., Perluigi M., and Sultana R. (2006). Oxidative stress in Alzheimer's disease brain: New insights from redox proteomics. Eur. J. Pharmacol. 545:39–50.
Calder P. C. (2004). n-3 Fatty acids, inflammation, and immunity - Relevance to postsurgical and critically ill patients. Lipids 39:1147–1161.

Calder P. C. (2005). Polyunsaturated fatty acids and inflammation. Biochem. Soc. Trans. 33: 423–427.

Calder P. C. (2006). n-3 Polyunsaturated fatty acids, inflammation, and inflammatory diseases. Am. J. Clin. Nutr. 83:1505S–1519S.

Calon F., Lim G. P., Yang F. S., Morihara T., Teter B., Ubeda O., Rostaing P., Triller A., Salem N. J., Ashe K. H., Frautschy S. A., and Cole G. M. (2004). Docosahexaenoic acid protects from dendritic pathology in an Alzheimer's disease mouse model. Neuron 43:633–645.

Chalon S. (2006). Omega-3 fatty acids and monoamine neurotransmission. Prostaglandins Leukot. Essent. Fatty Acids 75:259–269.

Chalon S., Delion-Vancassel S., Belzung C., Guilloteau D., Leguisquet A. M., Besnard J. C., and Durand G. (1998). Dietary fish oil affects monoaminergic neurotransmission and behavior in rats. J. Nutr. 128:2512–2519.

Champeil-Potokar G., Chaumontet C., Guesnet P., Lavialle M., and Denis I. (2006). Docosahexaenoic acid (22:6n-3) enrichment of membrane phospholipids increases gap junction coupling capacity in cultured astrocytes. Eur. J. Neurosci. 24:3084–3090.

Choi D. W. (1988). Glutamate neurotoxicity and diseases of the nervous system. Neuron 1: 628–634.

Colangelo V., Schurr J., Ball M. J., Pelaez R. P., Bazan N. G., and Lukiw W. J. (2002). Gene expression profiling of 12633 genes in Alzheimer hippocampal CA1: Transcription and neurotrophic factor down-regulation and up-regulation of apoptotic and pro-inflammatory signaling. J. Neurosci. Res. 70:462–473.

Cole G. M. and Frautschy S. A. (2006). Docosahexaenoic acid protects from amyloid and dendritic pathology in an Alzheimer's disease mouse model. Nutr. Health 18:249–259.

Cordain L., Eaton S. B., Sebastian A., Mann N., Lindeberg S., Watkins B. A., O'Keefe J. H., and Brand-Miller J. (2005). Origins and evolution of the Western diet: health implications for the 21st century. Am. J. Clin. Nutr. 81:341–354.

Correale J. and Villa A. (2004). The neuroprotective role of inflammation in nervous system injuries. J. Neurol. 251:1304–1316.

Craft J. M., Watterson D. M., and Van Eldik L. J. (2005). Neuroinflammation: a potential therapeutic target. Expert Opin. Ther. Targets 9:887–900.

Danbolt N. C. (2001). Glutamate uptake. Prog. Neurobiol. 65:1–105.

De Caterina R. and Massaro M. (2005). Omega-3 fatty acids and the regulation of expression of endothelial pro-atherogenic and pro-inflammatory genes. J. Membr. Biol. 206: 103–116.

Deckelbaum R. J., Worgall T. S., and Seo T. (2006). n-3 Fatty acids and gene expression. Am. J. Clin. Nutr. 83:1520S–1525S.

Denys A., Hichami A., and Khan N. A. (2005). n-3PUFAs modulate T-cell activation via protein kinase C-α and -ε and the NF-κ B signaling pathway. J. Lipid Res. 46:752–758.

Dwyer B. E., Takeda A., Zhu X. W., Perry G., and Smith M. A. (2005). Ferric cycle activity and Alzheimer disease. Curr. Neurovasc. Res. 2:261–267.

Esposito E., Rotilio D., Di Matteo V., Di Giulio C., Cacchio M., and Algeri S. (2002). A review of specific dietary antioxidants and the effects on biochemical mechanisms related to neurodegenerative processes. Neurobiol. Aging 23:719–735.

Facheris M., Beretta S., and Ferrarese C. (2004). Peripheral markers of oxidative stress and excitotoxicity in neurodegenerative disorders: Tools for diagnosis and therapy? J. Alzheimer's Dis. 6:177–184.

Farkas T., Kitajka K., Fodor E., Csengeri I., Lahdes E., Yeo Y. K., Krasznai Z., and Halver J. E. (2000). Docosahexaenoic acid-containing phospholipid molecular species in brains of vertebrates. Proc. Natl. Acad. Sci. USA 97:6362–6366.

Farooqui A. A. and Horrocks L. A. (1991). Excitatory amino acid receptors, neural membrane phospholipid metabolism and neurological disorders. Brain Res. Rev. 16:171–191.

Farooqui A. A. and Horrocks L. A. (1994). Excitotoxicity and neurological disorders: involvement of membrane phospholipids. Int. Rev. Neurobiol. 36:267–323.

Farooqui A. A. and Horrocks L. A. (2004). Beneficial effects of docosahexaenoic acid on health of the human brain. Agro Food Industry Hi-Tech 15:52–53.

Farooqui A. A. and Horrocks L. A. (2006a). Phospholipase A_2-generated lipid mediators in the brain: the good, the bad, and the ugly. Neuroscientist 12:245–260.

Farooqui A. A. and Horrocks L. A. (2007a). Glutamate and cytokine-mediated alterations of phospholipids in head injury and spinal cord trauma. In: Banik N. (ed.), *Brain and Spinal Cord Trauma.* Handbook of Neurochemistry, Lajtha, A. (ed.) Springer, New York, in press.

Farooqui A. A. and Horrocks L. A. (2007b). *Glycerophospholipids in the Brain: Phospholipases A_2 in Neurological Disorders*, pp.1–394. Springer, New York.

Farooqui A. A., Ong W. Y., Lu X. R., Halliwell B., and Horrocks L. A. (2001). Neurochemical consequences of kainate-induced toxicity in brain: involvement of arachidonic acid release and prevention of toxicity by phospholipase A_2 inhibitors. Brain Res. Rev. 38:61–78.

Farooqui A. A., Ong W. Y., and Horrocks L. A. (2004). Biochemical aspects of neurodegeneration in human brain: involvement of neural membrane phospholipids and phospholipases A_2. Neurochem. Res. 29:1961–1977.

Farooqui A. A., Ong W. Y., and Horrocks L. A. (2006). Inhibitors of brain phospholipase A_2 activity: Their neuropharmacological effects and therapeutic importance for the treatment of neurologic disorders. Pharmacol. Rev. 58:591–620.

Farooqui A. A., Horrocks L. A., and Farooqui T. (2007). Modulation of inflammation in brain: a matter of fat. J. Neurochem. 101:577–599.

Floyd R. A. and Hensley K. (2002). Oxidative stress in brain aging-Implications for therapeutics of neurodegenerative diseases. Neurobiol. Aging 23:795–807.

Freund-Levi Y., Eriksdotter-Jönhagen M., Cederholm T., Basun H., Faxén-Irving G., Garlind A., Vedin I., Vessby B., Wahlund L. O., and Palmblad J. (2006). ω-3 Fatty acid treatment in 174 patients with mild to moderate Alzheimer disease: OmegAD study - A randomized double-blind trial. Arch. Neurol. 63:1402–1408.

Frey R. S., Gao X., Javaid K., Siddiqui S. S., Rahman A., and Malik A. B. (2006). Phosphatidylinositol 3-kinase γ signaling through protein kinase Cζ induces NADPH oxidase-mediated oxidant generation and NF-κactivation in endothelial cells. J. Biol. Chem. 281: 16128–16138.

Gasparini L., Ongini E., and Wenk G. (2004). Non-steroidal anti-inflammatory drugs (NSAIDs) in Alzheimer's disease: old and new mechanisms of action. J. Neurochem. 91:521–536.

German J. B., Gillies L. A., Smilowitz J. T., Zivkovic A. M., and Watkins S. M. (2007). Lipidomics and lipid profiling in metabolomics. Curr. Opin. Lipidol. 18:66–71.

Gilgun-Sherki Y., Melamed E., and Offen D. (2001). Oxidative stress induced-neurodegenerative diseases: the need for antioxidants that penetrate the blood brain barrier. Neuropharmacology 40:959–975.

Gilgun-Sherki Y., Rosenbaum Z., Melamed E., and Offen D. (2002). Antioxidant therapy in acute central nervous system injury: current state. Pharmacol. Rev. 54:271–284.

Gilgun-Sherki Y., Melamed E., and Offen D. (2006). Anti-inflammatory drugs in the treatment of neurodegenerative diseases: current state. Curr. Pharmaceut. Design 12:3509–3519.

Graeber M. B. and Moran L. B. (2002). Mechanisms of cell death in neurodegenerative diseases: fashion, fiction, and facts. Brain Pathol. 12:385–390.

Grant W. B. (1997). Dietary links to Alzheimer's disease. Alz. Disease Rev. 2:42–55.

Hampel H., Teipel S. J., Alexander G. E., Pogarell O., Rapoport S. I., and Moller H. J. (2002). *In vivo* imaging of region and cell type specific neocortical neurodegeneration in Alzheimer's disease - Perspectives of MRI derived corpus callosum measurement for mapping disease progression and effects of therapy. Evidence from studies with MRI, EEG and PET. J. Neural Transm. 109:837–855.

Hashimoto M., Hossain S., Agdul H., and Shido O. (2005). Docosahexaenoic acid-induced amelioration on impairment of memory learning in amyloid β-infused rats relates to the decreases of amyloid β and cholesterol levels in detergent-insoluble membrane fractions. Biochim. Biophys. Acta Mol. Cell Biol. Lipids 1738:91–98.

Hodge W. G., Barnes D., Schachter H. M., Pan Y. I., Lowcock E. C., Zhang L., Sampson M., Morrison A., Tran K., Miguelez M., and Lewin G. (2006). The evidence for efficacy of omega-3 fatty acids in preventing or slowing the progression of retinitis pigmentosa: a systematic review. Can. J. Ophthalmol. 41:481–490.

Högyes E., Nyakas C., Kiliaan A., Farkas T., Penke B., and Luiten P. G. M. (2003). Neuroprotective effect of developmental docosahexaenoic acid supplement against excitotoxic brain damage in infant rats. Neuroscience 119:999–1012.

Horrocks L. A. and Farooqui A. A. (2004). Docosahexaenoic acid in the diet: its importance in maintenance and restoration of neural membrane function. Prostaglandins Leukot. Essent. Fatty Acids 70:361–372.

Horrocks L. A. and Yeo Y. K. (1999). Health benefits of docosahexaenoic acid (DHA). Pharmacol. Res. 40:211–225.

Hossain M. S., Hashimoto M., and Masumura S. (1998). Influence of docosahexaenoic acid on cerebral lipid peroxide level in aged rats with and without hypercholesterolemia. Neurosci. Lett. 244:157–160.

Hossain M. S., Hashimoto M., Gamoh S., and Masumura S. (1999). Antioxidative effects of docosahexaenoic acid in the cerebrum versus cerebellum and brainstem of aged hypercholesterolemic rats. J. Neurochem. 72:1133–1138.

Huang M. T., Ghai G., and Ho C. T. (2004). Inflammatory process and molecular targets for anti-inflammatory nutraceuticals. Compr. Rev. Food Sci. Food Safety 3:127–139.

Hunt A. N. and Postle A. D. (2006). Mass spectrometry determination of endonuclear phospholipid composition and dynamics. Methods 39:104–111.

Imbimbo B. P. (2004). The potential role of non-steroidal anti-inflammatory drugs in treating Alzheimer's disease. Expert Opin. Invest. Drugs 13:1469–1481.

Isbilen B., Fraser S. P., and Djamgoz M. B. A. (2006). Docosahexaenoic acid (omega-3) blocks voltage-gated sodium channel activity and migration of MDA-MB-231 human breast cancer cells. Int. J. Biochem. Cell Biol. 38:2173–2182.

Itokazu N., Ikegaya Y., Nishikawa M., and Matsuki N. (2000). Bidirectional actions of docosahexaenoic acid on hippocampal neurotransmissions *in vivo*. Brain Res. 862:211–216.

Jabaudon D., Scanziani M., Gahwiler B. H., and Gerber U. (2000). Acute decrease in net glutamate uptake during energy deprivation. Proc. Natl. Acad. Sci. USA 97:5610–5615.

Jellinger K. A. (2001). Cell death mechanisms in neurodegeneration. J. Cell. Mol. Med. 5:1–17.

Joardar A., Sen A. K., and Das S. (2006). Docosahexaenoic acid facilitates cell maturation and β-adrenergic transmission in astrocytes. J. Lipid Res. 47:571–581.

Johnson E. J. and Schaefer E. J. (2006). Potential role of dietary n-3 fatty acids in the prevention of dementia and macular degeneration. Am. J. Clin. Nutr. 83:1494S–1498S.

Juranek I. and Bezek S. (2005). Controversy of free radical hypothesis: reactive oxygen species - Cause or consequence of tissue injury? Gen. Physiol. Biophys. 24:263–278.

Kalmijn S., Van Boxtel M. P. J., Ocké M., Verschuren W. M. M., Kromhout D., and Launer L. J. (2004). Dietary intake of fatty acids and fish in relation to cognitive performance at middle age. Neurology 62:275–280.

Katchman A. N. and Hershkowitz N. (1993). Early anoxia-induced vesicular glutamate release results from mobilization of calcium from intracellular stores. J. Neurophysiol. 70:1–7.

Kawakita E., Hashimoto M., and Shido O. (2006). Docosahexaenoic acid promotes neurogenesis in vitro and *in vivo*. Neuroscience 139:991–997.

Kidd P. M. (2005). Neurodegeneration from mitochondrial insufficiency: nutrients, stem cells, growth factors, and prospects for brain rebuilding using integrative management. Altern. Med. Rev. 10:268–293.

Kim K. M., Jung B. H., Paeng K. J., Kim I., and Chung B. C. (2004). Increased urinary F_2-isoprostanes levels in the patients with Alzheimer's disease. Brain Res. Bull. 64: 47–51.

Kimelberg H. K. and Mongin A. A. (1998). Swelling-activated release of excitatory amino acids in the brain: relevance for pathophysiology. Contrib. Nephrol. 123:240–257.

King V. R., Huang W. L., Dyall S. C., Curran O. E., Priestley J. V., and Michael-Titus A. T. (2006). Omega-3 fatty acids improve recovery, whereas omega-6 fatty acids worsen outcome, after spinal cord injury in the adult rat. J. Neurosci. 26:4672–4680.

Kitajka K., Puskás L. G., Zvara A., Hackler L. J., Barceló-Coblijn G., Yeo Y. K., and Farkas T. (2002). The role of n-3 polyunsaturated fatty acids in brain: Modulation of rat brain gene expression by dietary n-3 fatty acids. Proc. Natl. Acad. Sci. USA 99:2619–2624.

Leist M. and Nicotera P. (1998). Apoptosis, excitotoxicity, and neuropathology. Exp. Cell Res. 239:183–201.

Leker R. R. and Shohami E. (2002). Cerebral ischemia and trauma - different etiologies yet similar mechanisms: neuroprotective opportunities. Brain Res. Rev. 39:55–73.

Lim W. S., Gammack J. K., Van Niekerk J., and Dangour A. D. (2006). Omega 3 fatty acid for the prevention of dementia. Cochrane Database of Systematic Reviews 2006:Art. No. CD005379. doi:10.1002/14651858.

Liu W., Liu R., Schreiber S. S., and Baudry M. (2001). Role of polyamine metabolism in kainic acid excitotoxicity in organotypic hippocampal slice cultures. J. Neurochem. 79: 976–984.

Liu H. T., Tashmukhamedov B. A., Inoue H., Okada Y., and Sabirov R. Z. (2006). Roles of two types of anion channels in glutamate release from mouse astrocytes under ischemic or osmotic stress. Glia 54:343–357.

Lu Y., Hong S., Gotlinger K., and Serhan C. N. (2006). Lipid mediator informatics and proteomics in inflammation-resolution. The Scientific World J. 6:589–614.

Lucas S. M., Rothwell N. J., and Gibson R. M. (2006). The role of inflammation in CNS injury and disease. Br. J. Pharmacol. 147(Suppl 1):S232–S240.

Luchsinger J. A., Tang M. X., Shea S., and Mayeux R. (2003). Antioxidant vitamin intake and risk of Alzheimer disease. Arch. Neurol. 60:203–208.

Luers G. H., Thiele S., Schad A., Volkl A., Yokota S., and Seitz J. (2006). Peroxisomes are present in murine spermatogonia and disappear during the course of spermatogenesis. Histochem. Cell Biol. 125:693–703.

Lukiw W. J. and Bazan N. G. (2006). Survival signalling in Alzheimer's disease. Biochem. Soc. Trans. 34:1277–1282.

Lukiw W. J., Cui J. G., Marcheselli V. L., Bodker M., Botkjaer A., Gotlinger K., Serhan C. N., and Bazan N. G. (2005). A role for docosahexaenoic acid-derived neuroprotectin D1 in neural cell survival and Alzheimer disease. J. Clin. Invest 115:2774–2783.

Marszalek J. R. and Lodish H. F. (2005). Docosahexaenoic acid, fatty acid-interacting proteins, and neuronal function: breastmilk and fish are good for you. Annu. Rev. Cell Dev. Biol. 21: 633–657.

Masters C. L., Cappai R., Barnham K. J., and Villemagne V. L. (2006). Molecular mechanisms for Alzheimer's disease: implications for neuroimaging and therapeutics. J. Neurochem. 97: 1700–1725.

Mayer K., Schaefer M. B., and Seeger W. (2006). Fish oil in the critically ill: from experimental to clinical data. Curr. Opin. Clin. Nutr. Metab. Care 9:140–148.

McIntosh T. K., Saatman K. E., Raghupathi R., Graham D. I., Smith D. H., Lee V. M., and Trojanowski J. Q. (1998). The molecular and cellular sequelae of experimental traumatic brain injury: pathogenetic mechanisms. Neuropathol. Appl. Neurobiol. 24:251–267.

Miggiano G. A. D. and De Sanctis R. (2006). Nutrigenomica: verso una dieta personalizzata [Nutritional genomics: toward a personalized diet]. Clin. Ter. 157:355–361.

Migliore L., Fontana I., Colognato R., Coppede F., Siciliano G., and Murri L. (2005). Searching for the role and the most suitable biomarkers of oxidative stress in Alzheimer's disease and in other neurodegenerative diseases. Neurobiol. Aging 26:587–595.

Miller A. A., Drummond G. R., and Sobey C. G. (2006). Novel isoforms of NADPH-oxidase in cerebral vascular control. Pharmacol. Ther. 111:928–948.

Milne S., Ivanova P., Forrester J., and Brown H. A. (2006). Lipidomics: An analysis of cellular lipids by ESI-MS. Methods 39:92–103.

Mori T. A. (2006). Omega-3 fatty acids and hypertension in humans. Clin. Exp. Pharmacol. Physiol. 33:842–846.

Mori T. A. and Beilin L. J. (2004). Omega-3 fatty acids and inflammation. Curr. Atheroscler. Rep. 6:461–467.

Morrow J. D. (2006). The isoprostanes - Unique products of arachidonate peroxidation: Their role as mediators of oxidant stress. Curr. Pharmaceut. Design 12:895–902.

Nakamura K., Kariyazono H., Komokata T., Hamada N., Sakata R., and Yamada K. (2005). Influence of preoperative administration of omega-3 fatty acid- enriched supplement on inflammatory and immune responses in patients undergoing major surgery for cancer. Nutrition 21:639–649.

Nalsen C., Vessby B., Berglund L., Uusitupa M., Hermansen K., Riccardi G., Rivellese A., Storlien L., Erkkila A., Yla-Herttuala S., Tapsell L., and Basu S. (2006). Dietary (n-3) fatty acids reduce plasma F_2-isoprostanes but not prostaglandin $F_{2\alpha}$ in healthy humans. J. Nutr. 136:1222–1228.

Newcomb R., Sun X. Y., Taylor L., Curthoys N., and Giffard R. G. (1997). Increased production of extracellular glutamate by the mitochondrial glutaminase following neuronal death. J. Biol. Chem. 272:11276–11282.

Nicotera P. and Leist M. (1997). Energy supply and the shape of death in neurons and lymphoid cells. Cell Death Differ. 4:435–442.

Nicotera P. and Lipton S. A. (1999). Excitotoxins in neuronal apoptosis and necrosis. J. Cereb. Blood Flow Metab 19:583–591.

Ohishi H., Shigemoto R., Nakanishi S., and Mizuno N. (1993). Distribution of the mRNA for a metabotropic glutamate receptor (mGluR3) in the rat brain: an in situ hybridization study. J. Comp. Neurol. 335:252–266.

Olney J. W., Fuller T., and de Gubareff T. (1979). Acute dendrotoxic changes in the hippocampus of kainate treated rats. Brain Res. 176:91–100.

Ordovas J. M. and Corella D. (2004). Nutritional genomics. Annu. Rev. Genomics Hum. Genet. 5:71–118.

Page G., Peeters M., Najimi M., Maloteaux J. M., and Hermans E. (2001). Modulation of the neuronal dopamine transporter activity by the metabotropic glutamate receptor mGluR5 in rat striatal synaptosomes through phosphorylation mediated processes. J. Neurochem. 76: 1282–1290.

Park E., Velumian A. A., and Fehlings M. G. (2004). The role of excitotoxicity in secondary mechanisms of spinal cord injury: a review with an emphasis on the implications for white matter degeneration. J. Neurotrauma 21:754–774.

Perluigi M., Poon H. F., Hensley K., Pierce W. M., Klein J. B., Calabrese V., De Marco C., and Butterfield D. A. (2005). Proteomic analysis of 4-hydroxy-2-nonenal-modified proteins in G93A-SOD1 transgenic mice - A model of familial amyotrophic lateral sclerosis. Free Radical Biol. Med. 38:960–968.

Phillis J. W., Horrocks L. A., and Farooqui A. A. (2006). Cyclooxygenases, lipoxygenases, and epoxygenases in CNS: their role and involvement in neurological disorders. Brain Res. Rev. 52:201–243.

Polidori M. C. (2004). Oxidative stress and risk factors for Alzheimer's disease: clues to prevention and therapy. J. Alzheimer's Dis. 6:185–191.

Prasad K. N., Cole W. C., Hovland A. R., Prasad K. C., Nahreini P., Kumar B., Edwards-Prasad J., and Andreatta C. P. (1999). Multiple antioxidants in the prevention and treatment of neurodegenerative disease: analysis of biologic rationale. Curr. Opin. Neurol. 12:761–770.

Puri B. K. (2005). Treatment of Huntington's disease with eicosapentaenoic acid. In: Yehuda S. and Mostofsky D. I. (eds.), *Nutrients, Stress and Medical Disorders.* Nutrition and Health (Series) Humana Press Inc, Totowa, pp. 279–286.

Puskás L. G., Kitajka K., Nyakas C., Barcelo-Coblijn G., and Farkas T. (2003). Short-term administration of omega 3 fatty acids from fish oil results in increased transthyretin transcription in old rat hippocampus. Proc. Natl. Acad. Sci. USA 100:1580–1585.

Rapoport S. I. (1999). *In vivo* fatty acid incorporation into brain phospholipids in relation to signal transduction and membrane remodeling. Neurochem. Res. 24:1403–1415.

Rapoport S. I. (2001). *In vivo* fatty acid incorporation into brain phospholipids in relation to plasma availability, signal transduction and membrane remodeling. J. Mol. Neurosci. 16: 243–261.

Rapoport S. I. (2005). *In vivo* approaches and rationale for quantifying kinetics and imaging brain lipid metabolic pathways. Prostaglandins Other Lipid Mediat. 77:185–196.

Reddy P. H. (2006). Mitochondrial oxidative damage in aging and Alzheimer's disease: Implications for mitochondrially targeted antioxidant therapeutics. J. Biomed. Biotechnol. 2006: Art. No. 31372, doi: 10.1155/JBB/2006/31372.

Riederer P. and Hoyer S. (2006). From benefit to damage. Glutamate and advanced glycation end products in Alzheimer brain. J. Neural Transm. 113:1671–1677.
Roettger V. and Lipton P. (1996). Mechanism of glutamate release from rat hippocampal slices during in vitro ischemia. Neuroscience 75:677–685.
Roland I., de Leval X., Evrard B., Pirotte B., Dogne J. M., and Delattre L. (2004). Modulation of the arachidonic cascade with omega 3 fatty acids or analogues: Potential therapeutic benefits. Mini-Rev. Medicin. Chem. 4:659–668.
Rossi D. J., Oshima T., and Attwell D. (2000). Glutamate release in severe brain ischaemia is mainly by reversed uptake. Nature 403:316–321.
Rothwell N. J. (1999). Annual review prize lecture cytokines - killers in the brain? J. Physiol. (London) 514:3–17.
Rubin B. B., Downey G. P., Koh A., Degousee N., Ghomashchi F., Nallan L., Stefanski E., Harkin D. W., Sun C. X., Smart B. P., Lindsay T. F., Cherepanov V., Vachon E., Kelvin D., Sadilek M., Brown G. E., Yaffe M. B., Plumb J., Grinstein S., Glogauer M., and Gelb M. H. (2005). Cytosolic phospholipase A_2-α is necessary for platelet-activating factor biosynthesis, efficient neutrophil-mediated bacterial killing, and the innate immune response to pulmonary infection-cPLA$_2$-α does not regulate neutrophil NADPH oxidase activity. J. Biol. Chem. 280: 7519–7529.
Samadi P., Gregoire L., Rouillard C., Bedard P. J., Di Paolo T., and Levesque D. (2006). Docosahexaenoic acid reduces levodopa-induced dyskinesias in 1-methyl-4-phenyl-1,2,3,6-tetrahydropyridine monkeys. Ann. Neurol. 59:282–288.
Sastry P. S. and Rao K. S. (2000). Apoptosis and the nervous system. J. Neurochem. 74:1–20.
Serhan C. N. (2005a). Mediator lipidomics. Prostaglandins Other Lipid Mediat. 77:4–14.
Serhan C. N. (2005b). Novel ω-3-derived local mediators in anti-inflammation and resolution. Pharmacol. Ther. 105:7–21.
Serhan C. N. (2006). Novel chemical mediators in the resolution of inflammation: Resolvins and protectins. Anesthesiol. Clinics North Am. 24:341–364.
Serhan C. N., Hong S., and Lu Y. (2006). Lipid mediator informatics-lipidomics: Novel pathways in mapping resolution. AAPS J. 8:E284–E297.
Simopoulos A. P. (2002a). Omega-3 fatty acids in inflammation and autoimmune diseases. J. Am. Coll. Nutr. 21:495–505.
Simopoulos A. P. (2002b). The importance of the ratio of omega-6/omega-3 essential fatty acids. Biomed. Pharmacother. 56:365–379.
Simopoulos A. P. (2004). Omega-3 fatty acids and antioxidants in edible wild plants. Biol. Res. 37:263–277.
Simopoulos A. P. (2006). Evolutionary aspects of diet, the omega-6/omega-3 ratio and genetic variation: nutritional implications for chronic diseases. Biomed. Pharmacother. 60:502–507.
Slusher B. S., Vornov J. J., Thomas A. G., Hurn P. D., Harukuni I., Bhardwaj A., Traystman R. J., Robinson M. B., Britton P., Lu X. C., Tortella F. C., Wozniak K. M., Yudkoff M., Potter B. M., and Jackson P. F. (1999). Selective inhibition of NAALADase, which converts NAAG to glutamate, reduces ischemic brain injury. Nat. Med. 5:1396–1402.
Smith M. A., Nunomura A., Lee H. G., Zhu X., Moreira P. I., Avila J., and Perry G. (2005). Chronological primacy of oxidative stress in Alzheimer disease. Neurobiol. Aging 26:579–580.
Soule J., Messaoudi E., and Bramham C. R. (2006). Brain-derived neurotrophic factor and control of synaptic consolidation in the adult brain. Biochem. Soc. Trans. 34:600–604.
Sun G. Y., Horrocks L. A., and Farooqui A. A. (2007). The role of NADPH oxidase and phospholipases A_2 in mediating oxidative and inflammatory responses in neurodegenerative diseases. J. Neurochem. in press.
Testa C. M., Standaert D. G., Young A. B., and Penney J. B., Jr. (1994). Metabotropic glutamate receptor mRNA expression in the basal ganglia of the rat. J. Neurosci. 14: 3005–3018.
Tsutsumi T., Yamauchi E., Suzuki E., Watanabe S., Kobayashi T., and Okuyama H. (1995). Effect of a high α-linolenate and high linoleate diet on membrane-associated enzyme activities in rat brain–modulation of Na^+, K^+-ATPase activity at suboptimal concentrations of ATP. Biol. Pharm. Bull. 18:664–670.

Tuz K., Peña-Segura C., Franco R., and Pasantes-Morales H. (2004). Depolarization, exocytosis and amino acid release evoked by hyposmolarity from cortical synaptosomes. Eur. J. Neurosci. 19:916–924.

Valenzuela A. and Nieto M. S. (2001). Docosahexaenoic acid (DHA) in fetal development and infant nutrition. Revista Med. Chile 129:1203–1211.

Wang Q., Yu S., Simonyi A., Sun G. Y., and Sun A. Y. (2005). Kainic acid-mediated excitotoxicity as a model for neurodegeneration. Mol. Neurobiol. 31:3–16.

Wang J. Y., Wen L. L., Huang Y. N., Chen Y. T., and Ku M. C. (2006). Dual effects of antioxidants in neurodegeneration: Direct neuroprotection against oxidative stress and indirect protection via suppression of glia-mediated inflammation. Curr. Pharmaceut. Design 12:3521–3533.

Watson A. D. (2006). Lipidomics: a global approach to lipid analysis in biological systems. J. Lipid Res. 47:2101–2111.

Wendum D., Svrcek M., Rigau V., Boelle P. Y., Sebbagh N., Parc R., Masliah J., Trugnan G., and Flejou J. F. (2003). COX-2, inflammatory secreted PLA2, and cytoplasmic PLA2 protein expression in small bowel adenocarcinomas compared with colorectal adenocarcinomas. Modern Pathol. 16:130–136.

Weylandt K. H. and Kang J. X. (2005). Rethinking lipid mediators. Lancet 366:618–620.

Wilde G. J. C., Pringle A. K., Wright P., and Iannotti F. (1997). Differential vulnerability of the CA1 and CA3 subfields of the hippocampus to superoxide and hydroxyl radicals in vitro. J. Neurochem. 69:883–886.

Willcox J. K., Ash S. L., and Catignani G. L. (2004). Antioxidants and prevention of chronic disease. Crit. Rev. Food Sci. Nutr. 44:275–295.

Williamson P. and Schlegel R. A. (2002). Transbilayer phospholipid movement and the clearance of apoptotic cells. Biochim. Biophys. Acta Mol. Cell Biol. Lipids 1585:53–63.

Wood P. L. (1998). *Neuroinflammation: Mechanisms and Management*. Humana Press, Totowa, New Jersey.

Yamamoto Y. and Gaynor R. B. (2004). IκB kinases: key regulators of the NF-κB pathway. Trends Biochem. Sci. 29:72–79.

Yamazaki R., Kusunoki N., Matsuzaki T., Hashimoto S., and Kawai S. (2002). Nonsteroidal anti-inflammatory drugs induce apoptosis in association with activation of peroxisome proliferator-activated receptor γ in rheumatoid synovial cells. J. Pharmacol. Exp. Ther. 302:18–25.

Yehuda S., Rabinovitz S., Carasso R. L., and Mostofsky D. I. (2002). The role of polyunsaturated fatty acids in restoring the aging neuronal membrane. Neurobiol. Aging 23:843–853.

Yehuda S., Rabinovitz S., and Mostofsky D. I. (2005). Essential fatty acids and the brain: from infancy to aging. Neurobiol. Aging 26(Suppl 1):98–102.

Yoshikawa K., Kita Y., Kishimoto K., and Shimizu T. (2006). Profiling of eicosanoid production in the rat hippocampus during kainic acid-induced seizure-Dual phase regulation and differential involvement of COX-1 and COX-2. J. Biol. Chem. 281:14663–14669.

Yoshikawa T., Sakaeda T., Sugawara T., Hirano K., and Stella V. J. (1999). A novel chemical delivery system for brain targeting. Adv. Drug Deliv. Rev. 36:255–275.

Zhang Q., Pangrsic T., Kreft M., Krzan M., Li N., Sul J. Y., Halassa M., Van Bockstaele E., Zorec R., and Haydon P. G. (2004). Fusion-related release of glutamate from astrocytes. J. Biol. Chem. 279:12724–12733.

Zhao J., Lopez A. L., Erichsen D., Herek S., Cotter R. L., Curthoys N. P., and Zheng J. (2004). Mitochondrial glutaminase enhances extracellular glutamate production in HIV-1-infected macrophages: linkage to HIV-1 associated dementia. J. Neurochem. 88:169–180.

Zhu X. W., Raina A. K., Perry G., and Smith M. A. (2004). Alzheimer's disease: the two-hit hypothesis. Lancet Neurol. 3:219–226.

Zimmer L., Delion-Vancassel S., Durand G., Guilloteau D., Bodard S., Besnard J. C., and Chalon S. (2000). Modification of dopamine neurotransmission in the nucleus accumbens of rats deficient in n-3 polyunsaturated fatty acids. J. Lipid Res. 41:32–40.

Index